# Conservation Behavior

*Applying Behavioral Ecology to Wildlife Conservation and Management*

Conservation behavior assists the investigation of species endangerment associated with managing animals impacted by anthropogenic activities. It employs a theoretical framework that examines the mechanisms, development, function and phylogeny of behavior variation in order to develop practical tools for preventing biodiversity loss and extinction.

Developed from a symposium held at the International Congress for Conservation Biology in 2011, this is the first book to offer an in-depth, logical framework that identifies three vital areas for understanding conservation behavior: anthropogenic threats to wildlife, conservation and management protocols, and indicators of anthropogenic threats. Bridging the gap between behavioral ecology and conservation biology, this volume ascertains key links between the fields, explores the theoretical foundations of these linkages, and connects them to practical wildlife management tools and concise applicable advice.

Adopting a clear and structured approach throughout, this book is a vital resource for graduate students, academic researchers, and wildlife managers.

ODED BERGER-TAL is a senior lecturer at the Mitrani Department of Desert Ecology of Ben Gurion University of the Negev, Israel. His research centers upon the integration of behavioral ecology into wildlife conservation and management.

DAVID SALTZ is a Professor of Conservation Biology at the Mitrani Department of Desert Ecology, and the director of the Swiss Institute for Desert Energy and Environmental Research of Ben Gurion University of the Negev, Israel. His research focuses on wildlife conservation and management.

# Conservation Biology

*Executive editor*
Alan Crowden – freelance book editor, UK

This series aims to present internationally significant contributions from leading researchers in particularly active areas of conservation biology. It focuses on topics where basic theory
is strong and where there are pressing problems for practical conservation. The series
includes both authored and edited volumes and adopts a direct and accessible style targeted
at interested undergraduates, postgraduates, researchers and university teachers.

1. *Conservation in a changing world*, edited by Georgina Mace, Andrew Balmford and
   Joshua Ginsberg o 521 63270 6 (hardcover), o 521 63445 8 (paperback)

2. *Behaviour and conservation*, edited by Morris Gosling and William Sutherland
   o 521 66230 3 (hardcover), o 521 66539 6 (paperback)

3. *Priorities for the conservation of mammalian diversity*, edited by Abigail Entwistle and
   Nigel Dunstone o 521 77279 6 (hardcover), o 521 77536 1 (paperback)

4. *Genetics, demography and viability of fragmented populations*, edited by Andrew G. Young
   and Geoffrey M. Clarke o 521 782074 (hardcover), o 521 794218 (paperback)

5. *Carnivore conservation*, edited by John L. Gittleman, Stephan M. Funk, David Macdonald
   and Robert K. Wayne o 521 66232 X (hardcover), o 521 66537 X (paperback)

6. *Conservation of exploited species*, edited by John D. Reynolds, Georgina M. Mace, Kent
   H. Redford and John G. Robinson o 521 78216 3 (hardcover), o 521 78733 5 (paperback)

7. *Conserving bird biodiversity*, edited by Ken Norris and Deborah J. Pain o 521 78340 2
   (hardcover), o 521 78949 4 (paperback)

8. *Reproductive science and integrated conservation*, edited by William V. Holt, Amanda
   R. Pickard, John C. Rodger and David E. Wildt o 521 81215 1 (hardcover), o 521 01110 8
   (paperback)

9. *People and wildlife, conflict or co-existence?*, edited by Rosie Woodroffe,
   Simon Thergood and Alan Rabinowitz o 521 82505 9 (hardcover), o 521 53203 5
   (paperback)

10. *Phylogeny and conservation*, edited by Andrew Purvis, John L. Gittleman and
    Thomas Brooks o 521 82502 4 (hardcover), o 521 53200 0 (paperback)

# Conservation Behavior

## Applying Behavioral Ecology to Wildlife Conservation and Management

*Edited by*

ODED BERGER-TAL
*Ben Gurion University of the Negev, Israel*
and
DAVID SALTZ
*Ben Gurion University of the Negev, Israel*

## CAMBRIDGE
### UNIVERSITY PRESS

Shaftesbury Road, Cambridge CB2 8EA, United Kingdom

One Liberty Plaza, 20th Floor, New York, NY 10006, USA

477 Williamstown Road, Port Melbourne, VIC 3207, Australia

314–321, 3rd Floor, Plot 3, Splendor Forum, Jasola District Centre, New Delhi – 110025, India

103 Penang Road, #05–06/07, Visioncrest Commercial, Singapore 238467

Cambridge University Press is part of Cambridge University Press & Assessment, a department of the University of Cambridge.

We share the University's mission to contribute to society through the pursuit of education, learning and research at the highest international levels of excellence.

www.cambridge.org
Information on this title: www.cambridge.org/9781107690417

© Cambridge University Press & Assessment 2016

First published 2016

A catalogue record for this publication is available from the British Library

Library of Congress Cataloging-in-Publication data
Berger-Tal, Oded, editor.
Conservation behavior : applying behavioral ecology to wildlife conservation
and management / edited by Oded Berger-Tal, Ben Gurion University, Israel,
and David Saltz, Ben Gurion University, Israel.
New York : Cambridge University Press, 2016. | Series: Conservation biology |
Includes index.
LCCN 2015042973 | ISBN 9781107040106
LCSH: Animal behavior. | Animal ecology. | Wildlife conservation.
LCC QL751 .C663 2016 | DDC 591.5–dc23
LC record available at http://lccn.loc.gov/2015042973

ISBN    978-1-107-04010-6    Hardback
ISBN    978-1-107-69041-7    Paperback

*This book is dedicated to my parents, Noa and Arieh,*
*for their inexhaustible love and support.*

OBT

*To the young conservation biology undergraduate and*
*graduate students around the world to whom I leave the*
*burden of repairing all the damage my generation has so*
*skillfully inflicted on this planet.*

DS

# Contents

## Contributors

BEN BELL, Centre for Biodiversity & Restoration Ecology, Victoria University of Wellington, New Zealand.

ODED BERGER-TAL, Mitrani Department of Desert Ecology, Jacob Blaustein Institutes for Desert Research, Ben-Gurion University of the Negev, Israel.

CARMEN BESSA-GOMEZ, AgroParisTech, ESE UMR 8079 – Université Paris-Sud, France.

DANIEL T. BLUMSTEIN, Department of Ecology and Evolutionary Biology, University of California Los Angeles, USA.

JOEL S. BROWN, Department of Biological Sciences, University of Illinois at Chicago, USA.

COLLEEN CASSADY ST. CLAIR, Department of Biological Sciences, University of Alberta, Canada.

ESTEBAN FERNÁNDEZ-JURICIC, Department of Biological Sciences, Purdue University, USA.

ROB FOUND, Department of Biological Sciences, University of Alberta, Canada.

ADITYA GANGADHARAN, Department of Biological Sciences, University of Alberta, Canada.

BURT P. KOTLER, Mitrani Department of Desert Ecology, Jacob Blaustein Institutes for Desert Research, Ben-Gurion University of the Negev, Israel.

DOUGLAS W. MORRIS, Department of Biology, Lakehead University, Canada.

MAUREEN MURRAY, Department of Biological Sciences, University of Alberta, Canada.

NORMAN OWEN-SMITH, Department of Animal, Plant and Environmental Sciences, University of Witwatersrand, South Africa.

DANIEL I. RUBENSTEIN, Department of Ecology and Evolutionary Biology, Princeton University, USA.

DAVID SALTZ, Mitrani Department of Desert Ecology, Jacob Blaustein Institutes for Desert Research, Ben-Gurion University of the Negev, Israel.

FRANÇOIS SARRAZIN, UPMC, CESCO UMR 7204 MNHN CNRS UPMC, France.

ZACHARY SCHAKNER, Department of Ecology and Evolutionary Biology, University of California Los Angeles, USA.

DEBRA M. SHIER, Applied Animal Ecology Division, San Diego Zoo Institute for Conservation Research, USA, & Department of Ecology and Evolutionary Biology, University of California Los Angeles, USA.

JOHN SWADDLE, Institute for Integrative Behavioral and Biodiversity Studies, Biology Department, College of William and Mary, USA.

# Prologue – don't feed the bear!

Like A. A. Milne's Winnie-the-Pooh, real bears love rich food, and just like Pooh, real bears strive to minimize the costs and maximize the benefits of obtaining that food. Like Pooh, real bears also don't realize the possible consequences. In Pooh's case, his attempts to get free food result in him becoming stuck in one of the entrances to Rabbit's den, after consuming all of Rabbit's honey, and becoming too fat to go back out the way he came in. In the real world, bears learn very quickly that humans can provide easy access to food resources that will increase their net energetic return, and start seeking out human activity and steal or beg for food.

"Don't feed the bears!" is a line commonly appearing on roadside notices in many US National Parks. It encompasses the realization that our actions may alter the behavior of the species around us; a realization that took many years to materialize.

Initially, the begging behavior of bears was considered amusing and the US Park Service actually encouraged this, so called, habituation. It was not long before problems concerning human safety began surfacing. Bears began actively seeking human contact and occasionally would become aggressive towards visitors who would not "share their lunch". In 1902, the Park Service outlawed the hand feeding of bears, but did not enforce it and the practice continued. The situation became increasingly dangerous for both humans and bears. Fatal attacks on humans became common and problem bears were shot. Finally, in 1970, the Park Service began enforcing the law and devised various methods to prevent bear access to anthropogenic food sources. These include raising awareness in humans, preventing the bears from accessing food (e.g. introducing bear-proof containers), and using bear deterrents and aversive behavioral conditioning to keep bears away. Behavioral conditioning relies on our understanding of how bears learn and how they react to novel stimuli, and behavioral ecologists continue to devise better and more effective methods that will allow wildlife managers to resolve human-bear conflicts in an efficient, non-lethal manner.

Avoiding the consequences that may follow human-wildlife contacts is but one example of how looking at the world from the animal's point of view can improve the way we conserve and manage wildlife. "Thinking like a mountain" is what conservation behavior is all about.

# Preface – the role of behavior in conservation biology

Conservation biology is an applied multidisciplinary science that often deals with crisis situations. Of the many sciences from which conservation biology draws, it relies most heavily on ecology and its various sub-disciplines (population biology and genetics, community ecology, landscape ecology, etc.). One of these sub-disciplines, behavioral ecology, began in the past two decades receiving particular attention regarding its role in conservation biology. Specifically, several books (e.g. Clemmons & Buchholz 1997, Caro 1998, Festa-Bianchet & Apollonio 2003, Blumstein & Fernández-Juricic E. 2010) and papers (e.g. Sutherland 1998, Caro 1999, Linklater 2004, Angeloni *et al.* 2008, Greggor *et al.* 2014) began focusing on the interface between conservation biology and behavioral ecology, arguing that the discipline of behavioral ecology is an important component of conservation theory and practice, but has not yet received the attention it deserves. Further published opinions claimed that, in contrast to other ecology sub-disciplines, behavioral ecology has little bearing on conservation (Caro 2007), while others argued that behavioral ecology is, and always was, an important component of conservation biology (Harcourt 1999 and Buchholz 2007, respectively). A survey of the literature by Angeloni *et al.* (2008) indicated that only ~5% of papers published in leading conservation journals included the term behavior (or its derivatives) in their title, and that there is no evidence of an increasing trend. Angeloni *et al.* (2008) concluded, based on these findings, that a gap exists between the two disciplines and that the importance of behavioral ecology to conservation has yet to be fully realized. More recently, Nelson (2014) made a similar analysis and reached the same conclusions. When one considers that similar debates never took place with regard to the role of other ecology sub-disciplines in conservation, this debate is somewhat intriguing. It is especially interesting since all the aforementioned papers appear to pose legitimate arguments backed by logic and data that underpin two basic points of contention: (1) is behavioral ecology an important factor in conservation thinking and decision-making? And (2) is conservation behavior (i.e. the application of animal behavior in conservation) a young discipline not yet receiving the attention it deserves?

Biodiversity is a pivotal issue in conservation biology. The logic is straightforward: Diversity is the engine that drives evolution and enables species to change as the world changes. If ecosystems are to continue to provide the services man needs, diversity must be maintained to enable adaptation to a rapidly changing globe. The study of the linkage between biodiversity and ecosystem functioning is considered a top priority in future conservation research (Sutherland *et al.* 2009). Thus, conservation biology focuses on preventing the loss of diversity (of all types) stemming from anthropogenic influences. The behavior of an animal is the outcome of the interaction between its genes and the environment (GXE) and fulfills the role of a mediator between these two elements. Thus, it is almost axiomatic that animal behavior is a component of biodiversity and should be considered in conservation biology. The ability of animals to respond to anthropogenic activity depends on their learning capabilities and their behavioral diversity. An inability to respond behaviorally may contribute to, and even be a direct cause of, extinction. Because behavior is the result of GXE interactions, changes in behavior can be used to assess anthropogenic impacts on the environment, and any conservation or development planning should consider the impact of such actions on the behavior of organisms. Thus, the importance of animal behavior in conservation is, for the most part, self-evident.

The realization that behavior is an important component of managing wild populations stems back to the earliest studies of threatened species long before conservation biology was a realm of science. In the first issue of the first scientific journal devoted to applied ecology – *The Journal of Wildlife Management* – two of the twelve (>15%) papers published: "Winter and spring studies of the sharp-tailed grouse in Utah" (Marshall & Jensen 1937) and "Goose nesting studies on Bear River migratory waterfowl refuge" (Williams & Marshal 1937), focused on behavior and its role in managing wild population. In his book *A Sand County Almanac*, first published in 1949, Aldo Leopold states (page 81) "Science knows little about home range: how big it is, at various seasons, what food and cover it must include, when and how it is defended against trespass, and whether ownership is an individual, family, or group affair. These are fundamentals of animal economics or ecology." So Leopold too realized that the behavior of wildlife is a fundamental part of their economy. Undergraduate programs in Wildlife Management dating back 40 years (Humboldt State University, California 1977 yearly catalog) typically offered two upper-division courses in wildlife ethology or behavioral ecology that focused on the relationship between animal behavior and management. Thus, the realization of the importance of behavioral ecology in managing wild populations dates back many years before conservation biology became an academic field. That said, the question remains as to whether it is receiving the attention it deserves?

Angeloni *et al.* (2008)-surveyed the conservation literature, focusing on three leading journals (*Conservation Biology*, *Biological Conservation* and *Ecological Applications*) over a one decade period (1996–2005) looking for the prevalence of the term "behav*" in the title or abstract of all papers published. They found behavioral issues occupy only 2–6% of the volume of published papers in leading conservation journals. Based on a similar survey we carried out (limited to one journal, *Conservation Biology*), this trend continues through 2009 and increases somewhat thereafter to ca. 8%. Although, at first glance, this seems low, one must consider that conservation biology is a multidisciplinary science covering many disciplines including: conservation genetics, population dynamics, community structure, ecosystem management, ecotoxicology – to name a few that are in essence ecological, and other non-ecological fields. There are several ways to evaluate this: First, one can check the prevalence of other sub-disciplines in the conservation literature – for example the occurrence of the term "genetic(s)." We checked its prevalence in *Conservation Biology* and found that from 1988 to 2003 "genetic*" was twice as prevalent in the titles and abstracts as "behav*" (on average 10.8 vs. 4.4%, respectively), but declined in the following decade, 2004–2013, to 8.5 versus 8.0%, respectively. Thus, an undoubtedly important sub-discipline such as genetics is no more prevalent in the conservation literature than behavior. We found a similar trend in the journal *Animal Conservation*,with considerably greater difference in the first decade the journal was published (1998–2008; 27.5 vs. 10.8%, respectively) but becoming closer between the years 2009–2013 (17.3 vs. 13.7%, respectively). The overall higher percentages in both these topics in an animal-oriented conservation journal is not surprising as it will not cover all realms a general conservation journal would (e.g. plant diversity and community structure). Another option is to compare animal-oriented journals in ecology to animal-oriented journals in conservation ecology – e.g. *Journal of Animal Ecology* versus *Animal Conservation*. In the 1998–2008 period, the percentage of papers in *Journal of Animal Ecology* with "genetic*" in the title or abstract was considerably lower than those with the term "behav*" (4.4 vs. 28.2%, respectively) but the gap shrank between the years 2009 and 2013 due to an increase in the prevalence of genetic studies (9.7 vs. 31.8%). The prevalence of behavior-oriented papers in this journal is just over twice as high as in *Animal Conservation*, reflecting a true difference in focus stemming from the inherent multidisciplinary character of conservation.

These numbers suggest that the volume of behavioral ecology papers within the realm of conservation biology is similar to that of other sub-disciplines (e.g. genetics), and cannot be expected to increase substantially (if at all) in a multidisciplinary field like conservation biology. If that is the case, then what spawned the debate regarding the prevalence of behavioral considerations in conservation in the first place? In his 2007 paper, Tim Caro (2007) claims that behavioral ecology *theory* and *paradigms* have little bearing on conservation. A recent survey of the literature suggested that conservation topics such as invasive species and climate change studies often consider foraging and dispersal behaviors. However, related behavior ecology theories – i.e. optimal foraging and ideal free distribution – are mentioned in only a small fraction (<<1%) of the papers (unpublished data). By contrast, genetic studies in conservation commonly refer to, and are driven by underlying theory (e.g. Hardy-Weinberg theorem, founder effects and genetic drift). Thus, the problem appears to be not the consideration of behavior in conservation, but rather placing conservation behavior studies within the theoretical behavioral ecology framework and paradigms.

\*\*\*

Frameworks are essential for the progression of science as they lend structure and layering, provide a linkage to theory and channel work in certain directions, spawning hypotheses and, later on, generalizations and paradigms. Without these, ad hoc explanations are provided to explain research findings and no rules evolve. Some realms of conservation behavior where key paradigms may evolve in the near future are, for example, the linkage of habituation and the landscape of fear, and the role of social structure in small/declining populations. Structure is not only vital for the progression of science; it also facilitates learning, as comprehension requires the organization of ideas relative to each other and to existing knowledge (Kintsch 1988). Frameworks are especially important in applied sciences dealing with crisis situations (such as conservation biology and medicine), because these sciences are goal oriented and involve frequent and vital decision-making processes. The frameworks provide focus and point to potential issues that need to be considered.

In 2011 we suggested a conceptual theoretical framework for the field of conservation behavior (Berger-Tal *et al.* 2011). The framework focuses on three main realms in which the role of behavior must be considered: anthropogenic threats to wildlife, conservation and management protocols, and indicators of anthropogenic threats or management success. These three elements form the backbone of the conservation behavior conceptual model and dictate the structure of this book.

This book is made up of four parts. The first part serves as an introduction to conservation behavior. In Chapter 1 we give a brief overview of the fields of conservation biology and behavioral ecology and introduce the conservation behavior framework. In Chapter 2 John Swaddle expands on the basic process through which animal behaviors and consequential responses to environmental changes are shaped – evolution. In Chapter 3 Schakner and Blumstein discuss another fundamental process that is vital to understanding how animals respond to human activity – the process of learning. The rest of the book closely follows the three themes of the conservation behavior framework.

Part II looks at how anthropogenic activities impact animal behavior and how these impacts are linked to demographic changes. In Chapter 4 we look at the consequences of not changing one's behavior in the face of a rapidly changing environment. In Chapter 5 Daniel Rubenstein provides the complementary view point and describes the possible consequences of altering one's behavior in response to a changing environment.

Part III considers the various uses of behavioral ecology in conservation and management planning. Esteban Fernández-Juricic provides an overview of the role of sensory ecology in behavioral-based management (Chapter 6). In Chapters 7 and 8 St. Clair *et al.* and Ben Bell discuss the use of behavioral knowledge for reserve design and management and for reintroductions, respectively. Bessa-Gomez and Sarrazin give a brief introduction to the use of behavior ecology in wildlife population modeling (Chapter 9). Lastly, in Chapter 10 Debra Shier considers how manipulating animal behavior can increase the success of captive breeding programs.

Part IV of the book deals with the use of behavior as a leading indicator either of anthropogenic threats to wildlife or of the success of management programs. In Chapter 11 Kotler *et al.* give a detailed overview and guidelines of how to use foraging behavior as a leading indicator for assessing populations' state. In Chapter 12 we look at how behavioral indicators can be used to gauge shifts in the community structure, assess ecosystem health and predict global changes.

The book is aimed at behavioral ecologists of all levels, especially those looking for ways to help with conservation of species, and at conservation practitioners and wildlife managers. Each non-introductory chapter ends with a section that focuses on giving concise and practical advice regarding the uses of behavioral theory and knowledge in management. The structured nature of the book also makes it an excellent basis for a conservation behavior course or for behavioral-oriented classes within a conservation biology course.

Behavior acts as a mediator between the animal and its environment. As such it rapidly varies over time and space and is a function of past experience and the genetic limits resulting from past selection. Behavior is therefore an important component of biodiversity, and like all other components of biodiversity should be regularly addressed when managing animal populations. We hope this book will help implement this concept and will serve as a basis for future development and improvements of the conservation behavior framework.

David Saltz and Oded Berger-Tal

## REFERENCES

Angeloni, L., Schlaepfer, M.A., Lawler, J.J. and Crooks, K.R. 2008. A reassessment of the interface between conservation and behavior. *Animal Behavior*, 75:731–737.

Berger-Tal, O., Polak, T., Oron, A. *et al.* 2011. Integrating animal behavior and conservation biology: a conceptual framework. *Behavioral Ecology*, 22:236–239.

Blumstein, D. T. and Fernández-Juricic, E. 2010. *A Primer of Conservation Behavior*. Sunderland: Sinauer Associates.

Buchholz, R. 2007. Behavioral biology: an effective and relevant conservation tool. *Trends in Ecology and Evolution*, 22:401–407.

Caro, T. 1998. The significance of behavioral ecology for conservation biology. In Caro, T. (ed.), *Behavioral Ecology and Conservation Biology*, pp. 3–26. Oxford: Oxford University Press.

Caro, T. 1999. The behavior-conservation interface. *Trends in Ecology and Evolution*, 14:366–369.

Caro, T. 2007. Behavior and conservation: a bridge too far? *Trends in Ecology and Evolution*, 22:394–400.

Clemmons, J.R. and Buchholz, R. 1997. *Behavioral Approaches to Conservation In The Wild*. Cambridge: Cambridge University Press.

Festa-Bianchet, M. and Apollonio, M. (Eds.) 2003. *Animal Behavior and Wildlife Conservation*. Washington, DC: Island Press.

Greggor, A.L., Clayton, N.S., Phalan, B. and Thornton, A. 2014. Comparative cognition for conservationists. *Trends in Ecology and Evolution*, 29:489–495.

Harcourt, A.H. 1999. The behavior–conservation interface. *Trends in Ecology and Evolution*, 14:490.

Kintsch, W. 1988. The role of knowledge in discourse comprehension: A constructive integration model. *Psychological Review*, 95:163–182.

Linklater, W.L. 2004. Wanted for conservation research: behavioral ecologists with a broader perspective. *Bioscience*, 54:352–360.

Marshall, W.H. and Jensen, M.S. 1937. Winter and spring studies of the sharp-tailed grouse in Utah. *Journal of Wildlife Management*, 1:87–99.

Nelson, X. J. 2014. Animal behavior can inform conservation policy, we just need to get on with the job – or can it? *Current Zoology*, 60:479–485.

Sutherland, W.J. 1998. The importance of behavioral studies in conservation biology. *Animal Behaviour*, 56:801–809.

Sutherland, W.J. *et al.* 2009. One hundred questions of importance to the conservation of global biological diversity. *Conservation Biology* 23:557–567.

Williams, C.S. and Marshall, W. H. 1937. Goose nesting studies on Bear River migratory waterfowl refuge. *Journal of Wildlife Management*, 1:77–86.

# Acknowledgments

The original conservation behavior framework which serves as the foundation for this book is the result of a thought-exercise that led to many long discussions during a conservation behavior course, held at the Jacob Blaustein Institutes for Desert Research, Ben-Gurion University of the Negev. We are in debt to Aya Oron, Tal Polak, Yael Lubin and Burt Kotler for their pivotal role in the framework's construction.

We deeply thank our contributors for lending their expertise to this book, investing so much of their time and energy into it, and patiently enduring our numerous, and sometimes frustrating, requests.

We are grateful to our reviewers: Peter Banks, Steve Beissinger, Luigi Boitani, Tamar Dayan, Paul Doherty, Clinton Francis, Wayne Getz, Andrea Griffin, Michael Heithaus, Todd Katzner, Yael Lubin, Misty McPhee, Bart Nolet, John Pearce, Guy Pe'er, Eloy Revilla, Bruce Roberson, Martin Schaefer, Kate Searle, Phillip Seddon, Tanya Shenk, Andy Sih, Ronald Swaisgood and Robert Young. Thank you for your excellent and insightful feedback.

We thank the editorial staff at Cambridge University Press, and especially Dominic Lewis, for their professionalism and for being incredibly patient with us as we discovered that editing this book is by far the most time-consuming endeavor we ever took upon ourselves.

Last but not least, we are forever grateful to Reut, Ahuvit, Maayan, Yonatan, Shira, Yael, Moria and Ariel, for being our anchor in crazy times, and for their everlasting encouragement and support.

# Part I

# The integration of two disciplines: conservation and behavioral ecology

# Introduction: the whys and the hows of conservation behavior

## ODED BERGER-TAL AND DAVID SALTZ

Our planet is changing at a startling pace. The rate of species extinction is alarmingly high (Barnosky *et al.* 2011) and unique ecosystems such as coral reefs and tropical forests are rapidly diminishing and disappearing. It is very clear that the only way to prevent, or at least slow down, this mass extinction, is by direct action. The science of conservation biology stands before the ongoing environmental crisis, offering some hope that through the implementation of our accumulating interdisciplinary scientific knowledge we can prevent, and even reverse, the decline of the diversity of life on Earth.

The behavior of an organism is, in a sense, the mediator between the organism and its environment and provides flexibility so the organisms can maintain a adequate fitness over a wider range of environmental conditions. This, of course, has limits, and under extreme changes the organism's behavior will fail to provide a sufficient buffer from the changing environment. Knowledge of a species' behavioral attributes provides, therefore, important insights into how anthropogenic actions (direct or indirect) will impact the species, and what actions can be taken to minimize this impact.

In this chapter we will start by giving a brief general overview of conservation biology's interdisciplinary foundations. Many excellent volumes have been dedicated to this field (e.g. Groom *et al.* 2006, Primack 2006, Hunter & Gibbs 2007), and they give a far more comprehensive picture of the history, practice and many challenges of conservation biology. However, we hope we provide enough background in the first part of this chapter to make our readers better understand the goals of conservation, and to have these goals stay in their minds, as they continue reading about the more specific aspects of using behavior in conservation. Before considering the

*Conservation Behavior: Applying Behavioral Ecology to Wildlife Conservation and Management*, eds. O. Berger-Tal and D. Saltz. Published by Cambridge University Press. © Cambridge University Press 2016.

role of behavior in conservation, we will first consider the roots of behavioral ecology, and then discuss the short history of conservation behavior – a field dedicated to the use of the knowledge of animal behavior in conservation biology. To conclude this introductory chapter, we will outline the principles of the conservation behavior framework that serves as the basis for the structure of this book.

Conservation biology has three objectives: (1) Documenting the extant biological diversity on Earth. (2) Locating, defining and investigating anthropogenic threats to biodiversity. (3) Developing and implementing practical approaches to reducing or eliminating these threats (Groom et al. 2006, Primack 2006). To achieve these objectives successfully, conservation biologists must understand the ultimate goals for protecting nature. Namely: What do we wish to conserve? Why do we wish to conserve it? And, how can we actually do it?

## 1.1   WHAT TO CONSERVE?

The answer is seemingly self-evident. We aim to conserve nature. But what is nature? A popular response would be – that which is not made or influenced by humans. However, humans are an important part of most ecosystems, and they have played a significant part in the shaping of these ecosystems, with many species evolving or co-evolving with humans. More than 83% of the Earth's land surface is directly influenced by anthropogenic activity (Sanderson et al. 2002), and it is logical to assume that much of the remaining 17% "pristine" landscape is indirectly influenced by humans through global processes such as climate change and pollution. Nevertheless, despite this fact, most of us have an inherent notion as to what nature is, and we could easily say that Yellowstone National Park is much more "natural" than downtown Los Angeles, for example. Therefore, we do not define nature by whether or not it is influenced by human activities, but rather by what is the magnitude of this influence. The less an area is *disturbed* by humans, the more natural it is.

For many years ecologists believed that nature is at a stable equilibrium and ecological systems can reach a stable steady state. According to this view, in order to conserve ecological systems, all we have to do is return the system to its steady state and protect it from further disturbances. However, we know now that nature is dynamic and constantly changing (Pickett et al. 1992), and that ultimately, the process that governs these changes across species, communities and ecosystems is evolution by natural selection (see Chapter 2 for a short review of evolutionary processes). Theodosius Dobzhansky's famous saying: "Nothing in biology makes sense except in the light of evolution" applies to conservation biology as well. Evolution is the basic axiom of biology and therefore we need to approach the question of what to conserve from an evolutionary point of view (Groom *et al.* 2006). This may be best understood through what is known as "Hutchinson's metaphor" that depicts nature as a performing art, namely "The ecological theater and the evolutionary play" (Hutchinson 1965). Human disturbances alter the course of the play by changing the evolutionary trajectory of many species, and in some cases, by stopping evolutionary processes altogether. Conservation biology aims to preserve the one process that keeps the play running and that is at the foundation of nature – evolution. More specifically, the aim of conservation is to prevent or minimize the foreclosure of evolutionary opportunities (Ehrlich 2001) as these opportunities provide the ecological systems (the theater) with the plasticity necessary to respond to unforeseen future changes.

## 1.2 WHY CONSERVE?

Ever since humans began to make use of nature's resources, altering their environment in the process, different philosophical and religious beliefs regarding the relationships between human societies and nature arose (Primack 2006). While there are many different reasons for conserving nature, they can roughly be divided into two main approaches:

(1) The intrinsic approach: According to this approach every organism has intrinsic value, and therefore a right to exist. This approach can be expanded to include the intrinsic value of ecosystems, evolutionary principles and nature in general (Rolston 1988). Since the current biodiversity crisis is the direct result of anthropogenic activity, nature conservation, in this context, is our ethical obligation.

(2) The utilitarian approach: Nature provides humans with irreplaceable and vital services (for example: carbon sequestration, nutrient cycling, water purification, crop pollination and many more). The monetary value assigned to these services has been estimated to be between 16 to 54 trillion US$ per year (Costanza *et al.* 1997). Thus, according to this approach, we conserve nature in order to continue and benefit from the various services it

provides. This approach also includes biomes and species that provide no known direct services to humans, both because they can interact with biomes and species that do provide services, and because they might provide important services in the future that we are yet to discover (e.g. industrial and medicinal plants in the rainforests of Madagascar [Rasoanaivo 1990]). The only ethical issue involved in this approach is the obligation of the human population to its future generations (Norton 2003), so the key focus is not a specific organism or ecosystem, but rather the sustainable use derived from them.

While nature conservation has been practiced in one form or another all over the world, in many ways the roots of modern conservation biology stem from the three main conservation philosophies that arose in North America in the late nineteenth to mid twentieth century (Callicott 1990).

(1) The romantic transcendental conservation ethic was derived from the writings of three prominent figures: Ralph Waldo Emerson, who in 1836 referred to nature as a temple in which people can achieve spiritual enlightenment; Henry David Thoreau, who believed that in wilderness lies the preservation of the world (Thoreau 1863); and John Muir, who advocated that natural areas have spiritual values that are superior to the material gain provided by their exploitation (Muir 1901). Thus, this philosophy represents an intrinsic and spiritual approach to nature, giving it a value in and of itself, apart from its value to humanity.

(2) The resource conservation ethic was preached by Gifford Pinchot, the first head of the US Forest Service, at the turn of the twentieth century. According to this utilitarian philosophy, nature is a collection of natural resources that are either: useful, useless or noxious to people. Useful resources should be conserved in a way that will ensure their sustainable use over time, while noxious resources should be removed (Pinchot 1947).

(3) The evolutionary land ethic was developed by Aldo Leopold in the mid twentieth century. This philosophy grew from Pinchot's resource conservation ethic, but with a realization of its scientific contradictions and inaccuracies. Leopold claimed that nature is not just a collection of individual resources but rather a complicated and integrated system of interdepended processes and components that function together like a fine Swiss watch (Leopold 1949).

Hence, we should strive to conserve even parts of nature that may seem unimportant to our eyes, since they may be vital to the long-term health of the system. According to this philosophy, the most important goal of conservation is to maintain the health of ecosystems and natural processes. Unlike Pinchot's anthropocentric approach that puts humans in the center of the natural world exploiting it for its purposes, the evolutionary land ethic is an ecocentric approach, considering humans as an integral part of the ecological community. It is Leopold's evolutionary land ethic that provides the philosophical foundation for modern conservation biology. However, whether this is a utilitarian, pragmatic approach to conservation where the only ethical consideration is our obligation to our children (Norton 2003, Minteer 2012) or whether it is founded on a belief in the intrinsic value of nature (Callicott 1999) is still being debated (Callicott *et al.* 2011). While most of Aldo Leopold's writings advocate the land ethic in a manner reflecting concern for the future existence of mankind, it is also clear that he considered nature and its components as having an intrinsic value (e.g. his description of a dying wolf in "Thinking Like a Mountain": *"We reached the old wolf in time to watch a fierce green fire dying in her eyes. I realized then, and have known ever since, that there was something new to me in those eyes – something known only to her and to the mountain. I was young then, and full of trigger-itch; I thought that because fewer wolves meant more deer, that no wolves would mean hunters' paradise. But after seeing the green fire die, I sensed that neither the wolf nor the mountain agreed with such a view,"* Sand County Almanac 1949). Thus, both the pragmatic school of thought focusing on the future well-being of mankind, and the intrinsic-value school of thought valuing nature for itself are valid approaches and, in fact, complement each other such that their joint consideration provides the optimal approach to support decision-making in conservation biology.

## 1.3 HOW TO CONSERVE?

In the second half of the twentieth century, with the acceleration of the biodiversity crisis and with ecosystems and species disappearing at an alarming rate throughout the world, it was becoming clear that there is a pressing need for an interdisciplinary approach that will bring together the growing number of people of different backgrounds and disciplines that were thinking and conducting research on conservation issues. In addition, the biodiversity crisis led to a series of legislations and agreements in the US and around the world (such as the US Endangered Species Act (ESA) in 1973 or the Convention on International Trade in Endangered Species of wild fauna and flora [CITES] in 1975), which increased the need for rigorous scientific input into conservation decision-making (Meine 2010). The first international conference on conservation biology was organized by Michael Soule and held in 1978 at the San Diego Wild Animal Park. Soon after the meeting, Soule and colleagues such as Paul Ehrlich and Jared Diamond began developing conservation biology as a discipline combining the practical experience of wildlife, forestry and fisheries management with the theoretical knowledge of population biology and biogeography (Primack 2006). A few years later Michael Soule wrote: "disciplines are not logical constructs; they are social crystallizations which occur when a group of people agree that association and discourse serve their interests. Conservation biology began when a critical mass of people agreed that they were conservation biologists" (Soule 1986).

Groom *et al.* (2006) define three guiding principles that are at the core of the science of conservation biology: The first principle states that *evolution* is the basic axiom that unites all of biology. Thus, conservation biology does not aim to stop evolutionary change and keep the status quo, but rather to conserve the ongoing evolutionary processes in order to ensure that populations may continue to respond to environmental change in an adaptive manner. The second principle states that the ecological world is *dynamic* and largely nonequilibrial. Therefore, we do not try to restore systems to some point of equilibrium, but rather to understand and preserve the nonequilibrial processes that maintain communities and ecosystems. The third principle is that *human* presence must be considered and included in conservation planning. Whether we like it or not, humans are an integral part of the ecological systems of our planet, and therefore any conservation attempt that does not take humans into consideration is doomed to fail.

We have established that we aim to conserve evolutionary and ecological processes. However, how can we preserve and protect something as intangible as evolution? The answer is biodiversity. It is this diversity that generates the evolutionary opportunities (Ehrlich 2001) and allows systems and organisms to change and adapt in response to a changing world. By preserving and protecting biodiversity, we are conserving the evolutionary process. There is a strong paradigm underlining this assumption: the loss of diversity reduces the ability of the system, and the populations and individuals it is comprised of, to respond to a change in the environment.

## 1.4 WHAT IS BIODIVERSITY?

The term "biological diversity" (or simply "biodiversity") has received many different definitions and interpretations, and although they are all variants of the same basic theme, the difference in content can have significant implications for determining conservation policy and action (Faith 2008). Gaston (1998) gives a detailed account of the different definitions for biodiversity and their consequences. For our purposes, we use the 1992 Convention on Biological Diversity definition, which is widely accepted among conservation biologists: "'Biological diversity' means the variability among living organisms from all sources including, *inter alia*, terrestrial, marine and other aquatic ecosystems and the ecological complexes of which they are part; this includes diversity within species, between species and of ecosystems" (Johnson 1993).

According to this definition, biodiversity must be considered on three hierarchical levels: genetic diversity, species diversity and ecosystem diversity. These levels of biodiversity are nested, i.e. higher levels enclose lower levels (Noss 1990). When an ecosystem is destroyed, all species within it are also destroyed, and with them all the genes that were stored in the DNA of all these species. A comprehensive approach to biodiversity must address the multiple levels of biodiversity on different spatial and temporal scales (Noss 1990).

### 1.4.1 Genetic diversity

Genetic diversity is the ultimate source of biodiversity at all levels. Without genetic variability there can be no selection, and therefore no evolutionary process. Low genetic variability decreases the ability of populations to adapt to environmental changes and increases their susceptibility to diseases. This, in turn, impacts the ability of the ecosystems that the species are a part of to respond to changes.

We can consider three levels of genetic diversity within a species: (1) Between populations of the same species – different populations can differ genetically because of different selection pressures, genetic drift and founder effect. Remote or isolated populations are of special importance to conservation, as their genetic composition may be unique, paving the way for the emergence of new species. (2) Within populations – different individuals within a given population differ from each other genetically.

A genetically diverse population will be more adaptable to changes and diseases than a genetically homogenous population. (3) Within individuals – since the genetic code is composed of pairs of alleles sampled from a population of one or more alleles, an individual can be homozygote or heterozygote in each allele, i.e., some individuals are more genetically diverse than others.

### 1.4.2 Species diversity

Species diversity includes all species found on our planet. It is what most people think of when they hear the term "biodiversity," and is the primary subject of the majority of conservation laws (such as ESA and CITES mentioned above). Most people understand the concept of species and of species diversity and best relate to such a tangible idea. Thus, species diversity is conceptually, legally and also practically the most considered and established form of biodiversity, and indeed natural scientists and conservation biologists have been focusing for many years on categorizing and studying different species.

Many conservation biologists measure species diversity by simply counting the number of different species within a community, a measure called species richness. This is a powerful and easy method to assess species diversity; however, it gives all species the same relative weight, regardless of how abundant they are. Other methods have been developed to include the relative frequencies of species within a community. Regardless, one of the more common methods of measuring species diversity is by separating it into geographical components (Whittaker 1960). Alpha diversity refers to local species diversity within one patch or habitat. Gamma diversity refers to regional species diversity – the diversity of species in a large collection of sites that make up a whole region of interest. Beta diversity links between the local and regional scales. It represents the rate of change in composition among sites within a region, and can be calculated as the gamma diversity for a region divided by the average alpha diversity for the sites in the region.

An important aspect of species diversity is biodisparity (Jablonski 1994), which is the range of morphologies or other attributes within a clade (a phylogenetic "branch"). If we aim to maximize the evolutionary potential of living organisms in nature, biodisparity can sometimes provide a better tool for setting conservation priorities than biodiversity (Jablonski 1995, Myers 1996), since some species display traits that are unique to their clade compared to other species whose traits are common (e.g. many of the traits of a panda are unique in nature compared for example to the traits of many fly species).

### 1.4.3  Ecosystem diversity

An ecosystem can be defined as a biological community (a group of species that occupy a particular place at a particular time and interact with each other) together with its associated physical and chemical environment (Primack 2006). While its definition is rather straightforward, ecosystems are still sometimes hard to define. This is because it is not always clear where one ecosystem starts and another ends. An ecosystem can be as small as a puddle of water or as big as a mountain range.

Ecosystems provide critical ecosystem services and support a broad array of biological functions (Groom *et al.* 2006). They are therefore a prime target for conservation, and concepts such as ecosystem health and ecosystem integrity have been developed to evaluate the species an ecosystem contains and the biotic and a biotic processes that take place within it. Furthermore, with the growing rate of species extinction scientists realized that it is all but impossible to create a management plan to protect each of the threatened species on our planet. Ecosystem management represents a more holistic approach to conservation which maintains that by keeping an ecosystem healthy (i.e. by preserving the ecological processes within an ecosystem) all species within it will also be protected.

## 1.5  ANTHROPOGENIC THREATS TO BIODIVERSITY

The biological diversity on this planet is increasingly threatened at all levels by human activities. The ultimate cause for the vast majority of these threats is the nearly exponential human population growth in the last century and a half (Ehrlich & Ehrlich 1990). Human population has increased from one billion in 1850 to well over seven billion in 2013 (USCB 2013), and is still increasing at an alarming rate. The need for natural resources to sustain this vast human population, as well as overconsumption of resources by some parts of the world's population, puts an enormous strain on all of Earth's ecosystems, and is the main driver of the various proximate causes for biodiversity decline. The most notable of these proximate causes are: *Habitat destruction, degradation and fragmentation; Overexploitation; Invasive species;* and *Climate change and pollution.*

### 1.5.1  Habitat destruction, degradation and fragmentation

Habitat destruction and degradation are the primary causes for the loss of biodiversity at all levels (Groom & Vynne 2006). Agricultural, industrial and urban development activities have directly transformed or destroyed most of our planet's habitats, and many other ecosystems have collapsed due to human activities. Such ecosystem collapses include, for example, the desertification of semi-arid ecosystems, the eutrophication of fresh water systems, and coral bleaching in coral reefs. In addition to the direct destruction of species and ecosystems, habitat loss also leads to habitat fragmentation, which has far-reaching consequences in nature conservation. Habitat fragmentation is the process where a large expanse of habitat is transformed into a number of smaller patches of smaller total area, isolated from each other by a matrix of habitats unlike the original (Wilcove *et al.* 1986, Fahrig 2003). Habitat fragmentation reduces biodiversity by reducing the total size of the habitat, reducing genetic flow over the landscape (e.g. reduced migration and dispersal), increasing edge effects within the habitat, crowding the habitat with animals displaced from areas that have been destroyed and so on (Groom & Vynne 2006). The fragmentation of habitats may splinter once-continuous populations into a metapopulation – local assemblages of breeding populations, separated by an inhospitable matrix, that have at least some migration among them (Levins 1969). Migration affects the local dynamics of the populations including the possibility of population reestablishment following extinction (Hanski & Simberloff 1997). The metapopulation approach to conservation puts a strong emphasis on connectivity between habitat fragments and the ability of individuals to migrate between populations through the inhabitable matrix, making the establishment of conservation corridors a commonly used strategy in wildlife management (Chetkiewicz *et al.* 2006).

### 1.5.2 Overexploitation

People have always hunted and harvested food and other resources to survive. However, the dramatic increase in human population combined with the drastic improvement in harvesting efficiency has led to an almost complete depletion of many species, such as large mammals and fish (Bennett *et al.* 2002). This reduction in density could lead to a series of detrimental processes such as vulnerability to demographic stochasticity, inbreeding and Allee effects (a positive correlation between population density and mean individual fitness [Allee 1931]), that in turn, may cause further population decline and even extinction.

Harvesting does not only affect the target species, but also other species that are taken accidently, opportunistically or interact (directly or indirectly) with the harvested species, resulting in a cascading effect (e.g. Estes *et al.* 1989, Terborgh *et al.* 1999). Marine fisheries, for example, are estimated to have an annual global by-catch weighing about 27 million tons, including many threatened species (Reynolds & Peres 2006). Non-sustainable exploitation may also damage the species' physical habitat, and harm other non-target species through cascading effects within communities and changes to the food webs (Reynolds & Peres 2006). In this way, the over-exploitation of sea otters (*Enhydra lutris*) in the Pacific coast of North America has completely transformed the marine kelp forest ecosystem (Estes *et al.* 1989). Otters prey on sea urchins, and their absence led to an urchin population explosion, which in turn led to the urchin overgrazing on kelp, devastating the kelp forests and creating "urchin barrens."

### 1.5.3 Invasive species

Human activities have distributed, either intentionally or unwittingly, a staggering number of species throughout the world, far beyond their native ranges (Wonham 2006). While in most cases the introduction of species into new habitats fails or has no deleterious effects, when introduced species do succeed in establishing themselves and impact local commu-nities they are considered invasive and the damage they create can be enormous, laying to waste entire ecosystems. One of the most famous examples of the devastating effects of invasive species is the introduction of the brown tree snake (*Bioga irregularis*) to the island of Guam. The snake had arrived on Guam shortly after World War II and within half a century has driven eight of eleven forest bird species to extinction (Wiles *et al.* 2003), as well as several species of reptiles, rodents and bats, radically changing the food web on the island (Wonham 2006). Invasive species can reduce native biodiversity through direct interactions with other species such as preda-tion, competition, parasitism and disease transfer, and through indirect effects such as by changing the trophic cascade, or by modifying the physical aspects of the habitat (Wootton 1994, Wonham 2006).

### 1.5.4 Climate change and pollution

Our planet's climate is changing rapidly. The Intergovernmental Panel on Climate Change has determined that the global rises in temperatures in the past 50 years were primarily due to global rises in anthropogenically pro-duced greenhouse gases and that the rate of temperature change will accelerate over the coming century (IPCC 2001). While we are just begin-ning to see the effects of climate change on biodiversity, the magnitude and severity of the expected impacts of climate change are overwhelming. Climate change is the one major threat that exists everywhere and cannot be reversed by local actions (Parmesan & Matthews 2006). The global increase in mean temperature is manifested locally in different ways: in some regions the yearly variance in temperature increases, making the winters colder and the summers hotter (Karl & Trenberth 2003). Precipitation has also increased globally, as well as the variance in precipitation, causing extreme droughts in some areas and major floods in others (Parmesan & Matthews 2006). These climatic changes affect biodiversity either directly – as many species find themselves maladapted to the new climate regime – or indirectly by altering and destroying existing ecosystems. Climate change also selects for a series of novel adaptations to the new environment, and thus may irrevocably alter the evolutionary trajectory for many species.

Similarly to climate change, pollution is a global phenomenon and various harmful substances make their way around the globe carried in the air, the water and the sediments. These pollutants can change the biogeochemical composition of ecosystems and interfere with a variety of ecosystem processes such as leaf-litter breakdown and nitrogen fixing (Woodward *et al.* 2012), or when coming in direct contact with animals and plants, they may reduce both survival and reproduction success.

## 1.6 BEHAVIOR AS A MEDIATOR BETWEEN THE ENVIRONMENT AND THE INDIVIDUAL

Behavioral traits are one of the products of genetic diversity. However, behavior is also highly influenced by the environment (which includes both the a biotic conditions and biotic interactions with other individuals). Thus, behavior lies at the center of the gene–environment interactions and as such it gives populations the evolutionary flexibility to withstand environmental changes. On the individual level, behavior serves as a crucial mediator between the environment and fitness, allowing individuals to withstand harsh conditions to which they are not physiologically adapted (e.g. harsh desert environments, Ward 2009). This makes behavior extremely important to conservation biologists as it also serves as a mediator between fitness and anthropogenic disturbances.

Behavior allows individuals to respond rapidly and effectively to environmental changes (Charmantier *et al.* 2008), and plays an important role in shaping the biophysical properties and functions of ecosystems (Schmitz *et al.* 2008, Luck *et al.* 2012). As such, it is not only an essential contributor to determining an organism's survival and reproductive success and, by extension, the growth and persistence of natural populations (Caro & Sherman 2013), but is also a major component of ecosystem diversity.

Environments change rapidly, and just as high genetic diversity allows populations to adapt to these changes on an evolutionary time scale, behavioral diversity allows populations to respond to environmental changes (including rapid ones) on an ecological time-scale. This has led scientists to call attention to the importance of behavioral diversity and the need to preserve it (Caro & Sherman 2012). Individual variation in behavior is an important component of behavioral biodiversity that stems from genetic variation and experience (which is a product of the variation in the landscape). We can exploit individual variation in behavior to improve conservation success; for example, in severely threatened populations steps can be taken to enhance certain behavior types to increase the viability of populations (see Box 1.1 by C. C. St. Clair). In general, behavioral diversity in a population could come about through either monomorphism of mixed strategies – all individuals within a population are behaviorally similar, and they all display behavioral plasticity, i.e. behavioral diversity is enclosed within the individuals; or polymorphism of pure strategies – individuals display pure and non-flexible behavioral strategies, but there are different behavioral morphs within the population, i.e. behavioral diversity is enclosed within the population. Understanding how the behavioral diversity of a population is formed is central for making the best management plan to preserve this diversity.

Several key conservation issues are predominantly behavioral in nature, and these relate to all of the major threats to biodiversity: Fragmentation – behavioral processes of habitat selection and movement determine how animals use landscapes and are thereby fundamental to the identification and evaluation of corridors within fragmented landscapes (Chetkiewicz *et al.* 2006, chapter 7); Overharvesting – populations that have been reduced to small numbers by over-exploitation commonly exhibit Allee effects, which further hasten the decline in the population's size. The mechanisms driving Allee effects are in many cases behavioral in nature and include, among many others, cooperative feeding, cooperative defense and mate choice (Chapter 4); Invasive species – the existence or absence of appropriate anti-predator behaviors is one of the major factors determining the susceptibility of native species to alien predators; Global changes – behavioral diversity may allow individuals to deal with harsh environmental conditions such as heat or water stress and allow populations to withstand large-scale changes to their environment (at least up to a certain level).

By now we hope that the importance of animal behavior to conservation has become apparent, and thus it is time to turn to the development of the field of behavioral ecology.

**Box 1.1:** **Exploiting individual variation for conservation**
COLLEEN CASSADY ST. CLAIR

In an ideal world, all of the behavioral types that exist in a population would be preserved as a deserving component of biodiversity. But conservation challenges do not occur in ideal worlds and some behavioral types will make it easier to retain the small, isolated and declining populations that require urgent conservation action. In many conservation contexts, the successful individuals are likely to exhibit higher levels of behavioral flexibility – or lower levels of rigidity – that are maintained or even increased through time. Rapid development of the theory underlying personality and corresponding behavioral assays make it easier than ever before to identify combinations of individuals and management actions that are more likely to advance conservation goals.

As a characteristic of individuals, behavioral flexibility is inversely analogous to the specialization that defines most endangered species. The undeniable effect of specialization on endangerment is apparent in every conservation textbook and in a casual review of the 45,000 species currently on the global red list (IUCN 2014). Yet the foundational literature for conservation action has always emphasized small and declining populations (Caughly 1994), which are actually consequences of specialization. Cause and consequence of rarity are confounded by genetic diversity, which tends to be lower for endangered species, small populations, and species with limited ranges (Frankham 1996), while imposing genetic deficits that make those populations even less viable over time (reviewed by Fox & Carroll 2008, Radwan et al. 2010). The vicious circle of rarity that revolves around specialization is usually only exacerbated by anthropogenic impacts that cause small populations to become increasingly isolated (Chapter 7).

Identifying and preserving the individuals in imperiled populations that exhibit the most flexibility in their behavior may counteract some of the effect of species-level specialization to reduce the rate of population decline. Put another way, the rapidity of human-induced environmental change usually creates strong directional selection for greater behavioral flexibility. All else being equal, less flexible individuals will typically be less likely to succeed in a changing environment and more likely to thwart conservation actions that only partially restore pre-existing habitats. Conservation triage has long been espoused as a logical method for allocating scarce conservation resources among populations (e.g. McDonald-Madden et al. 2008), species (e.g. Bottrill et al. 2008), ecosystems, and regions (e.g. Millar et al. 2007) of conservation concern. It is long past time for conservation biologists to consider applying the concepts of adaptive triage to individuals.

Conservation biologists have been slow to recognize the conservation commodity that is contained in individual variation of any sort, perhaps because the discipline is so focused on populations, does not much recognize the importance of selection – natural or artificial – and makes little use of behavioral

**Box 1.1:   (cont.)**

theory (this chapter). By contrast, behaviorists have always emphasized differences among individuals, which can sometimes be correlated as suites of recurring behavioral tendencies and generalized as behavioral types, personalities, or temperaments (reviewed by Sih *et al.* 2004, Reale 2007). Distinct behavioral types might be categorized as opposite ends of axes defined by more specific terms such as bold *versus* shy responses to novelty (Wolf & Weissing 2012), pro- *versus* asocial behavior (Wilson *et al.* 2009) or the more holistic proactive–reactive axis of coping styles (Koolhaas *et al.* 2010). However defined, there has been an explosion in attention to differences among individuals by behavioral ecologists.

The literature on personality, broadly defined, is now ripe for use by managers who might advance conservation goals by identifying and targeting particular individuals. Hundreds of papers have addressed personality in free-living, non-human species that include birds (e.g., Krajl-Fiser *et al.* 2010), mammals (e.g. Gartner & Powell 2012, Carter *et al.* 2012), fish (e.g. Adriaenssens & Johnsson 2013), and invertebrates (e.g. Jant *et al.* 2014). The concept of identifying particular individuals to promote conservation goals is analogous to the traditional identification of keystone species in ecosystems (Paine 1969) and the more recent recognition of keystone individuals in social groups (Modlmeier *et al.* 2014). By extension, keystone individuals in conservation contexts have disproportionate effects on population persistence, not only through their own survival and reproductive success, but through the way they model new behaviors that promote success by conspecifics in human-dominated landscapes.

Even with agreement that behavioral flexibility is a measurable attribute of individuals with the potential to exert positive effects on conservation outcomes, it will not be possible to apply it as uni-dimensional behavioral types across conservation contexts. For example, greater flexibility in behavior has been linked to both shy individuals (Verbeek *et al.* 1996, Koolhaas *et al.* 1999, Ciuti *et al.* 2012) and bold ones (e.g. Jones & Godin 2010, Verdolin & Harper 2013). Boldness itself can change with an individual's state, age, experience and context, in animals ranging from dumpling squid (*Euprymna tasmanica*; Sinn *et al.* 2008) to domestic dogs (Starling *et al.* 2013), but it may still be more changeable through life – a useful metric of flexibility – for some behavioral types than others. For example, female elk (*Cervus elaphus*) that decreased movement rates as they aged were more likely to survive in a hunted population (Cuiti *et al.* 2012).

Measures of more complex personality axes, such as coping style, will likely provide more robust measures of personality for use in conservation, but their application is still nuanced. As an example, proactive individuals with fast reaction times are more likely to be aggressive (Koolhaus *et al.* 1999), which should make them more prone to human–wildlife conflict, but the same tendency might protect them from chronic elevations of stress in a human-dominated landscape (Cockrem 2007). Single measurements of coping style are less likely to reveal this nuance than multiple measures. For example,

**Box 1.1: (cont.)**

individual ewes of bighorn sheep (*Ovis canadensis*) were highly consistent in the docility they exhibited when trapped, but that tendency did not correlate with reproductive status, age or their behavior in the field (Reale *et al.* 2000). By contrast, when dark-eyed juncos (*Junco hyemalis*) were subjected to a common garden experiment, those from an urban population were consistently more exploratory and less stressed when being approached by people than juncos from a wildland population (Atwell *et al.* 2012). With appropriate care to define and measure multiple variables in a framework of adaptive management (Salafsky *et al.* 2002), it could be possible to use animal personality to advance conservation goals while simultaneously advancing a better understanding of the role personality plays in the apparent trade-off between survival and reproduction (Quinn *et al.* 2011).

In addition to targeting the more flexible individuals within populations, managers could exploit behavioral flexibility as a conservation commodity of individuals, whatever their underlying temperament, by manipulating their flexibility to support greater coexistence with people. Some of the necessary techniques for changing behavior could be borrowed from existing interventions for species that exhibit too much or too little behavioral rigidity (Chapters 4 and 5, respectively). For example, bold fish that were exposed to superior competitors or watched shy demonstrators explore novel objects became shyer in their own behavior (Frost *et al.* 2007). Managers could use similar techniques to teach novel responses to anthropogenic stimuli or to change the direction and magnitude of previously learned associations (Chapter 3). In some cases, managers may strive to reduce the flexibility of individuals, for example, by using conditioned taste aversion, which can be used to protect both prey (e.g. Massei *et al.* 2002) and predators (e.g. O'Donnell *et al.* 2010). Managers might also counter the expression of flexibility that could otherwise cause some beneficial behaviors to be lost from a population (Chapter 5).

In sum, I suggest that flexibility in behavioral expression varies among individuals in virtually all populations and could be identified, exploited, augmented or suppressed both within and among individuals to support diverse conservation goals. Realizing the potential behavioral flexibility offers as a conservation commodity will require considerable flexibility by conservation managers themselves. This flexibility might be encouraged by behaviorists, provided they also have enough awareness of the conservation problem and management context to avoid suggesting manipulations that are intractable or even irresponsible. Finding this optimum will require a broad understanding of conservation problems (Chapter 1), an understanding of behavior as both a cause and consequence of selection (Chapter 2), a solid understanding of learning theory (Chapter 3), a thorough repertoire of behavior-based conservation action (this book), and on going study of behavior and physiology as interacting characteristics of individuals.

Box 1.1: (cont.)

## REFERENCES

Adriaenssens, B. and Johnsson, J.I. 2013. Natural selection, plasticity and the emergence of a behavioural syndrome in the wild. *Ecology Letters*, 16: 47–55.

Atwell, J.W., Cardoso, G.C., Whittaker, D.J. *et al.* 2012. Boldness behavior and stress physiology in a novel urban environment suggest rapid correlated evolutionary adaptation. *Behavioral Ecology*, 23: 960–969.

Bottrill, M.C., Joseph, L.N., Carwardine, J. *et al.* 2008. Is conservation triage just smart decision making? *Trends in Ecology & Evolution*, 23: 649–654.

Carter, A.J., Marshall, H.H., Heinsohn, R. and Cowlishaw, G. 2012. How not to measure boldness: novel object and antipredator responses are not the same in wild baboons. *Animal Behaviour*, 84: 603–609.

Caughley, G. 1994. Directions in conservation biology. *Journal of Animal Ecology*, 63: 215–244.

Ciuti, S., Muhly, T.B., Paton, D.G. *et al.* 2012. Human selection of elk behavioural traits in a landscape of fear. *Proceedings of the Royal Society B-Biological Sciences*, 279: 4407–4416.

Cockrem, J.F. 2007. Stress, corticosterone responses and avian personalities. *Journal of Ornithology*, 148: 169–178.

Fox, C.W. and Carroll, S.P. 2008. *Conservation Biology: Evolution in Action*. Oxford: Oxford University Press.

Frankham, R. 1995. Inbreeding and extinction: a threshold effect. *Conservation Biology*, 9: 792–799.

Frost, A.J., Winrow-Giffen, A., Ashley, P.J. and Sneddon, L.U. 2007. Plasticity in animal personality traits: does prior experience alter the degree of boldness? *Proceedings of the Royal Society B-Biological Sciences*, 274: 333–339.

Gartner, M.C. and Powell, D. 2012. Personality assessment in snow leopards (*Uncia uncia*). *Zoo Biology*, 31: 151–165.

IUCN 2014. *The IUCN Red List of Threatened Species v. 2013. 2.* www.iucnredlist.org [accessed April 01, 2014].

Jandt, J.M., Bengston, S., and Pinter-Wollman, N. 2014. Behavioural syndromes and social insects: personality at multiple levels. *Biological Reviews*, 89: 48–67.

Jones, K.A. and Godin, J.G.J. 2010. Are fast explorers slow reactors? Linking personality type and anti-predator behavior. *Proceedings of the Royal Society B-Biological Sciences*, 277: 625–632.

Koolhaaas, J.M., De Boer, S.F., Coppens, C.M. and Buwalda, B. 2010. Neuroendocrinology of coping styles: Towards understanding the biology of individual variation. *Frontiers in Neuroendocrinology*, 31: 207–321.

**Box 1.1:   (cont.)**

Koolhaas, J.M., Korte, S.M., De Boer, S.F. *et al.* 1999. Coping styles in animals: current status in behavior and stress-physiology. *Neuroscience and Biobehavioural Reviews,* **23**: 925–935.

Krajl-Fiser, S., Weib, B.M. and Kotrschal, K. 2010. Behavioural and physiological correlates of personality in greylag geese (*Anser anser*). *Journal of Ethology,* **28**: 363–370.

Massei, G., Lyon, A.J. and Cowan, D.R. 2002. Conditioned taste aversion can reduce egg predation by rats. *Journal of Wildlife Management,* **66**: 1134–1140.

McDonald-Madden, E., Baxter, P.W.J. and Possingham, H.P. 2008. Making robust decisions for conservation with restricted money and knowledge. *Journal of Applied Ecology,* **45**: 1630–1638.

Millar, C.I., Stephenson, N.L. and Stephens, S.L. 2007. Climate change and forests of the future: managing in the face of uncertainty. *Ecological Applications,* **17**: 2145–2151.

Modlmeier, A.P., Keiser, C.N., Watters, J.V., Sih, A. and Pruitt, J.N. 2014. The keystone individual concept: an ecological and evolutionary overview. *Animal Behaviour,* **89**: 53–62.

O'Donnell, S., Webb, J.K. and Shine, R. 2010. Conditioned taste aversion enhances the survival of an endangered predator imperilled by a toxic invader. *Journal of Applied Ecology,* **47**: 558–565.

Paine, R.T. 1969. A note on trophic complexity and community stability. *American Naturalist,* **103**: 91–93.

Quinn, J.L., Cole, E.F., Patrick, S.C. and Sheldon, B.C. 2011. Scale and state dependence of the relationship between personality and dispersal in a great tit population. *Journal of Animal Ecology,* **80**: 918–928.

Radwan, J., Biedrzycka, A. and Babik, W. 2010. Does reduced MHC diversity decrease viability of vertebrate populations? *Biological Conservation,* **143**: 537–544.

Reale, D., Reader, S.M., Sol, D., McDougall, P.T. and Dingemanse, N.J. 2007. Integrating animal temperament within ecology and evolution. *Biological Reviews* **82**: 291–318.

Reale, D., Gallant, B.Y., Leblanc, M. and Festa-Bianchet, M. 2000. Consistency of temperament in bighorn ewes and correlates with behaviour and life history. *Animal Behaviour,* **60**: 589–597.

Salafsky, N., Margoluis, R., Redford, K.H. and Robinson, J.G. 2002. Improving the practice of conservation: a conceptual framework and research agenda for conservation science. *Conservation Biology,* **16**: 1469–1479.

Sih, A., Bell, A. and Johnson, J.C. 2004. Behavioural syndromes: an ecological and evolutionary overview. *Trends in Ecology & Evolution,* **193**: 72–378.

Sinn, D.L., Gosling, S.D. and Moltschaniwskyj, N.A. 2008. Development of shy/bold behaviour in squid: context-specific phenotypes associated with developmental plasticity. *Animal Behaviour,* **75**: 433–442.

**Box 1.1:** (cont.)

Starling, M.J., Branson, N., Thomson, P.C. and McGreevy, P.D. 2013. Age, sex and reproductive status affect boldness in dogs. *Veterinary Journal*, 197: 868–872.

Verdolin, J.L. and Harper, J. 2013. Are shy individuals less behaviorally variable? Insights from a captive population of mouse lemurs. *Primates*, 54: 309–314.

Verbeek, M.E.M., Boon, A. and Drent, D.J. 1996. Exploration, aggressive behavior and dominance in pair-wise confrontations of juvenile male great tits. *Behaviour*, 133: 945–963.

Wilson, D.S., O'Brien, D.T. and Sesma, A. 2009. Human prosociality from an evolutionary perspective: variation and correlations at a city-wide scale. *Evolution and Human Behavior*, 30: 190–200.

Wolf, M. and Weissing, F. J. 2012. Animal personalities: consequences for ecology and evolution. *Trends in Ecology & Evolution*, 27: 452–461.

## 1.7 BEHAVIORAL ECOLOGY: STUDYING BEHAVIOR FROM AN EVOLUTIONARY PERSPECTIVE

### 1.7.1 Origins

Humans have been fascinated by animal behavior since ancient times and great thinkers, from Aristotle to Darwin, have repeatedly attempted to explain different behaviors displayed by animals. However, it was not until the first half of the twentieth century, under the influence of Darwin's theory of evolution by natural selection (Darwin 1859), that the study of animal behavior began to evolve as a structured science. At that time, there were at least three major groups of scientists studying animal behavior (Hogan & Bolhuis 2009): the comparative psychologists, based mostly in North America, and concentrating on controlled experiments in laboratory conditions; the evolutionary biologists, originally based in Britain, but later on in both North America and Europe, focusing on how variability in behavior can be understood through the principles of natural selection; and the ethologists, based in Europe, focusing mostly on field studies of natural history and on meticulous observations and descriptions of animal behavior. However, all of these groups were regarded as inferior in comparison to other fields of science such as psychology or chemistry (Hogan & Bolhuis 2009). It was Niko Tinbergen, one of the founders of the field of ethology (along with Konard Lorenz and Karl von Frisch whom he later shared a Nobel Prize with), that laid the foundations for the scientific study of animal behavior in his seminal article on the "four whys" of the biology of behavior (Tinbergen 1963). Tinbergen first posed a single, central and admittedly vague (in his own words) question: *Why do animals behave the way they do?* He then suggested four questions or approaches that can be used to address this central question: *causation, survival value, evolution* (the three of which were previously suggested by Huxley [1914]) and *ontogeny*.

### 1.7.2 Tinbergen's four questions

**Causation:** What are the immediate effects that external and internal factors have on the occurrence of behavior? This includes not only the physiological and neurological processes that lead to the behavior itself, but also the immediate motivation of the animal to behave the way it does (i.e. hunger, fear, etc.).

**Survival value:** This has been later termed as the *function* of behavior, as it asks what the function of a behavior to the survival of the animal (or its offspring) is.

**Evolution:** How did this behavior evolve? This question looks at the phylogeny of the species in order to understand the evolutionary history of the behavior.

**Ontogeny:** How did the behavior develop within the individual? This question investigates genetic predispositions and sensitive periods in the life of a developing organism, as both have a crucial role in determining behavior.

### 1.7.3 The rise of behavioral ecology

While it is easy to imagine behavioral ecology emerging from ethology with researchers from the field starting to put a greater emphasis on the evolutionary and functional questions of behavior, this is not the case (Cuthill 2009). Rather, the theoretical foundations of behavioral ecology came from ecology, evolutionary biology, and economics.

In 1966 Robert MacArthur and Eric Pianka combined economics and population biology to develop elegant mathematical theories to understand the choices that animals make when moving between resource patches (MacArthur & Pianka 1966), giving rise to optimal foraging theory. At the same time, more or less, Darwin's idea of sexual selection was revived as the idea of kin selection was introduced (Hamilton 1964, Maynard Smith 1964), followed by numerous studies looking at the evolution of sexual behavior and of sociality. By the early 1970s, researchers had started to employ game theory to understand the occurrences and frequency of behaviors within populations (Maynard Smith & Price 1973), and in 1975 Edward O. Wilson published the book *Sociobiology: The New Synthesis*, in which he linked population biology, evolutionary biology and behavior to address the evolution of social organization. All of these advances were a product of a new approach to behavior that was becoming more and more popular, exploring patterns of variation in behavior and employing an individual-based view of selection (Birkhead & Monaghan 2010).

Just as the study of animal behavior in the 1930s focused almost entirely on causal mechanisms and ignored the other three aspects of behavior, early behavioral ecologists focused entirely on functional explanations for behavior, ignoring mechanisms and treating animals as "scheming tacticians" weighing up the costs and benefits of every conceivable course of action and always choosing the best one (Krebs & Davies 1997). This set the discipline off to a rocky start and it suffered a great deal of criticism at first (Birkhead & Monaghan 2010). However, the field kept growing, successfully integrating function and mechanism, incorporating novel and advanced approaches to the study of behavior, raising new questions and opening new lines of research. These days behavioral ecology is a thriving field, greatly increasing our understanding of behavior and how it evolves in relation to ecological conditions (Davies *et al.* 2012).

### 1.7.4 Research methods in behavioral ecology

**Mathematical theory and models:** Behavioral ecology attempts to identify how behavior adapts an organism to its environment. This question can in many cases be investigated using theoretical mathematical models that can predict, under the assumptions of the particular model, the optimal behavior or array of behaviors, and their frequency in the population or community. All of the major theories in behavioral ecology, such as kin selection, ideal free distribution and optimal foraging theory, are the product of mathematical models, and although "all models are wrong," "some are useful" enabling us to better understand and predict the way animals behave.

**Experimentation:** By manipulating some aspect of the animal's behavior or some aspect of the environment while controlling for all other factors, researchers try to test the predictions of relevant theoretical models. Experiments can take place in the lab, where it is easier to control for unrelated factors, or in the field, where the setting is more ecologically relevant, but controlling for all factors is more challenging (and sometimes impossible).

**Genetic analysis:** The evolutionary approach to behavior assumes that behaviors are the outcome of natural selection, but by focusing on the phenotype alone (i.e. only on displayed behavior), we are making assumptions about the genetic mechanism of the evolutionary process without trying to verify them. Indeed, many times the relationships between genes and behavior are found to be much more complex than was originally assumed (e.g. behaviors can be the result of interactions between several genes), which if studied, can serve to better explain behavioral phenomena. The integration of genetic aspects into behavioral ecology had only begun in the last two decades, but is advancing rapidly, with new genetic techniques constantly developing (Cezilly *et al.* 2008).

**Comparative studies:** This method is based on the comparison of traits between species in order to investigate the adaptive value of a trait. This can be done on both the phenotypic and the genotypic levels. While the comparative approach is essentially correlative, making it more difficult to interpret the causalities of the relationships that are uncovered (Cezilly *et al.* 2008), it can be a very strong tool when complementing the other approaches to behavioral ecology.

## 1.8   THE INTEGRATION BETWEEN BEHAVIORAL ECOLOGY AND CONSERVATION BIOLOGY

Literature linking animal behavior to wildlife management precedes the establishment of the society of conservation biology (Geist & Walther 1974), and both ethology and behavioral ecology have been part of the wildlife management colloquium in many academic institutes for, at least, the past 40 years. However, despite intuitive awareness of the importance of animal behavior to conservation, many conservation practitioners rarely employ any knowledge from behavioral ecology in their management plans, or employ it too late into management programs, resulting in various setbacks, which in many cases can lead to the failure of management efforts (Knight 2001, Blumstein & Fernández-Juricic 2004).

At the turn of the twentieth century, interest in the use of behavioral ecology in conservation biology was rekindled, and a flurry of publications appeared, highlighting conservation issues and tools that can benefit from behavioral knowledge, and urging scientists from both disciplines to work together (a few examples are: Clemmons & Buchholz 1997, Caro 1998, Sutherland 1998, Caro 1999, Festa-Bianchet & Apollonio 2003, Linklater 2004). It was during this period that this "new" discipline received its catchy name: "Conservation Behavior." However, criticism on the usefulness of this budding field to conservation was soon to follow, and claims were made that theoretical advances in behavioral ecology have made little practical contribution to conservation biology (Caro 2007, also see preface to this book).

What might have limited the linkage between behavioral paradigms and conservation issues is that until recently there was no conceptual model organizing the field of conservation behavior. Conservation biology is a very complex interdisciplinary field and even simply underlining the different linkages between behavioral studies and conservation can result in complicated and perhaps hard to follow outputs (Caro 1998). With no underlying framework, every researcher invents the foundation for his or her own work, and the body of research becomes a random collection of observations with little structure. The vast majority of publications calling for the integration of behavioral ecology and conservation have simply highlighted case studies, and pointed towards areas in which behavioral ecology could be of use. This has definitely led to many specific studies in which mostly descriptive behavioral knowledge played a part in solving conservation problems, but has inhibited the contribution of theoretical advances in behavioral ecology to conservation. In an attempt to lend more structure to the field, Linklater (2004) and Buchholz (2007) suggested the use of Tinbergen's four questions to direct applied conservation research. While this approach has made notable accomplishments in certain areas (Moore *et al.* 2008), it has not succeeded in increasing the general impact of conservation behavior.

In 2011, Berger-Tal *et al.* suggested a new conceptual framework for conservation behavior. A good framework should be logical, parsimonious and hierarchical; and although tending to oversimplify, it enables better focused studies, identification of key research areas, and future research directions and development. The aim of this new framework was to lend structure to the evolving field of conservation behavior in order to define the goals of conservation behavior studies, sharpen our vision for what can be done and how, and set the stage for generating hypotheses and developing subfields within the discipline. For the rest of the chapter we will give a general description of the framework. The structure of this book closely follows the structure of

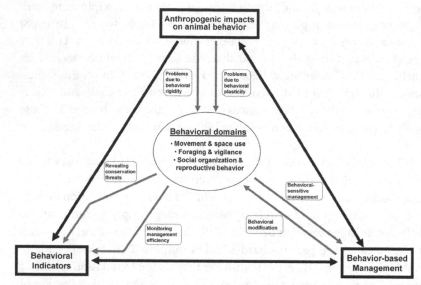

**Figure 1.1:** The conservation behavior framework is composed of three basic interrelated conservation themes: (1) Anthropogenic impacts on animal behavior; (2) behavior-based management; (3) behavioral indicators. The black arrows represent interactions between the conservation themes. Gray arrows represent the pathways that connect each theme to the behavioral domains.
Reprinted, with permission of Oxford University Press, from Berger-Tal *et al.* (2011).

the framework, with each chapter dedicated to a detailed exploration of one of the themes of the framework.

## 1.9 THE CONSERVATION BEHAVIOR FRAMEWORK

Behavior serves as a mediator between the environment (and anthropogenic disturbances) and individual fitness. As such, animals use behavior to regulate (or at least attempt to regulate) the effects of anthropogenic disturbances. There are three main ways in which wildlife managers can use the regulatory nature of behavior for conservation purposes. First, they can try to understand how the animals regulate anthropogenic disturbances behaviorally. Second, they can manipulate the behavioral regulator. And third, they can use behavior to learn about the disturbances (Figure 1.1). These three ways are the basis for the three basic themes that make up the conservation behavior framework (Figure 1.1), and by which conservation and behavior are linked. The themes are: (1) Direct and indirect *anthropogenic impacts on animal behavior* that, in turn, impact biodiversity;

(2) *behavior-based management*, representing the use and consideration of behavior in conservation practice; (3) *behavioral indicators*, which are behavioral changes that are potential pointers to processes that are of conservation concern.

The key element of behavioral ecology is the adaptive nature of behavior. Behavioral strategies in a population are the outcome of evolutionary processes that depend on the fitness of particular strategies under prevailing environmental conditions (Krebs & Davies 1997, Norris 2004). Behaviors should evolve to maximize the fitness of the individuals showing those behaviors (Owens 2006, Davies *et al.* 2012). Three key behavior domains are central to the attainment of high fitness in individuals of all species and are therefore of key concern in conservation (Berger-Tal *et al.* 2011): (1) Movement and space-use patterns. (2) Foraging and predator–prey related behaviors. (3) Social behavior and reproduction. All the different behaviors in the three domains affect survival and reproduction (hence recruitment), thus providing invaluable information on population and community dynamics. It is important to note that these behavior domains have been highlighted to make it easier for wildlife managers to recognize relevant (and sometimes crucial) behaviors. However, there could be a situation where the relevant behavior for conservation is not covered by these domains.

Within each of the three conservation behavior themes, we identified two focal pathways that define the way the theme relates to behavior.

### 1.9.1 Theme 1: Anthropogenic impacts on animal behavior

Anthropogenic impacts on animal behavior come about by direct human disturbances, such as overharvesting, fragmentation and nuisance disturbances, and by indirect disturbances, such as changes in community structure due to, for example, invasive species, the creation of ecological traps (Robertson & Hutto 2006) or Allee effects (Allee 1931). These disturbances can affect behavior-dependent animal fitness through two distinct pathways (Figure 1.1): First, as was already noted by Caughley (1994) in his classic "declining-population paradigm," an organism's population will decline when its behaviors are no longer adaptive for the changed environment that it inhabits. When humans alter the environment, the fitness values of existing behavioral strategies change. If fitness is reduced and the organism's behavioral strategies are not sufficiently "plastic" to respond to the environmental change or when an evolutionary response to the altered environment is slow relative to the rate of environmental change, the population will decline

(Norris 2004). This pathway therefore represents a conservation concern on an immediate ecological time scale, since the inability of the organism to change its behavioral strategies directly leads to a demographic decline of the organism's population. Second, and in contrast to the first, if behavior is plastic it may be altered by anthropogenic changes. Although this may be adaptive in the short term, the behavioral response may change other fitness-related behaviors, such as social structure or mating success, thus altering the evolutionary trajectory of the species or the ecosystems in question, which may drive the closure of evolutionary options (Ehrlich 2001), creating conservation concerns on a longer evolutionary time scale (Manor & Saltz 2003). Furthermore, a change in behavior of one species may alter the dynamics of an entire community or ecosystem (Wright *et al.* 2010). Chapters 4 and 5 of this volume will deal with rigid and plastic behavioral responses to anthropogenic changes, respectively.

### 1.9.2 Theme 2: Behavior-based management

Here too, we recognize two pathways incorporating animal behavior into active management for conservation. In the first, the species' behavior is considered in conservation decision-making and protocols. We term this pathway "behavior-sensitive management." Behavioral considerations may play a crucial role in a large variety of conservation-related areas, many of which will be described in detail in this volume (see Chapters 6–9). These areas include reserve design and corridor planning, wildlife epidemiology, reintroductions and translocations, population management and control of invasive species. The proximate goals of behavior-based management will usually have a strong demographic nature – whether they aim to stabilize or increase the numbers of small or declining populations or to control populations of invasive or pest species. The second pathway addresses cases where the animal's behavior itself is the cause for conservation concern, the proximate goal of the management efforts may be to change the behavior of the target population. This approach is commonly applied in training captive-bred individuals designated for reintroduction to become predator-savvy or better adapted to the conditions that await them in the wild, but can also be applied to train animals in the wild (e.g. teaching animals to use overpasses). Chapter 10 of this volume will be dedicated to such behavioral modifications.

### 1.9.3 Theme 3: Behavioral indicators

The various adaptive behaviors of organisms give us a great deal of information about the evolutionary forces shaping these behaviors, the environments that the organisms inhabit, and any recent changes to either the selection forces or the environment. Thus, we can use the behavior itself as an indicator of the organism's state as well as the state of its environment (Kotler et al. 2007). Such indicators can include a large array of behaviors from all three behavior domains. The two pathways in which behavioral indicators can be used in conservation are: (1) Behavioral indicators that reveal anthropogenic threats and provide an early warning to population decline or habitat degradation before numerical responses are evident. (2) Behavioral indicators used to monitor the effectiveness of management programs, or evaluate the success of a management program at its early stages, before population or ecosystem-level responses are evident, as a part of an adaptive management protocol (Holling 1978).

Indicators from both pathways can be further divided into two cases: The behaviors of animals can point to their own state and the state of their populations. We term these "direct behavioral indicators." The behaviors of animals can also serve as indicators to the state of other species in the community or to the health of the ecosystem. We term these "indirect behavioral indicators." Direct and indirect indicators are the topics of Chapters 11 and 12, respectively.

### 1.9.4 Linking between the themes

The three behavioral conservation themes are strongly linked. For example, anthropogenic impact on animal behavior may be detected using behavioral indicators and can suggest the need for behavior-sensitive management (Ikuta & Blumstein 2003, Zidon et al. 2009). Alternatively, behavior-sensitive active management can be evaluated using behavioral indicators, and this knowledge may modify the management plan accordingly. In many situations, the behavioral aspect of one theme may dictate a non-behavioral component of another theme (e.g. behavioral indicators may often be indicative of disturbances impacting non-behavioral population dynamics). One can view the three themes as entry points for behavioral ecologists aiming to use their knowledge and expertise of animal behavior in conservation. In each of the themes, new research should be based on previous knowledge of animal behavior as well as on pressing conservation concerns. In this way, conservation behavior can serve as a much-needed link between the ever-expanding knowledge in behavioral ecology and the more practical needs of conservation biologists. Although behavioral ecologists may, in many cases, address only one of the proposed three themes in any given research, the role of the conservation biologist is to consider and integrate all three themes into one adaptive management scheme.

## 1.10 SUMMARY

Our world is facing an environmental crisis of staggering proportions. The science of conservation biology stands before this crisis and aims to mitigate it by preserving evolutionary processes within functioning ecological settings. In order to achieve this goal, conservation biologists seek to conserve biodiversity at all its levels: genetic diversity, species diversity and ecosystem diversity. Behavioral diversity allows populations to respond rapidly and effectively to anthropogenic-induced environmental changes, and as such it plays a crucial role in maintaining biodiversity at all its levels. However, despite the fact that conservation biologists have been long aware of the important role of behavior in conservation, and despite the fact that the field of behavioral ecology has been developing rapidly and successfully in the past decades, the integration of behavioral ecology into conservation biology has been slow and relatively ineffective so far. What might have limited the linkage of behavioral paradigms and conservation is that until recently there was no conceptual model organizing the field of conservation behavior. This book is structured according to a new conservation behavior framework aimed at highlighting the linkages between behavior and conservation and allowing for better integration of behavioral ecology principles into practical conservation work. The framework is composed of three basic themes: anthropogenic impacts on animal behavior, behavioral-based management and behavioral indicators. The following chapters explore each of these themes in detail.

## REFERENCES

Allee, W.C. 1931. *Animal Aggregations, a Study in General Sociology.* Chicago: University of Chicago Press.

Angeloni, L., Schlaepfer, M.A., Lawler, J.J. and Crooks, K.R. 2008. A reassessment of the interface between conservation and behavior. *Animal Behaviour,* 75: 731–737.

Barnosky, A.D., Matzke, N., Tomiya, S. *et al.* 2011. Has the Earth's sixth mass extinction already arrived? *Nature,* 471: 51–57.

Bennett. E.L, Eves, H.E., Robinson, J.G. and Wilkie, D.S. 2002. Why is eating bushmeat a biodiversity crisis. *Conservation Biology in Practice,* 3: 28–29.

Berger-Tal, O., Polak, T., Oron, A. *et al.* 2011. Integrating animal behavior and conservation biology: a conceptual framework. *Behavioral Ecology,* 22: 236–239.

Birkhead, T.R. and Monaghan, P. 2010. Ingenious ideas: the history of behavioral ecology. In Westneat, D.F. and Fox, C.W. (eds.), *Evolutionary Behavioral Ecology*, pp 3–15. Oxford: Oxford University Press.

Blumstein, D.T. and Fernández-Juricic, E. 2004. The emergence of conservation behavior. *Conservation Biology*, 18: 1175–1177.

Buchholz, R. 2007. Behavioral biology: an effective and relevant conservation tool. *Trends in Ecology and Evolution*, 22: 401–407.

Caughley, G. 1994. Directions in conservation biology. *Journal of Animal Ecology*, 63: 215–244.

Callicott, J.B. 1990. Whither conservation ethics? *Conservation Biology*, 4: 15–20.

Callicott, J.B. 1999. *Beyond the Land Ethics: More Essays in Environmental Philosophy*. New York: State University of New York Press.

Callicott, J.B., Grove-Fanning, W., Rowland, J. *et al.* 2011. Reply to Norton: re: Aldo Leopold and pragmatism. *Environmental Values*, 20: 17–22.

Caro, T. 1998. The significance of behavioral ecology for conservation biology. In Caro, T. (ed.), *Behavioral Ecology and Conservation Biology*, pp 3–26. Oxford: Oxford University Press.

Caro, T. 1999. The behavior–conservation interface. *Trends in Ecology and Evolution*, 14: 366–369.

Caro, T. 2007. Behavior and conservation: a bridge too far? *Trends in Ecology andEvolution*, 22: 394–400.

Caro, T. and Sherman, P.W. 2012. Vanishing behaviors. *Conservation Letters*, 5: 159–166.

Caro, T. and Sherman, P.W. 2013. Eighteen reasons animal behaviourists avoid involvement in conservation. *Animal Behaviour*, 85: 305–312.

Cezilly, F., Danchin, E. and Giraldeau, L-A. 2008. Research methods in behavioural ecology. In Danchin, E., Giraldeau, L-A. and Cezilly, F. (eds.), *Behavioural Ecology*, pp. 55–96. Oxford: Oxford University Press.

Charmantier, A., McCleery, R.H., Cole, L.R. *et al.* 2008. Adaptive phenotypic plasticity in response to climate change in a wild bird population. *Science*, 320: 800–803.

Chetkiewicz, C.L.B., St. Claire, C.C. and Boyce, M.S. 2006. Corridors for conservation: integrating pattern and process. *Annual Review of Ecology, Evolution, and Systematics*, 37: 317–342.

Clemmons, J.R. and Buchholz, R. 1997. *Behavioral Approaches to Conservation in the Wild*. Cambridge: Cambridge University Press.

Costanza, R., d'Arge, R., de Groot, R. *et al.* 1997. The value of the world's ecosystem services and natural capital. *Nature*, 387: 253–260.

Cuthill, I. 2009. The study of function in behavioural ecology. In Bolhuis, J.J. and Verhulst, S. (eds.), *Tinbergen's Legacy: Function and Mechanism in Behavioral Biology*, pp. 107–126. Cambridge: Cambridge University Press.

Darwin, C. 1859. *On the Origins of Species by Means of Natural Selection*. London: John Murray.

Davies, N.B., Krebs, J.R. and West, S.A. 2012. *An Introduction to Behavioural Ecology*. Oxford: Wiley-Blackwell.

Ehrlich, P.R. 2001. Intervening in evolution: ethics and actions. *Proceedings of the National Academy of Sciences, USA*, 98: 5477–5480.

Ehrlich, P.R and Ehrlich, A.H. 1990. *The Population Explosion*. New York: Simon & Schuster.

Estes, J.A., Duggins, D.O. and Rathbun, G.B. 1989. The ecology of extinctions in kelp forest communities. *Conservation Biology*, 3: 252–264.

Fahrig, L. 2003. Effects of habitat fragmentation on biodiversity. *Annual Review of Ecology, Evolution, and Systematics*, 34: 487–515.

Faith, D.P. 2008. Biodiversity. In Zalta, E.N. (ed.), *The Stanford Encyclopedia of Philosophy*. URL: http://plato.stanford.edu/archives/fall2008/entries/biodiversity/.

Festa-Bianchet, M. and Apollonio, M. (eds.) 2003. *Animal Behavior and Wildlife Conservation*. Washington, DC: Island Press.

Gaston, K.J. 1998. Biodiversity. In Sutherland, W.J. (ed.), *Conservation Science and Action*, pp. 1–19. Oxford: Blackwell Science Ltd.

Geist, V. and Walther, F. 1974. *The Behavior of Ungulates and its Relation to Management*. IUCN, Morges, Switzerland.

Groom, M.J., Meffe, G.K. and Carroll, C.R. 2006. *Principles of Conservation Biology*. 4th edn. Sunderland: Sinauer Associates.

Groom, M.J. and Vynne, C.H. 2006. Habitat degradation and loss. In Groom, M.J., Meffe, G.K. and Carroll, C.R. (eds.), *Principles of conservation biology*. 4th edn., pp. 173–212. Sunderland: Sinauer Associates.

Hamilton, W.D. 1964. The genetical evolution of social behaviour. I. *Journal of Theoretical Biology*, 7: 1–16.

Hanski, I. and Simberloff, D. 1997. The metapopulation approach, its history, conceptual domain, and application to conservation. In Hanski, I. and Gilpin, M.E. (eds.), *Metapopulation Biology: Ecology, Genetics, and Evolution*, pp. 5–26. San Diego: San Diego Academic Press.

Hogan, J.A. and Bolhuis, J.J. 2009. Tinbergen's four questions and contemporary behavioral ecology. In Bolhuis, J.J. and Verhulst, S. (eds.), *Tinbergen's Legacy: Function and Mechanism in Behavioral Biology*, pp. 25–34. Cambridge: Cambridge University Press.

Holling, C.S. (ed.) 1978. *Adaptive Environmental Assessment and Management*. Chichester: John Wiley & Sons.

Hunter, M.L. and Gibbs, J. 2007. *Fundamentals of Conservation Biology*. 3rd edn. Oxford: Wiley-Blackwell.

Hutchinson, G.E. 1965. *The Ecological Theater and the Evolutionary Play*. New Haven: Yale University Press.

Huxley, J.S. 1914. The courtship-habits of the great crested Grebe (*Podiceps cristatus*); with an addition to the theory of sexual selection. *Proceedings of the Zoological Society of London*, 84: 491–562.

Ikuta, L.A. and Blumstein, D.T. 2003. Do fences protect birds from human disturbance? *Biological Conservation*, 112: 447–452.

IPCC. 2001. *Climate Change 2001: the Scientific Basis*. Intergovernmental Panel on Climate Change third assessment report. Cambridge: Cambridge University Press.

Jablonski, D. 1994. Extinctions in the fossil records. *Philosophical Transactions: Biological Sciences*, 344: 11–17.

Jablonski, D. 1995. Extinctions in the fossil record. In May, R.M. and Lawton, J.H. (eds.), *Extinction Rates*. Oxford: Oxford University Press.

Johnson, S.P. 1993. *The Earth Summit: The United Nations Conference on Environment and Development (UNCED)*. London: Graham and Trotman.

Karl, T.R. and Trenberth, K.E. 2003. Modern global climate change. *Science*, **302**: 1719–1723.

Knight, J. 2001. If they could talk to the animals. *Nature*, **414**: 246–247.

Kotler, B.P., Morris, D.W. and Brown, J.S. 2007. Behavioral indicators and conservation: wielding "the biologist's tricorder". *Israel Journal of Ecology and Evolution*, **53**: 237–244.

Krebs, J.R. and Davies, N.B. 1997. *Behavioural Ecology: An Evolutionary Approach*. 4th edn. Oxford: Blackwell Publishing.

Leopold, A. 1949. *A Sand County Almanac: And Sketches Here and There*. New York: Oxford University Press.

Levins, R. 1969. Some demographic and genetic consequences of environmental heterogeneity for biological control. *Bulletin of the Entomological Sciences of America*, **15**: 237–240.

Linklater, W.L. 2004. Wanted for conservation research: behavioral ecologists with a broader perspective. *Bioscience*, **54**: 352–360.

Luck, G.W., Lavorel, S., McIntyre, S. and Lumb, K. 2012. Improving the application of vertebrate trait-based frameworks to the study of ecosystem services. *Journal of Animal Ecology*, **81**: 1065–1076.

MacArthur, R.H. and Pianka, E.R. 1966. On optimal use of a patchy environment. *American Naturalist*, **100**: 603–609.

Manor, R. and Saltz, D. 2003. Impact of human nuisance disturbance on vigilance and group size of a social ungulate. *Ecological Applications*, **13**: 1830–1834.

Maynard Smith, J. 1964. Group selection and kin selection. *Nature*, **201**: 1145–1147.

Maynard Smith, J. and Price, G.R. 1973. The logic of animal conflict. *Nature*, **246**: 15–18.

Meine, C. 2010. Conservation biology: past and present. In: Sodhi, N.S. and Ehrlich, P.R. (eds.), *Conservation Biology for All*, pp. 7–26. Oxford: Oxford University Press.

Minteer, B.A. 2012. *Refounding Environmental Ethics: Pragmatism, Principle, and Practice*. Philadelphia: Temple University Press.

Moore, J.A., Bell, B.D. and Linklater, W.L. 2008. The debate on behavior in conservation: New Zealand integrates theory with practice. *Bioscience*, **58**: 454–459.

Muir, J. 1901. *Our National Parks*. Boston: Houghton Mifflin.

Myers, N. 1996. The biodiversity crisis and the future of evolution. *The Environmentalist*, **16**: 37–47.

Norris, K. 2004. Managing threatened species: the ecological toolbox, evolutionary theory and declining-population paradigm. *Journal of Applied Ecology*, **41**: 413–426.

Norton, B.G. 2003. *Searching for Sustainability: Interdisciplinary Essays in the Philosophy of Conservation Biology*. Cambridge: Cambridge University Press.

Noss, R.F. 1990. Indicators for monitoring biodiversity: a hierarchical approach. *Conservation Biology*, **4**: 355–364.

Owens, I.P.F. 2006. Where is behavioural ecology going? *Trends in Ecology and Evolution*, **21**: 356–361.

Parmesan, C. and Matthews, J. 2006. Biological impacts of climate change. In Groom, M.J., Meffe, G.K. and Carroll (eds.), *Principles of Conservation Biology*. 4th edn., pp. 333–374. Sunderland: Sinauer Associates.

Pickett, S.T.A., Parker, V.T. and Fieldler, P.L. 1992. The new paradigm in ecology: implications for conservation biology above the species level. In Fieldler, P.L. and Jain, S.K. (eds.), *Conservation Biology: The Theory and Practice of Nature Conservation Preservation and Management*, pp. 65–88. New York: Chapman and Hall.

Pinchot, G. 1947. *Breaking New Ground*. New York: Harcourt, Brace, and Co.

Primack, R.B. 2006. *Essentials of Conservation Biology*. 3rd edn. Sunderland: Sinauer Associates.

Rasoanaivo, P. 1990. Rain-forests of Madagascar – sources of industrial and medicinal-plants. *Ambio*, 19: 421–424.

Reynolds, J.D. and Peres, C.A. 2006. Overexploitation. In Groom, M.J., Meffe, G.K. and Carroll, C.R. (eds.), *Principles of Conservation Biology*. 4th edn., pp. 253–292. Sunderland: Sinauer Associates.

Robertson, B.A. and Hutto, R.L. 2006. A framework for understanding ecological traps and an evaluation of existing evidence. *Ecology*, 87: 1075–1085.

Rolston, H. 1988. Human-values and natural systems. *Society & Natural Resources*, 1: 271–283.

Sanderson, E.W., Jaiteh, M., Levy, M.A. *et al.* 2002. The human footprint and the last of the wild. *Bioscience*, 52: 891–904.

Schmitz, O.J., Grabowski, J.H. and Peckarsky, B.L. 2008. From individuals to ecosystem function: toward an integration of evolutionary and ecosystem ecology. *Ecology*, 89: 2436–2445.

Soule, M.E. 1986. Conservation biology and the "real world." In Soule M.E. (ed.), *Conservation Biology: The Science of Scarcity and Diversity*, pp 1–12. Sunderland: Sinauer Associates.

Sutherland, W.J. 1998. The importance of behavioral studies in conservation biology. *Animal Behaviour*, 56: 801–809.

Terborgh, J., Estes, J.A., Paquet, P. *et al.* 1999. The role of top carnivores in regulating terrestrial ecosystems. In Soule, M.E. and Terborgh, J. (eds.), *Continental Conservation*, pp. 39–64. Washington, DC: Island Press.

Tinbergen, N. 1963. On aims and methods in ethology. *Zeitschrift fur tierpsychologie*, 20: 410–433.

Thoreau, H.D. 1863. *Excursions*. Boston: Ticknor and Fields.

United States Census Bureau. 2013. www.census.gov/popclock/.

Ward, D. 2009. *The Biology of Deserts*. Oxford: Oxford University Press.

Whittaker, R.H. 1960. Vegetation of the Siskiyou mountains, Oregon and California. *Ecological Monographs*, 30: 279–338.

Wilcove, D.S., McLellan, C.H. and Dobson, A.P. 1986. Habitat fragmentation in the temperate zone. In Soule M.E. (ed.), *Conservation Biology: The Science of Scarcity and Diversity*, pp.237–256. Sunderland: Sinauer Associates.

Wiles, G.J., Bart, J., Beck, R.E. and Aguon, C.F. 2003. Impacts of the brown tree snake: patterns of decline and species persistence in Guam's avifauna. *Conservation Biology*, 17: 1350–1360.

Wilson, E.O. 1975. *Sociobiology. The New Synthesis*. Cambridge: Belknap Press.

Wonham, M. 2006. Species invasions. In Groom, M.J., Meffe, G.K. and Carroll (eds.), *Principles of Conservation Biology*. 4th edn., pp. 293–332. Sunderland: Sinauer Associates.

Woodward, G., Gessner, M.O., Giller, P.S. *et al.* 2012. Continental-scale effects of nutrient pollution on stream ecosystem functioning. *Science*, **336**: 1438–1440.

Wootton, J.T. 1994. The nature and consequences of indirect effects in ecological communities. *Annual Review of Ecology, Evolution, and Systematics*, **25**: 443–466.

Wright, J.T., Byers, J.E., Koukoumaftsis, L.P., Ralph, P.J. and Gribben, P.E. 2010. Native species behavior mitigates the impact of habitat-forming invasive seaweed. *Oecologia*, **163**: 527–534.

Zidon, R., Saltz, D., Shore, L.S. and Motro, U. 2009. Behavioral changes, stress, and survival following reintroduction of Persian fallow deer from two breeding facilities. *Conservation Biology*, **23**: 1026–1035.

# Evolution and conservation behavior

## JOHN P. SWADDLE

All behavior occurs within the theater of evolution and hence it would be remiss to ignore evolutionary forces and outcomes when considering how behavior and conservation interact with each other. This is particularly pertinent as conservation outcomes can play out over a long time period, increasing the probability that conservation of biota will be affected by evolutionary forces and the evolutionary responses of organisms and communities to such forces.

To interweave evolutionary thinking into an approach to conservation behavior I will first, briefly, review how evolution works and offer a primer on micro-evolutionary forces. From there I will build evidence for how human alteration of the environment influences evolutionary forces and biological outcomes. Importantly, I will also layer-on a view of how macro-evolutionary patterns relate to conservation outcomes over long time periods, framed in the context of modern threats to biodiversity.

This evolutionary foundation lets me argue that behavior can lead evolution and not just be the outcome of evolutionary forces. Hence, behavioral mechanisms can alter evolutionary trajectories over a surprisingly short time period and affect conservation outcomes; a view not commonly held even among evolutionary biologists. This view will be integrated back to this book's initial framework of how behavior and conservation intersect with each other, revising the Berger-Tal *et al.* (2011) model to explicitly include micro- and macro-evolutionary forces, patterns and outcomes.

## 2.1   A BRIEF PRIMER ON MICRO-EVOLUTIONARY FORCES

Evolution is classically defined as the change in allele frequencies in a population over time (Maynard Smith & Szathmáry 1995, Stearns &

*Conservation Behavior: Applying Behavioral Ecology to Wildlife Conservation and Management*, eds. O. Berger-Tal and D. Saltz. Published by Cambridge University Press. © Cambridge University Press 2016.

Hoekstra 2005, Swaddle 2010). What this definition does not readily admit to is that this is perhaps an overly gene-centric definition of micro-evolution, yet our concept of evolution is much larger than tracking gene variants over time. Evolution is really the change in heritable variation in populations (defined broadly, as I will explain later when tackling macro-evolutionary topics) over time. By heritable variation I mean differences in a population that can be passed on from one generation to another, which can occur by many mechanisms including traditional genetic inheritance, epigenetic processes, parental effects, and some would argue by cultural inheritance also. Box 2.1 provides an overview of the main principles of evolution by natural selection.

Without heritable variation evolutionary forces have nothing to act upon. With heritable difference among individual entities in a population, there are four major forces that can change the inheritance of this variation in the next generation: mutation, selection, drift and flow. Let us briefly consider each in turn.

Mutation is the direct alteration of the material that is inherited from one generation to another. This is most simply illustrated by genetic mutation where single nucleotides can be altered (Ingram 1957), chromosomes can be restructured (Finnegan 1989) or whole chromosomes can be deleted or duplicated (Zhang 2003). If these mutations occur in the germ line (e.g. to the chromosomes of sperm and egg cells) then they would immediately affect allelic frequencies in the next generation and result in an evolutionary change.

Selection occurs when one variant of a heritable phenotype confers a fitness advantage over other variants. Hence, in subsequent generations the higher fitness phenotype, with everything else held constant, will increase in relative frequency. This is the process of adaptation that most behaviorists and conservation scientists are very familiar with and is often the primary evolutionary force invoked to explain change in populations over time in response to environmental alterations (Arnold & Wade 1984, Bell 2013a, Fisher 1958).

Drift is a collective term for random processes that affect the genotypes and phenotypes of the next generation (Falconer & Mackay 1996, Lande 1976). Such processes can be genetic, such as the somewhat random division and recombination of parental chromosomes to form offspring (at least in sexually reproducing diploid organisms). It is not predictable which alleles will make it from the mother and the father into the viable zygote; there is randomness afoot. Hence, siblings are not identical even though they have the same parents. In the meiotic shuffle of genetic drift

**Box 2.1**

Charles Darwin and Alfred Russell Wallace authored the first formal papers about natural selection, which were read back-to-back to the Linnean Society of London on July 1, 1858. Because both scientists arrived at very similar conclusions independently of each other the Linnean Society felt it was appropriate to give both Darwin and Wallace equal billing. Neither Wallace nor Darwin were in attendance when the papers were read aloud.

Although Wallace generated very similar ideas it is Darwin who is most often credited with the formal origination of evolutionary theory, probably because some of his unpublished theories and mountains of evidence appear to pre-date Wallace's, but also because of the encyclopedic nature and impact of his famous book, *On the Origin of Species by Means of Natural Selection, or the Preservation of Favoured Races in the Struggle for Life* (Darwin, 1859). It is this latter subtitle of the *Origin of Species* that captures the essence of early evolutionary thinking. Darwinian selection is centered on four simple postulates. (1) Individuals within species are different to each other (i.e. they vary in phenotype). (2) These differences among individuals are at least partially passed on to their offspring (i.e. phenotypes are heritable and related to the genotype). (3) There is a "struggle for life" as more offspring are produced each generation than can possibly survive to reproduce. (4) Survival and reproduction are related to the variation among individuals so that some variants survive better and/or more frequently reproduce than others and are therefore more frequently expressed in the next generation. Hence, the principle of natural selection is elegantly simple: expression of a heritable trait is related to differential reproduction and survival. Individuals in a population who express the favored form of the trait will leave more descendants and those descendants will express the same (or similar) favored trait form, resulting in a change in trait frequency in the population over time. As the favored trait has a genetically heritable basis the population has evolved by natural selection because gene frequencies in the population have changed over time.

The four postulates of natural selection have been incorporated into the modern syntheses of evolution; the Darwin–Wallace framework for natural selection has been accompanied by the other recognized mechanisms of evolution: mutation, drift and gene flow (see main text). While neither Darwin nor Wallace knew anything of genes and how the favored variants were expressed, they accurately described the notion that some phenotypes helped individuals survive and reproduce and this would lead to a relative increase in those phenotypes (and their associated genotypes) in future generations. The principle of natural selection is routinely applied in many areas of science and society. For example, treatments of infectious diseases are formulated by understanding how the disease organisms are evolving in relation to the medical intervention (Gluckman *et al.*

Box 2.1 (cont.)

2009). In economics, the principles of game theory and economic competition are founded on natural selection mechanisms (Von Neumann & Morgenstern 2007). Evolution, initiated by the Darwin–Wallace formulation of natural selection, has become the central principle of the life sciences and underpins our understanding of biological function and diversity (Coyne 2009).

some alleles are lost by chance. Although drift is often explained in terms of genetic processes, it is relevant to conservation scientists to remember that drift also refers to organism- and population-level stochastic processes too. For example, if a fire sweeps through a region some organisms will die or be displaced simply because of their situational bad luck; others will survive.

In some ways flow is the simplest of all evolutionary mechanisms to understand and is highly relevant to conservation science. Flow refers to the emigration and immigration of heritable components from populations, often described in terms of gene flow (Falconer & Mackay 1996). As individuals move from one area to another and interbreed with the host population new allelic combinations arise because of the arrival of new individuals and their successful contributions to the resident gene pool. In general, gene flow is thought to homogenize genetic variation among populations, although there are examples where systematically biased gene flow can either continue to introduce new genetic variation to populations or create genetic differences among subpopulations (Postma & van Noordwijk 2005, Slatkin 1985).

As a final remark on micro-evolutionary processes it is important to stress that these forces are not mutually exclusive of each other; demonstration of one force does not mean that another is not occurring. In a rapidly changing environment, often the domain of conservation concerns, it is likely that all four mechanisms are working simultaneously.

## 2.2 HUMAN ALTERATION OF THE LANDSCAPE INFLUENCES EVOLUTIONARY FORCES AND CONSERVATION

In discussing evidence for whether and how anthropogenic alteration of habitats is affecting evolutionary processes and conservation status I have categorized studies according to how humans affect the micro-evolutionary mechanisms that were reviewed in the previous section.

### 2.2.1 Mutation

Most mutations are deleterious and we have known for a long time that various anthropogenically produced compounds and radioactive pollution have mutagenic effects through direct biochemical influence on organisms' cells (Møller 1998, Parsons 1990, Sagan 1989, Sankaranarayanan 1982). More recently, it has been suggested that any stress-inducing environmental pressure will increase the probability of mutation (Galhardo et al. 2007). This includes the behavioral environment of organisms that can induce stress, for example, through perceived predation risk or "fear" (Blumstein 2006), and this can trigger physiological cascades that could heighten the risk of mutation. Hence, human alteration of the landscape could fundamentally alter genetic and phenotypic variation (including behavior), and most of that induced variation is likely to be maladaptive (Sakai & Suzuki 1964).

Although human-introduced mutagens introduce new genetic variation, the selection against the majority of this variation should be strong. This means that human-affected rates of mutation are very unlikely to be evolutionarily beneficial. Nevertheless the mutations induced by anthropogenic pollution and environmental change can alter evolutionary trajectories as the mutagens introduce new genetic and phenotypic variation, and variation is the fuel for evolutionary processes. If populations cannot respond positively to the changing environment in contemporary ecological time through learning (see Chapter 3) and plasticity (see Chapter 5 and section 4 of this chapter) and/or over evolutionary time through adaptation, then populations are more likely to be at risk.

### 2.2.2 Selection

Humans have substantially altered the environment but we have yet to develop a robust theoretical understanding of how rapid environmental change will alter evolutionary trajectories of populations and species and, hence, their conservation status (Angilletta & Sears 2011, Cote & Reynolds 2012, Fraser et al. 2006, Hendry et al. 2011, Martinez-Abrain & Oro 2010, Norris 2004). It is commonplace to observe evolution as a slow process where behavioral traits are in part influenced by conserved gene clusters (Fitzpatrick et al. 2005) and retained even after selection pressures for those behavioral traits are removed (Blumstein 2002). However, we are gaining an increasing appreciation for how recent and dramatic anthropogenic alteration of the environment (sometimes termed Human Induced Rapid Environmental Change, HIREC) is affecting the fitness of populations of many taxa (French *et al.* 2011, Hendry *et al.* 2011, Kight & Swaddle 2007, Mergeay & Santamaria 2012, Palkovacs *et al.* 2012, Smith *et al.* 2008).

My intent is not to provide anything close to a comprehensive review of how HIREC is resulting in adaptive change in populations; here I offer some case studies synthesized from recent reviews (Coppack & Partecke 2006, Cote & Reynolds 2012, Palkovacs *et al.* 2012, Vander Wal *et al.* 2013). For example, it appears that disturbance regimes and environmental alterations are selecting for faster growth rates in California sea lions (*Zalophus californianus*) (French *et al.* 2011), smaller body sizes and changes in anti-predatory behaviors of Japanese mamushi snakes (*Gloydius blomhoffii*) (Sasaki *et al.* 2009), altered courtship strategies in male three-spined stickleback (*Gasterosteus aculeatus*) (Tuomainen *et al.* 2011), smaller herd size in mountain gazelle (*Gazella gazella*) (Manor & Saltz 2003) and could be selecting for increased physiological stress-coping mechanisms in some bird species (Jimenez *et al.* 2011). In laboratory experimental situations, designed to somewhat mimic more natural environmental change, it is abundantly clear that strong selective pressures can lead to rapid adaptation in a range of micro-organisms (Bataillon *et al.* 2013, Bell 2013a, Bell 2013b, Gomulkiewicz & Shaw 2013, Gonzalez & Bell 2013, Gonzalez *et al.* 2013, Kirkpatrick & Peischl 2013, Schiffers *et al.* 2013). However, it is also important to remember that intense selection is predicted to erode genetic variation and, hence, make populations susceptible to future environmental change and increase the risk of extinction in unpredictably changing environments (Miller *et al.* 2009, Reding *et al.* 2013), hence rapid adaptation to environmental pressures might not all be good news.

Although tempting to think that the selection pressures exerted by anthropogenic pressures are linear, it is possible that populations (and their traits) respond in non-linear ways to human-induced selection pressures. In other words, human-induced selection pressures could select for intermediate phenotypes (Kight & Swaddle 2007, Sagan 1989). Human disturbance can also bring about indirect selection pressures, for example by promoting invasive species that themselves exert selection pressures on the resident biota (Schlaepfer *et al.* 2005) or by altering the perceptual (e.g. acoustic) landscape and thereby altering whole communities and their suite of direct and indirect ecological and evolutionary relationships (Francis *et al.* 2009, Van Dyck 2012).

### 2.2.3 Drift

When humans disturb the environment we often create two processes that will likely affect genetic drift mechanisms. First, by altering habitats we tend to fragment populations, creating smaller effective population sizes in a series of sub-populations. When effective population size is reduced the effects of randomness (i.e. drift mechanisms) can be greatly enhanced (Falconer & Mackay 1996). Hence, habitat fragmentation will likely result in an increase in the magnitude of drift mechanisms and the further loss of genetic diversity.

Second, destruction of habitat and direct application (intentional or not) of extremely harmful chemicals can often result in large-scale mortality of individuals where there is no genetic variation for resisting the human disturbance. Therefore it is somewhat random who will survive such events. Consider deforestation to create human habitation. In such a case most of the biota in the forest will be destroyed, but perhaps a small pocket of the forest will survive. As populations will rarely have any genetic variance to assist with surviving such a catastrophic event there is little to no selection occurring on the population and which genotypes make it to the next generation is more associated with luck than adaptation. Hence, large-scale but incomplete destruction of suitable habitat can be a drift event that imposes an intense evolutionary pressure on a population.

One of the long-term predictions of drift is the loss of genetic variation. Hence, small populations are particularly vulnerable to the rapid loss of genetic variation (Cote & Reynolds 2012, Keogh 2009). One of the most striking conservation examples of drift resulting from a severe genetic bottleneck is that of the African cheetah (*Acinonyx jubatus jubatus*), which is almost genetically uniform at some loci because of the bottleneck the species experienced approximately 10,000 years ago (Menotti-Raymond & O'Brien 1993). Small founding populations also result in erosion of genetic variance due to drift. For example, the translocation of small founding populations of North Island saddlebacks (*Philesturnus rufusater*) to islands in New Zealand resulted in intense genetic drift and the dramatic loss of song diversity, which has implications for social and mating interactions in this species (Parker et al. 2012). Habitat fragmentation and the subsequent isolation of smaller populations may increase drift and not only increase extinction risks through ecological and population dynamic processes but also increase the probability of extinction in the longer term through the loss of genetic diversity (Cote & Reynolds 2012).

## 2.2.4 Gene flow

As gene flow is the movement of breeding individuals from one population to the other, substantial gene flow can help combat some of the negative effects of habitat fragmentation. It is possible that gene flow can rescue the loss of genetic diversity that often results from the isolation and reduced population sizes due to habitat fragmentation. Increased flow can introduce new genetic variance and thereby increase the evolutionary potential of populations. For example, fragmented populations of the lemon shark (*Negaprion brevirostris*) in the Bahamas were observed to have increased additive genetic variance, presumably because of an influx of breeding individuals from neighboring locations (DiBattista *et al.* 2011). In such cases, behaviorally mediated gene flow can help rescue populations in the short term but also increase the longer-term viability of populations because of increased genetic diversity, which allows for stronger evolutionary responses to subsequent environmental change (Kokko & Sutherland 2001). As gene flow most often results from the movement of individuals, the behavior of individuals is a prime determinant of this evolutionary mechanism.

Although gene flow can have positive ecological and evolutionary outcomes for fragmented populations, too much flow can have negative consequences. For example, if there is substantial movement of breeding individuals among all (sub)populations, then all of the populations will become genetically homogenized resulting in a relative decrease in genetic variance among populations and, therefore, a reduction in the evolutionary potential across populations. In such cases, gene flow may swamp any adaptation and slow down the rate at which populations could respond positively to environmental change. Conversely, too little gene flow, often due to physical barriers created by landscape alteration, can result in isolated populations that diverge from each other and suffer from a loss of genetic diversity, also reducing evolutionary potential (Hartmann *et al.* 2013). For these and other reasons authors have suggested that an intermediate degree of gene flow is optimal for many populations (Garant *et al.* 2007).

## 2.3 MACRO-EVOLUTIONARY PATTERNS RELATED TO CONSERVATION

Although extinction is the ultimate fate of all species, conservation scientists are trying to stem the tide and slow the rate of modern extinction so that we do not experience another global mass extinction event. In terms of evolutionary frameworks, extinction is the macro-evolutionary analogy of mortality and, hence, I argue that an appreciation for macro-evolutionary patterns and forces should be much better integrated into conservation frameworks that intend to take a longer view of species status. Importantly, there are species-level traits that influence the probability of extinction and some of these traits are at least in part the product of behavior (Harnik *et al.* 2012, Jablonski 1987, Reding *et al.* 2013). Hence, there is a need to understand what kinds of macro-evolutionary forces and traits are relevant to conservation science. Many behaviors will likely influence a species' susceptibility to anthropogenic activity, but to illustrate the possible relationships I have chosen to focus on one particular domain of behavior: reproductive behavior. Here, I give a brief overview of some reproductive traits that may be associated with indicators of species persistence and indicate why we might expect these traits to predict anthropogenic-related extinction risk.

Evolutionary theory often assumes that sexual selection pressures (i.e. selection for differential mating success) will cause traits to vary from their ecological optimum and thus adversely affect viability while maximizing mating success for each sex (Andersson 1994, Andersson & Simmons 2006, Shuster & Wade 2003). In this way, sexual selection and natural selection can represent antagonistic forces. Species under intense sexual selection pressures are thus thought to be "non-optimal" from an ecological standpoint and therefore more vulnerable to predation, parasitism, sexual competition and other factors that may ultimately result in population decline or extinction (Andersson 1994).

Several theoretical models have supported this fundamental prediction. For example, Tanaka (1996) showed that sexual selection could increase both natural and sexual selection loads (i.e. the fitness cost of losing genetic variance due to selection) and thus enhance population decline in populations undergoing an environmental change. Theoretical models have also predicted population declines in response to sexual selection when the costs of a trait are not paid by the same individuals who experience the benefits (Kokko & Brooks 2003). Other models have demonstrated the possibility of extinction in cases where male viability decreases with increasing female choice (Houle & Kondrashov 2002). What all these models have in common is that increased sexual selection pressures are predicted to be associated with increased risk of population extinction.

There is some empirical support for this prediction. For example, monochromatic bird populations (i.e. those presumed to be under less intense sexual selection) introduced to islands enjoy a higher success rate than similar dichromatic species that are under more intense sexual selection pressures (Cassey 2002, McLain *et al.* 1999, Sorci *et al.* 1998). This pattern is not universal as at least one study has failed to demonstrate such a connection (Donze *et al.* 2004). Dichromatic bird species in North America suffer from more frequent local extinctions than monochromatic species, although this local loss appears to be counter balanced by higher population turnover rates for the dichromatic species (Doherty *et al.* 2003). Such a population balancing act may explain why there is no apparent correlation between dichromatism or dimorphism metrics with population dynamics of European non-Passeriformes (Prinzing *et al.* 2002). One study has demonstrated that a macro-scale indicator of sexual selection is associated with increased extinction threat: Morrow and Pitcher (2003) showed that species descriptions of testis size, a measure of post-copulatory sexual selection by sperm competition, was correlated with the IUCN (International Union for Conservation of Nature) threat status of European birds; species with larger testes were more likely to be closer to the brink of extinction. In this same study the authors failed to find associations between threat status and measures of pre-copulatory sexual selection, such as plumage dichromatism. Therefore, the links between macro-evolutionary indicators of sexual selection and population extinction risk are not always evident. However, on balance, it is fair to say that a growing number of studies now report that more intense sexual selection among species is associated with increased extinction risk.

It should be noted that all of the studies summarized above are correlative. A recent experimental study strengthens the links between species-level sexual selection and extinction risk. Reding *et al.* (2013) evolved replicate populations of yeast (*Saccharomyces cerevisiae*) for approximately 250 generations under varying degrees of sexual selection. They showed that the populations under the most intense sexual selection were at a competitive disadvantage when transferred to a new growth medium, i.e. when they experienced a change of environmental conditions. More intense sexual selection resulted in greater risk of population extinction. Hence, overall, it appears that lineages with group- or species-level indicators of more intense sexual selection tend to be at greater extinction risk, especially under situations of environmental change.

The relevance of evolution to conservation outcomes is not just manifest through the presence of particular behavioral trait states.

Past niche conservatism can help predict capacity for bird species to adapt to a rapidly changing environment (Lavergne *et al.* 2013). In other words, the evolutionary history of a species can constrain the subsequent evolutionary response of that lineage – species are often phylogenetically constrained in the ways they can respond to environmental change. Some lineages are less able to persist in the face of environmental change because their common ancestors were also less likely to persist. In addition, as noted earlier, the current evolutionary regime can affect conservation outcomes. For example, sexual selection can impose evolutionary forces that make it more difficult for populations to adapt to a changing environment (Reding *et al.* 2013). Hence, past and current rates of evolution can be used as predictors of future capacity for evolutionary rescue and, therefore, it is important to understand the evolutionary history of lineages in determining how management practices and future environmental change could maximize population viability and persistence.

These remarks point to the importance of both paleontology and phylogenetic reconstruction to conservation science. Paleontology can let us understand the form and function of shared ancestors and the rate of change of traits (largely phenotypes) over time. Such information helps to predict how traits (and species) can evolve as the capacity to produce these phenotypes will often be conserved within a lineage. Molecular phylogenetic reconstruction can directly tell us about the genetic capacity for evolutionary change within lineages, as well as the rate of molecular genetic change over time. In addition, the emergence of functional genomics and proteomic studies could give us a molecular view of how groups of organisms share the capacity to produce traits that may assist with population (or species) persistence.

Thus far there has been relatively little emphasis on integrating a macro-evolutionary view into behavior conservation. The published literature is heavily biased toward proximate measurement of the consequences of individual-level behavioral change on metrics that could affect population dynamics. This is, perhaps, not surprising as behavioral ecologists tend to focus on individual-level processes and conservation scientists are worried about the here-and-now. However, in reframing how behavior and conservation inform each other I believe it behooves us to pay greater attention to group- and species-level behavioral traits and processes and integrate evolutionary histories into our consideration of how management practices will ultimately affect the conservation status of species. After all, conservationists are at least partly focused on ultimate questions that play out over deep evolutionary time: will species go extinct, and why?

## 2.4 IS BEHAVIOR A FOLLOWER OR LEADER OF EVOLUTIONARY CHANGE?

It is commonplace to think of behavior being the resultant product of an evolved adaptive strategy. There is plenty of support for this view, as indicated elsewhere in this volume and many other texts that describe the evolution of behavioral strategies (Alcock 2013, Andersson 1994, Fitzpatrick et al. 2005, Shuster & Wade 2003). However, it is possible, perhaps probable, that the plasticity and flexibility of behavior (both of which I will define below) in concert with mechanisms of genetic accommodation and assimilation can drive evolutionary change and not simply be the resultant change from other evolutionary mechanisms (Badyaev 2005, Baldwin 1896, Chevin et al. 2013, Duckworth 2009, Greenberg et al. 2012, Lind & Johansson 2011, Mery & Burns 2010, Moczek et al. 2011, West-Eberhard 2003, Wund 2012). Under a situation of genetic assimilation the environmental cues needed to produce a phenotype are no longer required because a mutation arises to fix a developmental pathway on the trajectory that was previously determined by environmental stimulation (Baldwin 1896, Uller & Helantera 2011, West-Eberhard 2003). In other words, as many others have, I posit that behavioral plasticity can lead genetic change (Badyaev 2005, Badyaev 2009, Baldwin 1896, Uller & Helantera 2011, West-Eberhard 2003) and, therefore, increase the relevance of behavioral ecology to understanding how populations will evolve to environmental change – an issue of pressing concern to any conservation scientist. Before launching into this argument I want to reposition the term "plasticity" in such a way that allows behavioral variation to have even broader influences on evolutionary processes.

Phenotypic plasticity refers to the ability of a genotype to produce different phenotypes (e.g. variation in morphology, physiology, or behavior) when developing in different environmental conditions. For example, a butterfly developing in hot dry conditions might develop fundamentally different wing coloration patterns than the same butterfly (i.e. the same genotype) developing in cool wet conditions (Brakefield et al. 1998). Phenotypic plasticity is the property of a genotype, can evolve separately to average trait value (Ghalambor et al. 2007) and is well defined in the evolutionary literature. Behavioral traits can show developmental plasticity in that a genotype can produce different behaviors if the organism grows up in one environment versus another. However, that concept of behavioral plasticity is fundamentally different to the behavioral changes that you might observe as an individual temporarily alters behavior in response to environmental and internal cues (Dingemanse *et al.* 2012, Dingemanse *et al.* 2010). Therefore, I will use the term "behavioral flexibility" (Bretman *et al.* 2012) to refer to these rapid and transient changes in behavior that most behavioral ecologists and conservation scientists study, but are often confused with behavioral plasticity (which means something quite different to evolutionary biologists). To reiterate, behavioral plasticity is the property of a genotype and refers to the production of different behavioral phenotypes across an environmental gradient; whereas behavioral flexibility is the property of an individual in a particular environment that results in a (often temporary) change in behavior due to the conditions of the biotic and abiotic environment at that point in time.

To explain how behavior can lead evolutionary genetic change it is important to elucidate how both behavioral plasticity and flexibility can facilitate rapid evolutionary change in populations, in positive (adaptive) and negative (maladaptive) ways. Populations are often described as occupying space on an adaptive landscape, which is a theoretical landscape in which populations are attempting to get to a point of greater fitness – often referred to as a peak on an adaptive landscape (Arnold *et al.* 2001). If a population occupies an adaptive peak then it enjoys high fitness and productivity. If a population is stuck in an adaptive trough then it has traits that are very poorly adapted to current environmental conditions and the population will likely not sustain itself for very long. I will construct my whole argument for the importance of behavioral traits in leading evolutionary change in the context of a traditional "reaction norm" view of phenotypic plasticity (Chevin *et al.* 2013, Dingemanse *et al.* 2010) and its integration with an adaptive landscape.

A reaction norm visualizes the production of phenotypes by a genotype across an environmental gradient (Figure 2.1). A non-zero slope indicates phenotypic plasticity and, in the case of our discussion, behavioral plasticity. A flat line (slope of zero) indicates behavioral inflexibility and rigidity. As a hypothetical example that may help frame the context of my argument we could think about the environmental gradient ($x$-axis) representing the density of predators and the behavioral phenotype ($y$-axis) representing the amount of vigilance behavior. Specifically, I propose to expand our

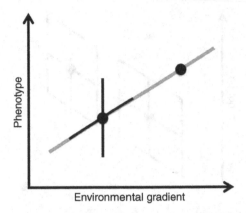

**Figure 2.1:** An example of a behavioral reaction norm showing the phenotypic response of a genotype that is expressed across an environmental gradient. The slope of the line indicates phenotypic plasticity in the expression of the phenotype. The vertical line around the left-most point on the line indicates additional phenotypic variation due to behavioral flexibility and developmental instability.

view of additional sources of behavioral (phenotypic) variation that need to be accounted for. Explicitly there is additional behavioral variation resulting from (a) developmental instability (Swaddle 2011); and (b) behavioral flexibility (Shuster & Arnold 2007). The addition of these two sources of behavioral variation results in an "error bar" in the plane of the y-axis on the traditional reaction norm; represented by a vertical line bisecting the left-most point on Figure 2.1. At any one point along the environmental gradient (the x-axis in Figure 2.1) an individual could display a range of behaviors. Some of this variation is because the genotype that individuals possess is most likely to produce a particular amount of vigilance behavior (the part of variation due to behavioral plasticity, i.e. the position along the sloped line in Figure 2.1), while some of the variation is because the individual could temporarily react to stimuli in the environment and alter its vigilance behavior (i.e. demonstrate behavioral flexibility). Additionally, the environment could cause stress that disrupts developmental processes resulting in further variation in vigilance behavior, which can be described as developmental instability (i.e. the inability of a genotype to buffer developmental processes against the current environmental conditions [Møller & Swaddle 1997]).

The addition of behavioral flexibility (and developmental instability) creates a "plasticity space" that can be occupied by an individual organism (which possesses its own unique genotype). For example, the plasticity

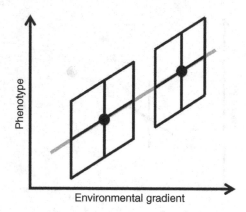

**Figure 2.2:** An example of a behavioral reaction norm indicating the plasticity spaces associated with individual expression of behavior. Each dot on the line represents an individual and the parallelogram around the dot indicates the range of behaviors that could be expressed by the individual.

space can be drawn around two samples (two points on the environmental gradient in Figure 2.2), where an individual shows variation in vigilance behavior (differences in the values on the y-axis) and could also move along the environmental gradient (left to right on the x-axis) by relocating itself in areas of fewer or more predators, or experiencing a change in the surrounding environment by predators entering or leaving the area.

It is possible that not all phenotypes (i.e. levels of vigilance in our example) are equally likely, so subparts of the range of possible behavioral phenotypes could also be associated with a range of likelihoods (or probabilities) depending on the behaviors associated with the genotype and the environment an individual finds itself in. For example, an individual exposed to a specific level of predation risk (i.e. a specific point on the environmental gradient) may be more vigilant most of the time, though there may be some occasions where the same individual may have to reduce vigilance behaviors perhaps to find food or attend to a mate even though the level of predation risk (i.e. the environmental gradient) has not changed. Thus, for a given point on the reaction norm the behavioral response may vary in a bivariate manner around this point due to variation in the landscape and the animal's state and experience (Figure 2.3) forming a probabilistic plasticity space. With knowledge of how organisms move, how they manipulate their own biotic and abiotic environment, how the environment itself is changing, how plastic the behavior is across the environmental gradient and how flexible their behavior is at any point on the gradient, this

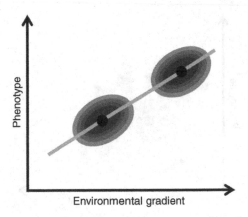

**Figure 2.3:** An example of a behavioral reaction norm where the plasticity spaces shown in Figure 2.2 are converted into probabilities of expressing a behavior. The darker colors indicate behaviors that are more likely to be expressed.

form of reaction norm could be operationalized and parameterized in a realistic manner. For illustrative purposes these probabilistic plasticity spaces are drawn as ovals, but they could take on other (irregular) shapes. This framework clearly illustrates the importance of bringing together information about how behaviors are expressed in different environments; in other words, I am not only stressing the importance of evolutionary biology to conservation science but we also need to understand the development and mechanistic control of behaviors. Conservation behavior, as a field, requires these levels of integration to offer useful understanding of how environmental change and behavioral plasticity and flexibility are linked.

With this new framework, it is also possible to put many individuals (genotypes) and their associated probabilistic plasticity spaces on the graph (Figure 2.4), building to a population representation of behavioral plasticity and flexibility, further extending the concept of the behavioral reaction norm (Dingemanse *et al.* 2010). In our example this would mean that for some individuals vigilance is increasing with the density of predators, for others individuals, vigilance is decreasing. Probabilities of the production of vigilance behaviors by individuals in the population could then be used to render a population probability measure of behavioral expression that accounts for gene-by-environment interaction (the slope of the line on the reaction norm), developmental instability and behavioral flexibility (shorter-term changes in individual behavior at one point on the environmental gradient).

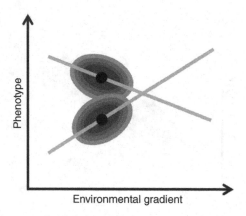

**Figure 2.4:** An example of a behavioral reaction norm showing two genotypes and their associated plasticity spaces. Note that the two genotypes have inverse slopes across the environmental gradient and cross each other.

Now that we have a framework that can explain the expression of behavior in relation to individual genotypes and variation in the environment we can link this to evolutionary processes. Explicitly, I will link the extended behavioral reaction norm to adaptation because many conservation scientists and behavioral ecologists alike are intrinsically interested in population productivity. To make this link we can compare plasticity spaces against a selection gradient (i.e. how the amount of vigilance behavior is related to fitness differences) to calculate probabilities of certain behavioral phenotypes and their associated genotypes being under selection. This could be achieved by adding a third axis to create an adaptive landscape (Arnold *et al.* 2001) that indicates the relative fitness value of particular amounts of vigilance behaviors. Such a framework would be immediately valuable in showing how populations will respond to environmental change in contemporary ecological time; for example, if there was an influx of predators to the area. It would also be possible to model an axis of phenotypic (behavioral) ontogeny to analyze how phenotypes (and their associated selection pressures) change over the development of individuals so that the age demographics of a population can be incorporated into predicting how populations will adapt to rapid environmental change. In the context of our example, vigilance behaviors in response to predation risk may change with age and so the age structure of the population would need to be accounted for as we model population responses.

As a further consideration of how behavioral mechanisms could directly influence evolution within this expanded view of an adaptive landscape it is important to recognize that individuals can modify their biotic and abiotic environment and/or move to new environments; hence, shifting the gradients of the adaptive landscape itself through behavioral modification. In our example, individuals could seek out areas with a lower density of predators and, hence, occupy a fundamentally different part of the reaction norm and be exposed to different selection pressures. This is yet another way in which behavior can lead evolutionary processes.

If we can now model how selection pressures relate to behavioral phenotypes and the probability of these behaviors being associated with genotypes, we can calculate population evolutionary responses if we can also understand the inheritance of phenotypes. I am stressing the inheritance of phenotypes (rather than genotypes) here as I think it will be important to expand upon our usual (quantitative genetic) concept of inheritance to include epigenetic and behavioral mechanisms of inheritance (Badyaev 2009, Bonduriansky 2013, Duckworth 2013, Jablonka 2013, Jablonka & Lamb 1995, Ledón-Rettig et al. 2013, Snell-Rood 2013). A behavior can be inherited by cultural (e.g. learning and copying) mechanisms (Dugatkin 2007, Kirkpatrick & Dugatkin 1994, Laland 1994, Swaddle et al. 2005, Uehara et al. 2005, White & Galef 2000) as well as through additive genetic variance ($h^2$) and epigenetic mechanisms (Duckworth 2013, Jablonka 2013, Jablonka & Lamb 1995, Ledón-Rettig et al. 2013, Snell-Rood 2013). Essentially, each value in the plasticity space would also be assigned an inheritance value so that the selection on phenotypes (i.e. vigilance behaviors) can be scaled to evolutionary responses in both genotypes and subsequent phenotypes. Computationally, I propose that we map the probabilistic plasticity space on to a fitness landscape and multiply this by location- (and phenotype-) specific inheritance values to calculate the probability of particular genotypes and phenotypes being represented in the next generation. This method may be computationally challenging but could be akin to multiplying layers of maps by each other and so could be visualized and computed within a GIS (Geographic Information Systems) framework, that brings together genotype, phenotype (i.e. vigilance behavior in our example), an adaptive landscape (i.e. how vigilance is linked to fitness) and probabilities of inheritance of vigilance behaviors (through genetic, epigenetic and cultural mechanisms).

There are at least three broad consequences of taking this new approach of explicitly incorporating behavioral plasticity and flexibility into modeling evolutionary outcomes, all of which are related to conservation. First, our revised concept of the behavioral reaction norm, and the concomitant definition of plasticity space associated with behaviors, may let us understand how increased behavioral variation can help (or hinder) populations as environmental conditions change. When the environment changes populations will often be challenged with a new suite of evolutionary selection pressures; populations that can rapidly adapt to the new environmental conditions will be favored. In cases where the heritable phenotypic variance afforded by behavioral plasticity and flexibility is aligned with these new selective pressures (i.e. aligned in the direction of an adaptive peak) it is likely that the populations will adapt more quickly and populations will be more likely to persist in the face of the changing environment (Badyaev 2009, Ghalambor *et al.* 2007). If the behavioral variation resulting from plasticity and behavioral flexibility is not aligned with an adaptive state and lowers population fitness then this additional form of variation will likely harm the long-term prospects of the population (Ghalambor *et al.* 2007).

Second, the existence of plasticity spaces (associated with genotypes) can result in the same selection pressure potentially selecting for different genotypes at different points along the environmental gradient. For example, if the same selection pressure for expression of the amount of vigilance behavior (dashed line; Figure 2.5) occurs in two environments (Env1 and Env2, indicated by arrows on Figure 2.5) and the plasticity spaces for the expression of the behavior inverts over the range of environments, then there will be a fundamental change in the probability of selection on the two genotypes ($G_1$ and $G_2$; Figure 2.5). In the example illustrated in Figure 2.5, $G_1$ is more strongly selected for in Env1 but $G_2$ is more strongly selected in Env2. The direction of selection acting on genotypes has changed based on a shift in the environment alone – the selection pressure acting on vigilance behavior has not changed. This phenomenon is particularly important for conservation when the environment shifts dramatically, such as in environments affected by fires, flooding or biological invasions. Behaviors that appear adaptive in one environmental condition (e.g. pre-invasion by existing-predators) may be relatively maladaptive in another environmental condition (e.g. post-invasion by new predators), meaning that a population that has adapted to one environmental state may be evolutionarily stuck in a low-fitness state if the environment changes quickly.

Third, within an environment (Env1 for example), selection will favor $G_2$ over $G_1$ by a calculable probability. However, as the plasticity spaces of $G_2$ and $G_1$ overlap and intersect with the selection regime (dashed line), there is a probability that $G_1$ can be selected over $G_2$ without changing the selection pressure, nor the genotypes present in the population, nor the point in the environmental gradient where the

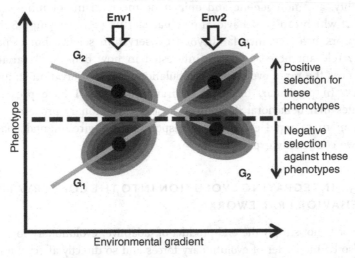

**Figure 2.5:** The same behavioral reaction norm indicated in Figure 2.4 with addition of a dotted line indicating selection for particular phenotypes (and their associated genotypes). Phenotypes above the dotted line are selected for; whereas phenotypes below the line are selected against. Env1 and Env2 indicate two levels on the environmental gradient ($x$-axis), and $G_1$ and $G_2$ indicate the two genotypes associated with the two lines plotted on the figure.

phenotypes are expressed. I predict that there will be inherent variation (that is not precisely predictable from one time point to another but can be calculated as a probability over a longer time period) in what behavioral state will render higher fitness in a population. Hence, selection can be reversed because of inclusion of our broadened view of the behavioral reaction norm and its associated plasticity space. This potential inversion of selection on genotypes can maintain variation in populations, which is essential to understanding the future capacity of evolution in populations as they experience further environmental change. In terms of conservation, these predictions are likely to be good news for population persistence as the varying selection pressures will help to maintain adaptive variation in populations. I envisage that the greatest challenge to population persistence will occur when the population is already small in size and the variation in selection pressures could tip an unlucky population in to a low-fitness state and increase the probability of local extinction.

This discussion lets us reexamine evolutionary processes, which is relevant to how conservation plays out in the field. Explicitly, we need to account for non-genetic sources of variation, which include behavioral flexibility, and non-genetic and epigenetic mechanisms of inheritance, some of which can be behavioral (such as learning and copying). Such a framework has relevance far beyond conservation science but is particularly relevant here as we are interested in how behavioral variation influences the rapid evolution of populations and conservation status of species. In particular, the kind of framework that I have presented indicates that behavioral flexibility can lead evolutionary processes and result in more rapid evolutionary responses to environmental change than we commonly expect.

## 2.5 INTEGRATING EVOLUTION INTO THE CONSERVATION BEHAVIOR FRAMEWORK

Behavior is not simply the end-product of (adaptive) evolutionary forces; it can also be the leader of evolutionary forces and so directly affect the likelihood of persistence of a population and species. To understand how behavior and conservation status of populations and species interact we must add an evolutionary lens to the framework initially proposed by Berger-Tal et al. (2011). In particular, I stress the importance of understanding how the environment moderates development of the phenotype and how this resultant (flexible or inflexible) phenotype then affects fitness and the niche space occupied by particular species (Angilletta & Sears 2011, Duckworth 2009, Hodson et al. 2010, Kozak & Boughman 2012, Sutherland et al. 2013, Yamamichi et al. 2011). Importantly, behavior can mediate all of these relationships and therefore should play a central role in understanding how environmental change and conservation relate over short (Jimenez et al. 2011, Montague et al. 2012, Tuomainen & Candolin 2011) and long time periods (Buchholz 2007, Caro & Sherman 2011, Mery & Burns 2010, Sih et al. 2010).

Specifically, I have supplemented the initial framework proposed by Berger-Tal et al. (2011), which tends to focus on ecological mechanisms and patterns, by adding a layer of evolutionary processes (Figure 2.6). Visually, this evolutionary layer is intended to float above the Berger-Tal et al. framework and is indicated by hashed gray arrows. At this evolutionary level the connections between the anthropogenic impacts on animal behavior and the behavioral domains (i.e. movement and space use, foraging and vigilance, and social organization and reproductive behavior) are mediated by behavioral (in)flexibility (Figure 2.6). My framing of behavioral flexibility allows for large changes in behavior or, conversely, inflexible and fixed behavior, both of which could affect immediate ecological and longer-term evolutionary responses to environmental change (Duckworth 2009).

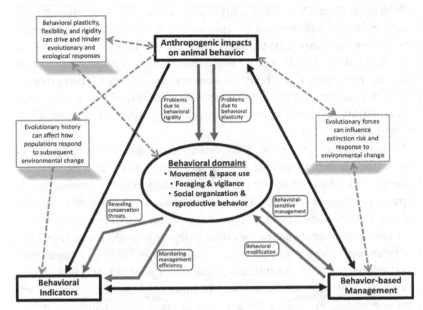

**Figure 2.6:** An adaptation of the Berger-Tal *et al.* (2011) framework that integrates micro- and macro-evolutionary processes into a holistic view of conservation behavior. The dashed line connections add an evolutionary layer to the original framework.

Although much of the discussion has focused on changes in behavior (through plasticity and flexibility) it is also possible that behavior is relatively fixed and invariable and this too can cause substantial problems for the conservation of populations (see Chapter 4). For example, species often select nesting and foraging habitats based on sensory cues and these general preferences seem rather fixed and rigid. Hence, unsuitable breeding and foraging habitats that have been substantially altered by humans but that possess these cues, often for aesthetic purposes (e.g. exotic fruit trees or manicured lawns), can be ecological and evolutionary traps for some species (Kokko & Sutherland 2001, Schlaepfer *et al.* 2002). In these cases, a lack of behavioral flexibility and plasticity and a slow rate of behavioral evolutionary change can doom species to occupying highly unsuitable areas.

I have made these links between the anthropogenic impacts on behavior and the behavioral domains bi-directional as some behaviors, such as the movement of individuals, can shift individuals into a new plasticity space (see section 4) and fundamentally change how human disturbance alters the adaptive value of particular behaviors.

Behavioral plasticity and flexibility, I propose, are the main mediators of how populations will respond to rapid environmental change over ecological and evolutionary time periods. Integrating this pivotal role of behavior allows us to visualize how evolutionary and ecological studies of behavior can both contribute to conservation science. Importantly, I want to reiterate that behavioral plasticity and flexibility can alter rates of adaptation to environmental change (Ghalambor et al. 2007), as explained in section 4.

Taking a longer-term view of the evolutionary history of organisms as well as the current evolutionary pressures experienced by a population, I have also added an evolutionary bi-directional link between anthropogenic impacts on animal behavior and behavior-based management. By this addition I claim that we need to account for evolutionary forces in considering management strategies as the current degree of selection on reproductive behaviors can also influence the immediate probability of population loss (Reding et al. 2013). Additionally, most management practices will have an evolutionary consequence and thereby influence how humans are affecting the evolution of behaviors. Such consequences may include the reduction of effective population size (which would affect drift), the introduction or separation of populations (which would affect flow), and/or environmental change that affects the adaptive value of particular behaviors (which would affect selection). Hence, we need an evolutionary lens on how behavioral management strategies interact with populations' responses to environmental change, ultimately affecting the conservation status of populations through behavioral ecology.

The evolutionary history of a lineage, such as the intensity of sexual selection, can influence extinction risk from contemporary environmental change (see section 3); hence, there is also a role for evolutionary thinking as we diagnose the links between anthropogenic impacts on animal behavior and behavioral indicators of conservation status. To this end I have added an evolutionary connection between these two elements of the Berger-Tal et al. (2011) framework.

Overall, I see behavior as a key mediator of how populations respond to environmental change as behavior can alter evolutionary forces and also lead evolutionary change. By adding an evolutionary view to the emerging Berger-Tal et al. (2011) framework I present ways in which we can blend ecological, developmental and evolutionary thinking to view behaviors as dynamic and forceful traits that influence the conservation status of populations. This integrative framework can also lead to practical considerations of how we can design effective conservation and management practices that take into account the evolutionary history and evolutionary potential of populations.

## REFERENCES

Alcock, J 2013. *Animal Behavior: An Evolutionary Approach*. 10th edn. New York: Sinauer Associates, Inc.

Andersson, M. 1994. *Sexual Selection*. Princeton, NJ: Princeton University Press.

Andersson, M. and Simmons, L.W. 2006. Sexual selection and mate choice. *Trends in Ecology and Evolution*, 21: 296–302.

Angilletta, M.J. and Sears, M.W. 2011. Coordinating theoretical and empirical efforts to understand the linkages between organisms and environments. *Integrative and Comparative Biology*, 51: 653–661.

Arnold, S., Pfrender, M. and Jones, A. 2001. The adaptive landscape as a conceptual bridge between micro- and macroevolution. *Genetica*, 112–113: 9–32.

Arnold, S.J. and Wade, M.J. 1984. On the measurement of natural and sexual selection: Theory. *Evolution*, 38: 709–719.

Badyaev, A.V. 2005. Stress-induced variation in evolution: from behavioural plasticity to genetic assimilation. *Proceedings of the Royal Society B-Biological Sciences*, 272: 877–886.

Badyaev, A.V. 2009. Evolutionary significance of phenotypic accommodation in novel environments: an empirical test of the Baldwin effect. *Philosophical Transactions of the Royal Society B-Biological Sciences*, 364: 1125–1141.

Baldwin, J.M. 1896. A new factor in evolution. *The American Naturalist*, 30: 441–451.

Bataillon, T., Joyce, P. and Sniegowski, P. 2013. As it happens: current directions in experimental evolution. *Biology Letters*, 9: 20120945.

Bell, G. 2013a. Evolutionary rescue and the limits of adaptation. *Philosophical Transactions of the Royal Society B: Biological Sciences*, 368: 20120080.

Bell, G. 2013b. Evolutionary rescue of a green alga kept in the dark. *Biology Letters*, 9: 20120823.

Berger-Tal, O., Polak, T., Oron, A. *et al.* 2011. Integrating animal behavior and conservation biology: a conceptual framework. *Behavioral Ecology*, 22: 236–239.

Blumstein, D.T. 2002. Moving to suburbia: ontogenetic and evolutionary consequences of life on predator-free islands. *Journal of Biogeography*, 29: 685–692.

Blumstein, D.T. 2006. Developing an evolutionary ecology of fear: how life history and natural history traits affect disturbance tolerance in birds. *Animal Behaviour*, 71: 389–399.

Bonduriansky, R. 2013. Nongenetic inheritance for behavioral ecologists. *Behavioral Ecology*, 24: 326–327.

Brakefield, P.M., Kesbeke. F. and Koch, P.B. 1998. The regulation of phenotypic plasticity of eyespots in the butterfly *Bicyclus anynana*. *The American Naturalist*, 152: 853–860.

Bretman, A., Westmancoat, J.D., Gage, M.J.G. and Chapman, T. 2012. Individual plastic responses by males to rivals reveal mismatches between behaviour and fitness outcomes. *Proceedings of the Royal Society B-Biological Sciences*, 279: 2868–2876.

Buchholz, R. 2007. Behavioural biology: an effective and relevant conservation tool. *Trends in Ecology & Evolution*, 22: 401–407.

Caro, T. and Sherman, P.W. 2011. Endangered species and a threatened discipline: behavioural ecology. *Trends in Ecology & Evolution*, 26: 111–118.

Cassey, P. 2002. Life history and ecology influences establishment success of introduced land birds. *Biological Journal of the Linnean Society*, 76: 465–480.

Chevin, L-M., Gallet, R., Gomulkiewicz, R., Holt, R.D. and Fellous, S. 2013. Phenotypic plasticity in evolutionary rescue experiments. *Philosophical Transactions of the Royal Society B: Biological Sciences*, 368: 20120089.

Coppack, T. and Partecke, J. 2006. The urbanization of birds: from behavioral plasticity to adaptive evolution. *Journal of Ornithology*, 147: 284–284.

Cote, I.M. and Reynolds, J.D. 2012. Meta-analysis at the intersection of evolutionary ecology and conservation. *Evolutionary Ecology*, 26: 1237–1252.

Coyne, J.A. 2009. *Why Evolution is True*. New York: Penguin.

Darwin, C. 1859. *On the Origin of Species by Means of Natural Selection*. London: John Murray.

DiBattista, J.D., Feldheim, K.A., Garant, D., Gruber, S.H. and Hendry, A.P. 2011. Anthropogenic disturbance and evolutionary parameters: a lemon shark population experiencing habitat loss. *Evolutionary Applications*, 4: 1–17.

Dingemanse, N.J., Barber, I., Wright, J. and Brommer, J.E. 2012. Quantitative genetics of behavioural reaction norms: genetic correlations between personality and behavioural plasticity vary across stickleback populations. *Journal of Evolutionary Biology*, 25: 485–496.

Dingemanse, N.J., Kazem, A.J.N., Réale, D. and Wright, J. 2010. Behavioural reaction norms: animal personality meets individual plasticity. *Trends in Ecology & Evolution*, 25: 81–89.

Doherty, P.F., Sorci, G., Royle, J.A. *et al.* 2003. Sexual selection affects local extinction and turnover in bird communities. *Proceedings of the National Academy of Sciences of the United States of America*, 100: 5858–5862.

Donze, J., Moulton, M.P., Labisky, R.F. and Jetz, W. 2004. Sexual plumage differences and the outcome of game bird (Aves: Galliformes) introductions on oceanic islands. *Evolutionary Ecology Research*, 6: 595–606.

Duckworth, R.A. 2009. The role of behavior in evolution: a search for mechanism. *Evolutionary Ecology*, 23: 513–531.

Duckworth, R.A. 2013. Epigenetic inheritance systems act as a bridge between ecological and evolutionary timescales. *Behavioral Ecology*, 24: 327–328.

Dugatkin, L.A. 2007. Developmental environment, cultural transmission, and mate choice copying. *Naturwissenschaften*, 94: 651–656.

Falconer, D.S. and Mackay, T.F.C. 1996. *Introduction to Quantitative Genetics*, 4th edn. Essex, UK: Longman.

Finnegan, D.J. 1989. Eukaryotic transposable elements and genome evolution. *Trends in Genetics*, 5: 103–107.

Fisher, R.A. 1958. *The Genetical Theory of Natural Selection.*, 2nd edn. New York: Dover.

Fitzpatrick, M.J., Ben-Shahar, Y., Smid, H.M. *et al.* 2005. Candidate genes for behavioural ecology. *Trends in Ecology & Evolution*, 20: 96–104.

Francis, C.D., Ortega, C.P. and Cruz, A. 2009. Noise pollution changes avian communities and species interactions. *Current Biology*, 19: 1415–1419.

Fraser, D.J., Coon, T., Prince, M.R., Dion, R. and Bernatchez, L. 2006. Integrating traditional and evolutionary knowledge in biodiversity conservation: a population level case study. *Ecology and Society*, 11.

French, S.S., Gonzalez-Suarez, M., Young, J.K., Durham, S. and Gerber, L.R. 2011. Human disturbance influences reproductive success and growth rate in California sea lions (*Zalophus californianus*). *PLoS ONE*, 6.

Galhardo, R.S., Hastings, P. and Rosenberg, S.M. 2007. Mutation as a stress response and the regulation of evolvability. *Critical Reviews in Biochemistry and Molecular Biology*, 42: 399–435.

Garant, D., Forde, S.E. and Hendry, A.P. 2007. The multifarious effects of dispersal and gene flow on contemporary adaptation. *Functional Ecology*, 21: 434–443.

Ghalambor, C.K., McKay, J.K., Carroll, S.P. and Reznick, D.N. 2007. Adaptive versus non-adaptive phenotypic plasticity and the potential for contemporary adaptation in new environments. *Functional Ecology*, 21: 394–407.

Gluckman, P., Beedle, A. and Hanson, M. 2009. *Principles of Evolutionary Medicine.* Oxford: Oxford University Press.

Gomulkiewicz, R. and Shaw, R.G. 2013. Evolutionary rescue beyond the models. *Philosophical Transactions of the Royal Society B: Biological Sciences*, 368: 20120093.

Gonzalez, A. and Bell, G. 2013. Evolutionary rescue and adaptation to abrupt environmental change depends upon the history of stress. *Philosophical Transactions of the Royal Society B: Biological Sciences*, 368: 20120079.

Gonzalez, A., Ronce, O., Ferriere, R. and Hochberg, M.E. 2013. Evolutionary rescue: an emerging focus at the intersection between ecology and evolution. *Philosophical Transactions of the Royal Society B: Biological Sciences*, 368: 20120404.

Greenberg, J.K., Xia, J., Zhou, X. *et al.* 2012. Behavioral plasticity in honey bees is associated with differences in brain microRNA transcriptome. *Genes Brain and Behavior*, 11: 660–670.

Harnik, P.G., Lotze, H.K., Anderson, S.C. *et al.* 2012. Extinctions in ancient and modern seas. *Trends in Ecology & Evolution*, 27: 608–617.

Hartmann, S.A., Steyer, K., Kraus, R.H.S., Segelbacher, G. and Nowak, C. 2013. Potential barriers to gene flow in the endangered European wildcat (*Felis silvestris*). *Conservation Genetics*, 14: 413–426.

Hendry, A.P., Kinnison, M.T., Heino, M. *et al.* 2011. Evolutionary principles and their practical application. *Evolutionary Applications*, 4: 159–183.

Hodson, J., Fortin, D., LeBlanc, M.L. and Belanger, L. 2010. An appraisal of the fitness consequences of forest disturbance for wildlife using habitat selection theory. *Oecologia*, 164: 73–86.

Houle, D. and Kondrashov, A.S. 2002. Coevolution of costly mate choice and condition-dependent display of good genes. *Proceedings of the Royal Society of London Series B-Biological Sciences*, 269: 97–104.

Ingram, V.M. 1957. Gene mutations in human haemoglobin: the chemical difference between normal and sickle cell haemoglobin. *Nature*, 180: 326–328.

Jablonka, E. 2013. Behavioral epigenetics in ecological context. *Behavioral Ecology*, 24: 325–326.

Jablonka, E. and Lamb, M.J. 1995. *Epigenetic Inheritance and Evolution: The Lamarckian Dimension*. Oxford: Oxford University Press.

Jablonski, D. 1987. Heritability at the species level: analysis of geographic ranges of cretaceous mollusks. *Science (New York, N. Y.)*, **238**: 360–363.

Jimenez, G., Lemus, J.A., Melendez, L., Blanco, G. and Laiolo, P. 2011. Dampened behavioral and physiological responses mediate birds' association with humans. *Biological Conservation*, **144**: 1702–1711.

Keogh, J.S. 2009. Evolutionary, behavioural and molecular ecology must meet to achieve long-term conservation goals. *Molecular Ecology*, **18**: 3761–3762.

Kight, C.R. and Swaddle, J.P. 2007. Associations of anthropogenic activity and disturbance with fitness metrics of eastern bluebirds (*Sialia sialis*). *Biological Conservation*, **138**: 189–197.

Kirkpatrick, M. and Dugatkin, L.A. 1994. Sexual selection and the evolutionary effects of copying mate choice. *Behavioral Ecology and Sociobiology*, **34**: 443–449.

Kirkpatrick, M. and Peischl, S. 2013. Evolutionary rescue by beneficial mutations in environments that change in space and time. *Philosophical Transactions of the Royal Society B: Biological Sciences*, **368**: 20120082.

Kokko, H. and Brooks, R. 2003. Sexy to die for? Sexual selection and the risk of extinction. *Annales Zoologici Fennici*, **40**: 207–219.

Kokko, H. and Sutherland, W.J. 2001. Ecological traps in changing environments: ecological and evolutionary consequences of a behaviourally mediated Allee effect. *Evolutionary Ecology Research*, **3**: 537–551.

Kozak, G.M. and Boughman, J.W. 2012. Plastic responses to parents and predators lead to divergent shoaling behaviour in sticklebacks. *Journal of Evolutionary Biology*, **25**: 759–769.

Laland, K.N. 1994. Sexual selection with a culturally transmitted mating preference. *Theoretical Population Biology*, **45**: 1–15.

Lande, R. 1976. Natural selection and random genetic drift in phenotypic evolution. *Evolution*, **30**: 314–334.

Lavergne, S. Evans, M.E.K., Burfield, I.J., Jiguet, F. and Thuiller, W. 2013. Are species' responses to global change predicted by past niche evolution? *Philosophical Transactions of the Royal Society B: Biological Sciences*, **368**: 20120091.

Ledón-Rettig, C.C., Richards, C.L. and Martin, L.B. 2013. A place for behavior in ecological epigenetics. *Behavioral Ecology*, **24**: 329–330.

Lind, M.I. and Johansson, F. 2011. Testing the role of phenotypic plasticity for local adaptation: growth and development in time-constrained *Rana temporaria* populations. *Journal of Evolutionary Biology*, **24**: 2696–2704.

Manor, R. and Saltz, D. 2003. Impact of human nuisance disturbance on vigilance and group size of a social ungulate. *Ecological Applications*, **13**: 1830–1834.

Martinez-Abrain, A. and Oro, D. 2010. Applied conservation services of the evolutionary theory. *Evolutionary Ecology*, **24**: 1381–1392.

Maynard Smith, J. and Szathmáry, E. 1995. *The Major Transitions in Evolution*. San Francisco: W. H. Freeman.

McLain, D.K., Moulton, M.P. and Sanderson, J.G. 1999. Sexual selection and extinction: The fate of plumage-dimorphic and plumage-monomorphic birds introduced onto islands. *Evolutionary Ecology Research*, **1**: 549–565.

Menotti-Raymond, M. and O'Brien, S.J. 1993. Dating the genetic bottleneck of the African cheetah. *Proceedings of the National Academy of Sciences*, 90: 3172–3176.

Mergeay, J. and Santamaria, L. 2012. Evolution and biodiversity: the evolutionary basis of biodiversity and its potential for adaptation to global change. *Evolutionary Applications*, 5: 103–106.

Mery, F. and Burns, J.G. 2010. Behavioural plasticity: an interaction between evolution and experience. *Evolutionary Ecology*, 24: 571–583.

Miller, K.A., Nelson, N.J., Smith, H.G. and Moore, J.A. 2009. How do reproductive skew and founder group size affect genetic diversity in reintroduced populations? *Molecular Ecology*, 18: 3792–3802.

Moczek, A.P., Sultan, S., Foster, S. *et al.* 2011. The role of developmental plasticity in evolutionary innovation. *Proceedings of the Royal Society B-Biological Sciences* 278: 2705–2713.

Møller, A.P. 1998. Developmental instability of plants and radiation from Chernobyl. *Oikos*, 81: 444–448.

Møller, A.P. and Swaddle, J.P. 1997. *Asymmetry, Developmental Stability and Evolution*. Oxford: Oxford University Press.

Montague, M.J., Danek-Gontard, M. and Kunc, H.P. 2012. Phenotypic plasticity affects the response of a sexually selected trait to anthropogenic noise. *Behavioral Ecology*, 24: 343–348.

Morrow, E.H. and Pitcher, T.E. 2003. Sexual selection and the risk of extinction in birds. *Proceedings of the Royal Society of London Series B-Biological Sciences*, 270: 1793–1799.

Norris, K. 2004. Managing threatened species: the ecological toolbox, evolutionary theory and declining-population paradigm. *Journal of Applied Ecology*, 41: 413–426.

Palkovacs, E.P., Kinnison, M.T., Correa, C., Dalton, C.M. and Hendry, A.P. 2012. Fates beyond traits: ecological consequences of human-induced trait change. *Evolutionary Applications*, 5: 183–191.

Parker, K.A., Anderson, M.J., Jenkins, P.F. and Brunton, D.H. 2012. The effects of translocation-induced isolation and fragmentation on the cultural evolution of bird song. *Ecology Letters*, 15: 778–785.

Parsons, P.A. 1990. Extreme environmental stress: asymmetry, metabolic cost and conservation. In *Evolutionary Genetics of* Drosophila (Barker J.S.F., ed.). New York: Plenum; 75–86.

Postma, E. and van Noordwijk, A.J. 2005. Gene flow maintains a large genetic difference in clutch size at a small spatial scale. *Nature*, 433: 65–68.

Prinzing, A., Brandle, M., Pfeifer, R. and Brandl, R. 2002. Does sexual selection influence population trends in European birds? *Evolutionary Ecology Research*, 4: 49–60.

Reding, L.P., Murphy, H.A. and Swaddle, J.P. 2013. Sexual selection hinders adaptation in experimental populations of yeast. *Biology Letters*, 9: 20121202.

Sagan, L.A. 1989. On radiation, paradigms, and hormesis. *Science*, 245: 574, 621.

Sakai, K-I. and Suzuki, A. 1964. Induced mutation and pleiotropy of genes responsible for quantitative characters in rice. *Radiation Botany*, 4: 141–151.

Sankaranarayanan, K. 1982. *Genetic Effects of Ionizing Radiation in Multi-Cellular Eukaryotes and the Assessment of Genetic Radiation Hazards in Man.* Amsterdam: Elsevier.

Sasaki, K.Fox, S.F. and Duvall, D. 2009. Rapid evolution in the wild: changes in body size, life-history traits, and behavior in hunted populations of the Japanese mamushi snake. *Conservation Biology,* 23: 93–102.

Schiffers, K., Bourne, E.C., Lavergne, S., Thuiller, W. and Travis, J.M.J. 2013. Limited evolutionary rescue of locally adapted populations facing climate change. *Philosophical Transactions of the Royal Society B: Biological Sciences,* 368: 20120083.

Schlaepfer, M.A., Runge, M.C. and Sherman, P.W. 2002. Ecological and evolutionary traps. *Trends in Ecology & Evolution,* 17: 474–480.

Schlaepfer, M.A., Sherman, P.W., Blossey, B. and Runge, M.C. 2005. Introduced species as evolutionary traps. *Ecology Letters,* 8: 241–246.

Shuster, S.M. and Arnold, E.M. 2007. The effect of females on male–male competition in the isopod, *Paracerceis sculpta*: a reaction norm approach to behavioral plasticity. *Journal of Crustacean Biology,* 27: 417–424.

Shuster, S.M. and Wade, M.J. 2003. *Mating Systems and Strategies.* Princeton, NJ: Princeton University Press.

Sih, A., Stamps, J., Yang, L.H., McElreath, R. and Ramenofsky, M. 2010. Behavior as a key component of integrative biology in a human-altered world. *Integrative and Comparative Biology,* 50: 934–944.

Slatkin, M. 1985. Gene flow in natural populations. *Annual Review of Ecology and Systematics,* 16: 393–430.

Smith, T.B., Mila, B., Grether, G.F. *et al.* 2008. Evolutionary consequences of human disturbance in a rainforest bird species from Central Africa. *Molecular Ecology,* 17: 58–71.

Snell-Rood, E. 2013. The importance of epigenetics for behavioral ecologists (and vice versa). *Behavioral Ecology,* 24: 328–329.

Sorci, G., Møller, A.P. and Clobert, J. 1998. Plumage dichromatism of birds predicts introduction success in New Zealand. *Journal of Animal Ecology,* 67: 263–269.

Stearns, S.C. and Hoekstra, R.F. 2005. *Evolution,* 2nd edn. London: Oxford University Press.

Sutherland, W.J., Freckleton, R.P., Godfray, H.C.J. *et al.* 2013. Identification of 100 fundamental ecological questions. *Journal of Ecology,* 101: 58–67.

Swaddle, J.P. 2010. Evolution. In *Encyclopedia of Earth* (Duffy, J.E., J.C, eds.). Washington, DC: National Council for Science and the Environment.

Swaddle, J.P. 2011. Assessing the developmental stress hypothesis in the context of a reaction norm. *Behavioral Ecology,* 22: 13–14.

Swaddle, J.P., Cathey, M.G., Correll, M. and Hodkinson, B.P. 2005. Socially transmitted mate preferences in a monogamous bird: a non-genetic mechanism of sexual selection. *Proceedings of the Royal Society of London B-Biological Sciences,* 272: 1053–1058.

Tanaka, Y. 1996. Sexual selection enhances population extinction in a changing environment. *Journal of Theoretical Biology,* 180: 197–206.

Tuomainen, U. and Candolin, U. 2011. Behavioural responses to human-induced environmental change. *Biological Reviews,* 86: 640–657.

Tuomainen, U., Sylvin, E. and Candolin, U. 2011. Adaptive phenotypic differentiation of courtship in response to recent anthropogenic disturbance. *Evolutionary Ecology Research*, 13: 697–710.

Uehara, T., Yokomizo, H. and Iwasa, Y. 2005. Mate-choice copying as Bayesian decision making. *American Naturalist*, 165: 403–410.

Uller, T. and Helantera, H. 2011. When are genes "leaders" or "followers" in evolution? *Trends in Ecology & Evolution*, 26: 435–436.

Van Dyck, H. 2012. Changing organisms in rapidly changing anthropogenic landscapes: the significance of the Umwelt'-concept and functional habitat for animal conservation. *Evolutionary Applications*, 5: 144–153.

Vander Wal, E., Garant, D., Festa-Bianchet, M. and Pelletier, F. 2013. Evolutionary rescue in vertebrates: evidence, applications and uncertainty. *Philosophical Transactions of the Royal Society B: Biological Sciences*, 368: 20120090.

Von Neumann, J. and Morgenstern, O. 2007. *Theory of Games and Economic Behavior* (commemorative edition). Princeton, NJ: Princeton University Press.

West-Eberhard, M.J. 2003. *Developmental Plasticity and Evolution*. Oxford University Press.

White, D.J. and Galef, B.G. Jr. 2000. "Culture" in quail: social influences on mate choices of female *Coturnix japonica. Animal Behaviour*, 59: 975–979.

Wund, M.A. 2012. Assessing the impacts of phenotypic plasticity on evolution. *Integrative and Comparative Biology*, 52: 5–15.

Yamamichi, M., Yoshida, T. and Sasaki, A. 2011. Comparing the effects of rapid evolution and phenotypic plasticity on predator–prey dynamics. *American Naturalist*, 178: 287–304.

Zhang, J. 2003. Evolution by gene duplication: an update. *Trends in Ecology & Evolution*, 18: 292–298.

# Learning and conservation behavior: an introduction and overview

## ZACHARY SCHAKNER AND DANIEL T. BLUMSTEIN

### 3.1 CONCEPTUAL BACKGROUND

Learning is a key aspect of behavior that may greatly enhance the survival and fecundity of animals, especially in a changing environment. Wildlife conservation problems often involve increasing the population of threatened or endangered species, decreasing the population of species deemed over abundant or encouraging animals to move to or from certain areas. Learning is an example of reversible plasticity (for review see Dukas 2009), which typically remains open to change throughout life. Old associations can be replaced, relearned and reinstated, facilitating behavioral modifications across an individual's lifetime. Because learning is potentially demographically important, and because it can be used to modify individual's behavior, it may therefore be an important tool for conservation behaviorists (Blumstein & Fernández-Juricic 2010). Our aim in this chapter is to introduce the fundamentals of learning that will later be developed and applied in subsequent chapters.

Animal learning theory defines learning as experience that elicits a change in behavior (Rescorla 1988, Heyes 1994). There are three basic mechanisms, or types of experiences, that underlie animal learning. The simplest learning process is non-associative because it involves an individual's experience with a single stimulus. During this process, exposure to the single stimulus results in a change in the magnitude of response upon subsequent exposures to that stimulus. If the response increases, the process is called *sensitization*; if the response decreases, the process is called *habituation*. More complex associative learning mechanisms involve a change in behavior as a result of experience with two stimuli through *Pavlovian conditioning* (also referred to as *classical conditioning*),

*Conservation Behavior: Applying Behavioral Ecology to Wildlife Conservation and Management*, eds. O. Berger-Tal and D. Saltz. Published by Cambridge University Press. © Cambridge University Press 2016.

or the relationship between a subject's own behavior in response to a stimulus, which is called *instrumental conditioning*. Finally, learning can also occur as a result of interactions or observations with other individuals through *social learning*, but it is currently unclear whether social learning actually represents separate learning mechanisms than individual learning (Heyes 1994). Below we will describe these in more detail and outline the conditions that influence them. Later we will explain how knowledge of mechanisms of learning can be applied to wildlife management and conservation.

### 3.1.1 Non-associative learning: habituation and sensitization

#### 3.1.1.1 What is it?

Single-stimulus learning is the simplest learning process and involves a change in the frequency or intensity of response to a stimulus. Non-associative, single-stimulus learning involving a reduction of a behavioral response to repeated exposure to stimuli that is not due to sensory fatigue is called habituation (Groves & Thompson 1970). Unlike generalized sensory adaptation or motor fatigue (which would exhibit generalized responses within a modality to stimuli), habituation is characterized by stimulus specificity, which can be tested by showing responsiveness to novel stimuli (Rankin *et al.* 2009). This specificity suggests the function of habituation is to filter harmless stimuli from novel stimuli (Rankin *et al.* 2009). In contrast to habituation, heightened responsiveness after repeated exposure is termed sensitization. According to the *dual process theory of habituation*, an observed behavior after repeated exposure to a stimulus represents the sum of the two underlying learning processes of habituation and sensitization (Groves & Thompson 1970).

#### 3.1.1.2 Conditions influencing habituation

Generally, simple parameters such as intensity, modality and frequency influence single-stimulus learning in animals. More frequent exposure typically results in quicker or more pronounced habituation (Groves & Thompson 1970, Rankin *et al.* 2009). Correspondingly, repeated exposure to less intense stimuli results in a response decrement, whereas repeated exposure to higher intensity stimuli may either elicit no habituation or may result in sensitization (Groves & Thompson 1970, Rankin *et al.* 2009). After becoming habituated, withholding the stimulus results in a partial recovery in responsiveness, a process termed *stimulus recovery*. Response decrement exhibits specificity within a modality, which can be demonstrated by restored responsiveness to novel stimuli. During the course of habituation, the presentation of another, strong stimulus results in *dishabituation*, or restored responsiveness to a previously habituated stimulus. These behavioral characteristics of habituation have been clearly described in Groves and Thompson (1970), and since refined in Rankin *et al.* (2009).

### 3.1.2 Pavlovian conditioning

#### 3.1.2.1 What is it?

Pavlovian learning is seen when individuals learn the relationship between two stimuli; it is also called classical conditioning and, broadly, is one type of associative learning (Mackintosh 1974, Dickinson 1980, Rescorla 1988). In this type of learning, a biologically relevant stimulus, called the *Unconditioned Stimulus* (abbreviated US) is preceded by another stimulus, the *Conditioned Stimulus* (abbreviated CS). According to contemporary animal learning theory successful classical conditioning depends upon the contingency between the CS and US. This contingency can be positive, meaning that the US reliably follows the CS, or negative, meaning the CS reliably signals the absence of US. As a result of this pairing, animals are able to learn the relations between the two stimuli and generate an adaptive response (Dickinson 1980, Shettleworth 2010).

The capacity to learn about the relationship between two stimuli, such as sounds preceding the presence of a predator, or taste cues associated with edible food, is functional because it guides how an animal can adaptively respond to exogenous stimuli as well as anticipate future events (Domjan 2005, Shettleworth 2010). *Pavlovian fear conditioning* is an associative form of learning in which individuals are exposed to an aversive stimulus (US) paired with an innocuous stimulus (CS) (Fanselow 1984, Grillon 2008, Fanselow and Ponnusamy 2008). Once conditioning has occurred, exposure to the unconditioned stimulus generates fear reactions to the conditioned stimulus. For example, by learning the cues that predict a predator attack, prey are able to modify their behavior and reduce the probability of death (Domjan 2005). From this functional learning perspective, learning about the relationship between two stimuli influences the adaptive decision-making process and can modify an individual's behavioral response (Hollis 1982).

### 3.1.2.2 Conditions influencing Pavlovian conditioning

Functionally, there are particular conditions in which animals are able to learn patterns or relationships between stimuli in the natural world. The temporal relationship between two stimuli influences the conditioning process. Generally, a CS that precedes a US in time leads to more robust conditioning (Domjan & Burkhard 1986, Rescorla 1988). This is intuitive because in nature it is adaptive to learn the cues that precede consequences (i.e. certain tastes may precede sickness, or alarm calls are likely to precede predator presence).

The Rescorla-Wagner model (RW) is a generally accepted model for predicting the behavioral consequences and conditions driving associative learning between a CS and a US (Rescorla & Wagner 1972). According to the model, learning occurs as a result of the difference between what an animal expects to happen versus what happens. The RW model suggests that all learning curves are similar and asymptotic (Figure 3.1). For example, the first pairing of a CS (e.g. a neutral tone) followed by a US (e.g. a shock) is surprising, and results in a significant amount of learning (Figure 3.1). After subsequent pairings, the amount that is learned decreases because the US is less surprising when it follows the CS, resulting in a negatively accelerating curve. At the asymptote, the past experience with the CS/US pairing means that the CS accurately predicts the US, and thus little more is learned. According to the model, learning curves may differ in their slope, which is determined by the values of the rate parameters (i.e. magnitude of US or CS and US salience). In other words, some relationships can be learned more quickly than others (e.g., taste aversion or fear conditioning). The model can be used to help understand differences between species (Trimmer *et al.* 2012) and help explain differences in the speed of learning. For instance, the value of alpha (the CS learning rate) for auditory cues may be higher in one species than another, which will then lead to the former learning more quickly than the latter when an auditory cue signals something like the imminent delivery of food. The RW model produces idealized learning curves during controlled conditions. In the wild, differences in parameter values across species may explain observed patterns of learning in different situations, although this requires further study.

Conditioning also depends on the nature and relationship of the stimuli being paired. Conditioning experiments confirm that learning particular combinations of stimuli can be especially effective. For example, pigeons form effective associations when auditory cues are the CS preceding a shock and visual cues precede food (Shapiro *et al.* 1980). Taste aversion learning is

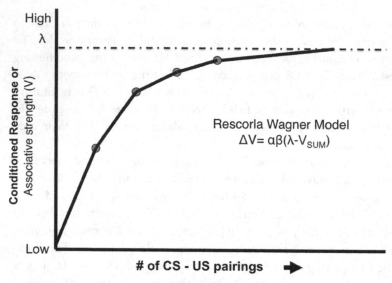

**Figure 3.1:** The Rescorla-Wagner model of learning. On the *y*-axis is the performance, which represents underlying learning (associative strength). The change in predictive value of a CS, $\Delta V$ is a result of the discrepancy between what is expected versus what actually happens ($\lambda$-$V_{SUM}$). and are learning rate parameters that correspond to salience of the $\alpha$CS and $\beta$US, and $V_{SUM\ I}$ is the sum of current associative strengths for all the CSs present. During the first few trials of CS/US (*x*-axis), the associative strength is large because the US is surprising. With subsequent trials, however, the associative strength decreases because it becomes less surprising. At the asymptote, the CS predicts the US with certainty, thus there is nothing more to be learned.

a well-known example of selective associations. In Garcia and Koelling's (1966) experiment, rats with two cues, taste CS and audiovisual CS, were then exposed to a nausea-inducing US or shock US. Shocked individuals associated the shock with the audiovisual cue and poisoned individuals associated the sickness with the taste cue (Garcia & Koelling 1966). There is also evidence of preparedness (Öhman & Mineka 2002), or evolved predispositions to associate particular stimuli (Griffin & Evans 2003). Animals form rapid associations between ecologically relevant CSs and certain aversive USs compared to fear-irrelevant CSs (Mineka & Öhman 2002).Examples include fearful responses to foxes (*Vulpes vulpes*), but not goats (*Capra hircus*), by tammar wallabies (*Macropus eugenii*) (Griffin *et al.* 2001), or fear responses to snakes but not flowers by primates (including humans) (Öhman & Mineka 2002). These, and many other examples (Domjan 2005) have suggested animals are predisposed to learn the relationships between evolutionarily relevant stimuli; findings which help develop an ecologically relevant perspective on general learning theory.

### 3.1.3 Instrumental conditioning
#### 3.1.3.1 What is it?
In instrumental conditioning, the animal learns a relationship between an operant behavior and the consequence of that behavior, and behavioral frequencies are adjusted accordingly (Thorndike & Bruce 1911, Domjan & Burkhard 1986). This is a second type of associative learning. During conditioning, a stimulus, typically termed a *reinforcer* influences the likelihood of a response. Thus, behaviors followed by positive consequences will increase in occurrence, whereas behaviors followed by negative consequences will decrease. Functionally, instrumental conditioning is a mechanism that enables individuals to modify, shape or create complex patterns of behavior.

#### 3.1.3.2 Conditions influencing instrumental conditioning
The rate of instrumental conditioning is influenced by the reinforcer type, the reinforcement schedule and the nature of the response (Domjan & Burkhard 1986). Generally, positive reinforcers, such as food or water, increase the frequency of a behavioral response, whereas negative reinforcers, such as shock or other pain-inducing events, decrease the frequency of a behavioral response. Similar to Pavlovian conditioning, instrumental conditioning depends on the temporal association between the reinforcer and response as well as on the contingency between the response and occurrence of the reinforcer. Finally, instrumental conditioning is limited by the degree that reinforced behaviors fit into an animal's natural behavior patterns, as well as the *belongingness* (the fit between the animal's behavior and stimuli used to reinforce them – e.g. Shettleworth 1975).

### 3.1.4 Social learning
#### 3.1.4.1 What is it?
We use Hoppitt & Laland's (2008) definition of social learning as "the process through which one individual influences the behavior of another individual in a manner that increases the probability that the observer learns" (further reviewed in Heyes 1994, Galef & Laland 2005). Research has emphasized the adaptive value of social learning (Laland 2004, Rendell et al. 2010, Heyes 2012). Social learning can function as a multiplier, since new traits can spread more quickly socially than by individual learning alone. There is evidence that asocial and social learning rely on the same

underlying associative and non-associative mechanisms (Heyes 1994, 2012).

For instance, stimulus enhancement has been proposed as a form of single-stimulus social learning. It occurs when a demonstrator's presence exposes an observer to a stimulus, resulting in an increase or decrease in responsiveness in the observer's interaction with that stimulus (Heyes 1994). From this perspective, stimulus enhancement may sensitize or habituate a response to a stimulus following an observer's interaction with a stimulus. For example, Heyes *et al.* (2000) found that rats observing conspecifics pressing a lever increased the probability of the observer interacting with that lever. It should be noted, however, that it is difficult to rule out associative learning in many cases of stimulus enhancement, because individuals may be learning to associate a location or stimulus with a reward (Hoppit & Laland 2008).

Observational conditioning is another form of social learning that, in this case, involves associative learning. Learning occurs when an observer's exposure to a demonstrator enables it to learn the relationship between two stimuli. For example, classic work by Mineka and Cook (1984) on rhesus monkeys (*Macaca mulata*) showed that naïve monkeys, when exposed to videos of wild monkeys responding fearfully to snakes, quickly learned the relationship between the fear response and the snake stimulus. In this experiment, the demonstrator monkeys' fear response is believed to be a US and associative learning occurs when paired with the snake, a CS.

Finally, observational learning occurs when an observer's experience with a demonstrator facilitates the observer's learning of a stimulus and response. For example, Akins and Zentall's (1996, 1998) work on Japanese quail (*Coturnix japonica*) used a two-action test to show observer quail learn to peck or step on a treadle based upon the demonstrator's action and the observed reward for that specific action.

### 3.1.4.2 Conditions influencing social learning

If we assume that asocial and social learning are governed by the same underlying fundamental learning mechanisms (Heyes 1994), then the conditions for both will be similar but with an added condition for social learning: the presence of conspecific or traces of conspecific stimuli. The components of an individual's social milieu provide opportunities for individuals to interact with and learn from conspecifics or traces of conspecifics (Coussi-Korbel & Fragaszy 1995). Variables such as rank, age, familiarity and social group size can influence how and from whom individuals learn, and this is termed directed social learning (Coussi-Korbel & Fragaszy 1995, Swaney *et al.* 2001, Nunn *et al.* 2009).

A given individual is not equally exposed to all animals in space and time (Coussi-Korbel & Fragaszy 1995), so there is some uncertainty as to who can and should learn socially. Network-based diffusion analysis (NBDA) uses formal network statistics to test for social learning in social groups (Hoppit *et al.* 2010). NBDA tracks the passage of information along established social networks in animal groups (Franz & Nunn 2009) because behaviors are expected to be transmitted across existing social connections.

While social learning may be potentially adaptive, like many traits, whether or not animals that learn specific things enhance their fitness may depend on the specific situation. For instance, social learning may lead to animals acting upon outdated information because they copied demonstrators who had learned something that is no longer valuable, and making the outdated behavioral response last longer within the population despite being less valuable (for review, see Laland 2004, Rendell *et al.* 2010).

## 3.2 LEARNING AND CONSERVATION: HOW KNOWLEDGE OF LEARNING MECHANISMS MAY HELP SOLVE CONSERVATION PROBLEMS

With this fundamental review of learning behind us, we shall now go on to highlight some important conservation questions that will be enhanced by the study of learning. Many conservation and management problems can benefit from mechanistic insights into how animals respond to stimuli and learn about biologically important events. We organize this section according to the three conservation behavior themes (Berger-Tal *et al.* 2011, Chapter 1).

### 3.2.1 Theme 1. Anthropogenic impacts on behavior

3.2.1.1 What constrains animal learning in response to anthropogenic change?

Anthropogenic change may increase environmental variation and may create novel environments that animals may have not experienced before (Sih *et al.* 2011). When faced with variable environments, learning is an adaptive mechanism that permits individuals to acquire predictive information from local conditions to generate adaptive behavioral responses (Shettleworth 2010). But, there are constraints on learning, and there is variation in how species respond to environmental change (Sol *et al.* 2002). We divide constraints to learning into internal and external. Internal constraints are largely cognitive, while external constraints include the rapidity of the stimulus exposure, the magnitude of the consequence and its consistency over time.

**Internal constraints** Differences in underlying input mechanisms, such as a species' perceptual abilities, the attention an individual can allocate to a task, or an individual's motivation (Macphail & Barlow 1985, Shettleworth 2010, Heyes 2012) are likely to influence learning capacities.

Non-detectable stimuli can't be learned. Some anthropogenic stimuli may simply not be detected, such as glass windows by birds. An animal that relies on vision might not learn to avoid a highway, compared to an animal with acute hearing that is disturbed by distant sounds. Sensory disturbances vary (Lowry *et al.* 2011) and so does the combination of sensory modalities during association formation (taste precedes sickness, sound precedes pain, etc.).

The Rescorla-Wagner model predicts that novel or surprising unconditioned stimuli will be more effective at strengthening CS–US associations than those whose occurrence is not surprising. But this may be a double-edged sword to managers. Novel foraging resources, such as crops or fishing lines, can be attractive, highly rewarding and lead to accelerated learning of nuisance behaviors. By contrast, other novel anthropogenic disturbance stimuli are perceived as threatening, resulting in accelerated avoidance (Frid & Dill 2002). Thinking about stimuli with respect to their potential RW learning rate parameter values (such as salience, suprisingness, belongingness) may be a fruitful way to categorize anthropogenic stimuli, particularly if the goal is to train animals to selectively make associations or to train animals to selectively avoid resources.

Motivational mechanisms mediate an individual's tolerance for conspecifics, heterospecifics (including humans) or other potentially novel stimuli. Ultimately, motivational mechanisms will influence the stimuli an animal encounters, and how effectively they are conditioned. Neophobia is defined as a propensity to avoid novel stimuli (Greenberg 2003). Differences in neophobia may underlie the propensity to learn (Sol 2013). For example, there is evidence that urban zenaida doves (*Zenaida aurita*) that experience a highly dynamic environment, become less neophobic, learn faster and are more inclined to learn from conspecific demonstrators than less urbanized doves (Carlier & Lefebvre 1997, Seferta *et al.* 2001). Motivational mechanisms can also vary intraspecifically due to personality differences in boldness/shyness (Shettleworth 2010, Sih & Giudice 2012). In numerous species, such as guppies (*Poecilia reticulata*), trout, (*Oncorhynchus mykiss*) and black-capped chickadees (*Poecile atricapillus*), bolder or more exploratory individuals learn a conditioning or discrimination task more quickly than shy individuals (Dugatkin & Alfieri 2003, Sneddon 2003, Guillette *et al.* 2009).

Comparative studies in birds and mammals suggest that the correlations between brain size, learning and overall behavioral flexibility enable species to respond to novel ecological challenges (Sol *et al.* 2002, 2008). Generating behavioral flexibility through learning may enable individuals to modify, copy or create novel anti-predator responses (Berger *et al.* 2001), prey choice (Estes *et al.* 1998), or habitat selection (Doligez *et al.* 2002). Comparative approaches suggest that species with larger brains (relative to body size) have enhanced survival in novel, disturbed or dynamic environments (Sol et al. 2005, 2007, 2008; Amiel *et al.* 2011). Thus, while behavioral plasticity, generated by learning, is widespread in nature, there is variation in the degree to which animals can learn to respond to the new situations that are generated by anthropogenic change, and relative brain size is a rough index of this flexibility. Managers should be sensitive to this variation and future research should identify other correlates of flexibility. It is important to note, however, that the effect of brain size on behavioral complexity remains highly debatable (Healy & Rowe 2007), and no study that we know of has looked at the influence of brain size on the effectiveness of different learning mechanisms.

In some species, there may be a sensitive time period during which most learning about a particular biologically important process occurs (Hogan & Bolhuis 2005). The classic example is filial imprinting in precocial birds (Lorenz 1970). However, there is also strong evidence of sensitive periods for habitat (Davis & Stamps 2004) and sexual preferences (Bateson 1978). More generally, however, individuals at different life stages may be more or less likely to learn (Dukas 2008). Hawkins *et al.* (2008) demonstrated age dependent learning of predator cues in hatchery-reared salmon. Their results suggest heightened receptivity to learning predator cues during the life history stage at which juveniles would be undergoing a habitat shift and thus are particularly sensitive toward predation. Such variation in the ability to learn may allow young, but not older, animals to learn appropriate responses in an anthropogenically disturbed environment.

Learning during sensitive periods can be via individual associative learning mechanisms or via social learning from parents. These so-called parental effects may be obligatory for survival in some species. However, parental effects can also act as multipliers, spreading maladaptive behaviors through populations. For instance, wild black bear (*Ursus americanus*) cubs raised by garbage-pilfering sows were significantly more likely to rely on human resources (Mazur & Seher 2008).

Finally, managers should be mindful of sensitive periods to optimize reintroductions/translocations. For instance, if animals are to be moved to a new environment, pre-exposure to that environment (or certain characteristics of that environment, such as food sources) during a sensitive period may be essential for successful establishment. Much work remains to be done to provide concrete examples that can help inform management.

**External constraints** Learning is adaptive because it enables individuals to track environmental variation. We know that the type of reinforcer, the temporal relationship between the reinforcer and the consequence, and the magnitude of the consequence will all constrain the rate of learning (Shettleworth 2010). Positive reinforcers, such as food, safety or conspecifics tend to increase behavioral responses. Negative reinforcers, such as painful, noxious or distracting stimuli, may only require a single exposure to create long-term learning (Rau & Fanselow 2009).

Learning can only occur if the rate of learning is faster than the rate of environmental change (Johnston 1982). If anthropogenic change is too rapid, learning cannot occur and individuals in a population will be unable to modify their behavior and behaviorally track the changes. In such cases, given sufficient additive genetic variation, there will be strong selection against those animals with an inadequate behavioral response.

The magnitude of the consequence, the speed (rapidity) at which a stimulus reaches its full magnitude, and its consistency over time (anthropogenic noise, for instance, may cycle over 24 hours) will also influence learning. An event or stimulus that is always lethal will prevent any learning from occurring, whereas highly profitable food sources (such as crops or garbage cans), or painful/nearly lethal encounters, may stimulate rapid and complete learning after one or a few exposures. Intense stimuli with a rapid onset elicit startle responses (Yeomans *et al.* 2002). In organisms vulnerable to high-intensity acoustic stimuli, such as sea turtles or cetaceans, rapid onset exposures (seismic airgun arrays or sonar) may lead to sensitization of avoidance responses (Gotz & Janik 2011, DeRuiter & Doukara 2012).

### 3.2.1.2 Anthropogenic impacts on behavior: can we develop an evolutionary ecology of habituation?

A fundamental question in wildlife conservation and management concerns the causes and consequences of habituation and sensitization. Why do some species habituate, while others sensitize to anthropogenic stimuli? The "life–dinner principle" suggests that for a prey species, the costs of getting predated far outweigh the costs of missing a meal (Dawkins & Krebs 1979). From a life–dinner principle perspective, there is an asymmetry between the fitness costs of failing to detect a predator (Type 1 error) and over-reacting to non-threatening stimuli (Type 2 error).

Habituation to non-threatening stimuli is somewhat expected since anxiety or stress from over-generalized threat recognition may be costly in terms of energy or time allocated to unnecessary defenses (Blanchard 2008). We therefore expect animals to show an initial heightened response, followed by rapid habituation to repeated unreinforced exposures of even potentially threatening stimuli (Groves & Thompson 1970). Habituation is thus a mechanism to reduce the costs of false alarms (Thorpe 1956, Shalter 1984).

Remarkably, given how long we have known about mechanistic processes involved in habituation and sensitization (Groves & Thompson 1970), little is known about habituation in the wild, or what we will refer to as the evolutionary ecology of habituation. Perhaps this is in part because habituation has been extensively investigated under controlled experimental conditions. By contrast, in nature, an organism's environment is noisy and filled with threatening and non-threatening stimuli that occur in a variety of different contextual situations. To deal with this uncertainty, there is evidence that habituation under natural conditions is quite selective and enables individuals to learn what is not threatening (Deecke et al. 2002, Hemmi & Merkle 2009, Raderschall et al. 2011). In a series of studies of anti-predator responses in wild hermit crabs, Hemmi (2011) demonstrated that habituated responses are recovered when the same predator stimulus is presented at a different distance or angle. Similar to laboratory investigations of dishabituation, this study shows that in the wild even small changes in stimulus presentation can result in recovered responsiveness. Correspondingly, selective habituation is hypothesized to be the mechanism by which harbor seals (*Phoca vitulina*) discriminate between threatening and non-threatening killer whale (*Orcinus orca*) vocalizations (Deecke et al. 2002). Harbor seals responded with flight to playback of vocalizations from local marine mammal-eating killer whales and novel fish-eating killer whales, but not local fish-eating killer whales. These results suggest that the seals habituated to non-threatening local fish-eating killer whales, but were fearful to unknown vocalizations. This specificity of habituation makes sense in terms of the fundamental characteristics of habituation described in our introduction and illustrates its evolutionary context.

Ultimately, to develop a natural history of habituation we will need to understand what sorts of stimuli in nature lead to habituation and then understand what life history and natural history features are correlated with habituation or sensitization. As a step towards this, (Li *et al.* 2011) developed a mixed-modeling statistical approach to identify how different anthropogenic stimuli (people, people on bicycles, people in cars) influenced flight initiation distance decisions in yellow-bellied marmots (*Marmota flaviventris*). Flight initiation distance (FID) is a particularly sensitive assay for how animals respond to approaching threats, and animals repeatedly exposed to humans often tolerate closer approaches before fleeing.

The nature, spatio-temporal pattern and context of exposure to stimuli influence the rate of habituation and whether sensitization occurs. For example, yellow-eyed penguins (*Megadyptes antipodes*) show sensitized stress responses to tourists in Sandfly Bay (Ellenberg *et al.* 2009). The authors suggested that the unpredictable and abrupt behavior of tourists that ran, shouted and chased penguins prevented habituation and facilitated sensitization. During exposure to threatening stimuli, animals assess the type and risk of the threat, as well the contextual cues (whether or not escape was possible) and used these factors to generate an appropriate response (Blanchard 2008, Blanchard *et al.* 2011). Risk assessment studies using laboratory rats show that an individual's response is the result of the type and distance of threat, and the local environment, to produce the adaptive response (Blanchard 2008, Blanchard *et al.* 2011). In the wild, whether an animal habituates or not is likely to be influenced both by the immediate environment (for instance, is a safe place to escape available?) and its own locomotor abilities (can it escape?).

Species and individuals within species may vary in how quickly they habituate as a result of personality or sex differences (Rodríguez-Prieto *et al.* 2010a). In humans, personality traits such as extroversion and impulsivity are correlated with a faster startle habituation response (LaRowe 2006). This suggests that over time there will be a non-random distribution of personalities in response to anthropogenic disturbance. Thus, we can predict that more tolerant species or individuals will be able to colonize more disturbed areas (Carrete & Tella 2010).

Habitat availability may be another factor that influences the likelihood of habitation or sensitization. Blumstein (2013) proposed the "contiguous habitat hypothesis" to explain why some Southern California birds habituated while others sensitized. The contiguous habitat hypothesis predicts that species that find themselves in highly fragmented and rare habitats will be more likely to habituate to increased human disturbance. This might result from a process of sorting whereby individuals and species that were unable to tolerate increased disturbance have been eliminated while those that tolerated disturbance persisted in the patches. The net result would be that "tolerant" species will be found in this highly patchy habitat while those in more contiguous habitat might be more variable and indeed might respond to increased disturbance by sensitizing. If generally true, the hypothesis suggests that the opportunity to move within habitat patches will be more often associated with sensitization than situations where animals are so constrained that they have no other choices than habituation.

### 3.2.1.3 Novel mismatches between cues and fitness: is learning important?

Individuals may naturally learn to identify cues that help them detect suitable habitats in which they historically have had relatively high survival or reproductive success (reviewed in Davis & Stamps 2004, Stamps & Swaisgood 2007). In some circumstances, individuals may select suboptimal habitats because of a mismatch between the cues they evolved to evaluate and novel fitness consequences associated with those cues; this is referred to as an ecological trap (reviewed by Schlaepfer *et al.* 2002, Sih *et al.* 2011, chapter 4). Whether ecological traps are more or less likely in species that learn about their habitat (or other biologically important characteristics) is an open question. For instance, animals that disperse may rely on learning cues from their natal habitat to help them develop a template by which they can evaluate habitat quality and determine where to settle while dispersing (Davis & Stamps 2004). The degree that animals learn would influence how those cues can be manipulated.

We expect that associative learning mechanisms (e.g. Pavlovian and instrumental associative learning) should enable individuals to select suitable habitats if learning is a mechanism underlying habitat selection. Even if learning is not a natural mechanism, it might be possible to generate positive experiences to train animals to use a desired habitat and/or negative experiences to train animals to avoid a particular habitat. Stimuli such as tastes, smells or visual cues can give information on relative forage quality or risk of predation that will influence animal decisions.

Extensive work on learning and life skill training in hatchery-reared fish represents an important application of learning theory that has translated to applied value. Hatchery fish that learn life skills such as predator recognition, prey handling and foraging locations exhibit enhanced post-release survival (reviewed in Brown & Laland 2001, Brown *et al.* 2003, Hawkins *et al.* 2008). Additionally, social learning can act as a multiplier of these skills, facilitating quicker learning and transmission, which is more efficient for the aquacul-turist whose aim is to produce animals that will survive upon release.

### 3.2.2 Theme 2. Behavior-based management: training for conservation

Knowledge of learning mechanisms is also of use to managers who wish to modify animal behavior. Training animals with basic learning mechanisms may help repel animals from human resources, attract them to particular habitats/regions or generate basic survival skills to enhance survival during translocations/reintroductions.

#### 3.2.2.1 Teaching attraction

Animal learning principles can provide general rules on how animals can be taught specific behaviors or attraction to habitats as well as the conditions under which they may not be able to be taught. Positive reinforcers can be used to attract an animal, locate food source or increase the frequency of a particular behavior. Stimuli used for positive reinforcement include food, shade, odors, shelter or access to conspecifics. These stimuli can be manipulated to facilitate the learning of habitat preferences. Preferences can be taught via Pavlovian conditioning where the taste is associated with food quality or via instrumental means where, for example, animals are trained to use tunnels beneath freeways. Additionally, conspecific or hetero-specific stimuli can act as positive reinforcers during food source localiza-tion or habitat selection (for review see Avarguès-Weber *et al.* 2013). The constraints to learning mentioned above similarly apply – there may be certain critical periods for learning to develop certain preferences.

#### 3.2.2.2 Teaching avoidance

The creation of novel concentrations of resources, such as crops, garbage cans, fishing lines and domesticated livestock, provide motivation for animals to learn to exploit those resources, resulting in human/wildlife conflict. Since anthropogenic resources, such as fishing lines or crops, can reduce the costs compared to natural foraging, the motivation to form the association between humans and food reward is not only high, but learning is expected to occur quickly (Schakner & Blumstein 2013). Once learned, the association is difficult to break and thus management efforts require foresight and a preventative mindset. Since learning to acquire human resources involves associative mechanisms, there are points in the learning process that management efforts should target to be most effective in teaching avoidance: pre association formation, during association forma-tion and post association formation.

Animals require a contingency to form an association between two stimuli or stimulus/response (Rescorla 1968). In the wild, animals can learn the association between human resources and the cues that reliably precede them. For example, marine mammals, such as sperm whales (*Physeter macrocephalus*), have learned to associate vessel sounds (CS) with a food reward (US: fish on line) (Thode et al. 2007). In order to form that association, the vessel sounds must reliably predict the food reward. Therefore, the most effective management of depredation is preventing animals from learning the depredative behavior in the first place by reducing the contingency between stimulus and reward. This can be accomplished by decoupling the spatio-temporal overlap between potential depredators and the human resources. For example, in the Gulf of Alaska, demersal longline fisheries management shifted from a 10-day derby-style fishing season (vessels catch a year's quota in a set period of time) to an 8-month-long individual fishing quota regime. As a result of the extended overlap between sperm whales and fishing vessels in space and time with the new quota fishing regime, there was ample opportunity for the animals to learn to exploit the resources and the whales are now attracted to boats setting and hauling in lines, which results in a loss of valuable fish (Hill et al. 1999). A lesson from this case study is that foresight may be necessary to prevent learning from occurring in the first place.

When innovators initially learn to depredate or crop raid, social learning can have a multiplier effect by spreading behaviors through populations quickly (Lefevbre 1995). In elephants (*Loxodonta africana*), for example, network analysis of crop raiders has demonstrated that the behavior appears to be socially learned through social networks (Chiyo et al. 2012). Correspondingly, social learning is believed to underlie the diffusion of depredation in sperm whales, killer whales and pilfering black bears (Whitehead 2004, Mazur & Seher 2008, Schakner et al. 2014). In these cases, it is important to know both the identity of innovators (age/sex) and the pattern of diffusion. This knowledge is useful to stop the spread of the behavior and for targeted repellents or removals of individuals.

Once the association between humans and food reinforcers has formed, management efforts rely on raising the cost to the individual depredator. Because the association is difficult to extinguish, management efforts must rely on forming new negative associations or on decoupling the contingency between humans and reward. Deterrents and repellents produce noxious, aversive or painful stimuli to prevent animals from interacting with human habitat or resources (Ramp et al. 2011). Here we suggest that associative learning may produce long-term learned avoidance.

During painful encounters, animals rapidly learn the cues, context or local conditions that are associated with that danger. This learning mechanism, i.e. fear conditioning, enables animals to learn from, respond to and detect danger. Repellents, therefore, should capitalize on insights from the fear conditioning literature to generate avoidance. The use of painful stimuli such as rubber bullets or electric shocks are widespread for eliciting avoidance, but their effectiveness can be short term or impractical, and this raises ethical issues (e.g. is it ethical to continue to do something that's both painful and ineffective?). However, painful deterrent stimuli may be an integral part of a fear-conditioning program. Once conditioning has occurred, exposure to the conditioned stimulus generates fear reactions.

During painful encounters, an animal's unconditioned response is different from the conditioned response. For example, rats exposed to shock (US) react with a burst of motor activity. In contrast, rats exposed to a stimulus that predicts shock (CS such as context or experimenter) evoke behavioral responses such as fleeing, hyper-vigilance or freezing. This suggests a conditioning approach may offer promise, especially if the conditioned response to the target CS is avoidance.

What cues animals pick up on to avoid an area remains an open question. For instance, it is known that animals learn to avoid environments, stimuli or conditions that are correlated with a decrease in fitness (i.e. death; Lima & Dill 1990, Frid & Dill 2002). Habitats, however, contain a suite of stimuli such as landscape features, conspecifics, heterospecifics and background sounds. During an aversive event (a predator attack), individuals likely associate features of the environment (such as open space or shadows) as well as other cues (such as predator scents). According to the Rescorla-Wagner model, contextual stimuli compete with the CS to predict the US. In contrast to simplified experimental conditions of context (a cage), the natural world is full of stimuli, and thus the animal may make associations between competing contextual cues and salient predator cues, This means that managers should use conditional stimuli that are obvious, discriminable and detectable, preceding the biologically relevant aversive stimuli, when designing and implementing repellents. If habitat avoidance is the goal, diffuse CS stimuli, such as a strobe or sound, can be implemented (Table 3.1).

**A checklist for US and CS selection** Effective deterrence relies on stimuli that are both aversive enough to cause rapid fear conditioning, and sufficiently aversive to prevent rapid habituation. To accomplish this, managers must tailor deterrent stimuli toward species-specific sensory modalities and sensory sensitivities. For example, sound is a fundamental channel for

**Table 3.1.** *Advantages and disadvantages of a variety of stimuli that can be used as both conditioned and unconditioned stimuli for management-based training.*

| Stimulus | Advantages | Disadvantages |
|---|---|---|
| **Conditioned stimulus** | | |
| Sound (e.g. Neutral tone) | Localized transmission | Non-target species impact |
| Light (e.g. Strobe light) | Discriminable | Limited to night or dark locations |
| Object (e.g. Flag or person) | Useful for place avoidance | Difficult to associate object with US |
| Chemosensory (e.g. Taste or scent) | Salient cue for food aversion | Limited to nauseating US |
| **Unconditioned stimulus** | | |
| Pain (e.g. Electric shock) | Long-lasting associations after few exposures | Can cause physical damage |
| Distracting (e.g. White noise) | Wide-ranging | Impact non-target species |
| Ecologically relevant stimuli (e.g. Predator cue) | Species-specific | Rapid habituation |
| Frightening stimuli (e.g. Looming, novel or abrupt stimuli) | Can elicit fear responses | Rapid habituation |
| Nauseating (e.g. LiCl) | One trial learning | Unwanted prey avoidance |

communication, foraging and predator detection in marine mammals and this makes it a useful modality in which to develop acoustic deterrents (Jefferson & Curry 1996). However, the input of aversive acoustic stimuli can impact non-target species, which should be considered during the development and implementation of acoustic deterrents (T. Gotz, pers. comm.). Deterrents can be modulated to match a species' sensory sensitivity while still being outside non-target animals' sensory range. Unconditioned stimuli that elicit pain must be practical as well and not cause permanent damage to the depredator. Finally, in social species, fearful responses by conspecifics can serve as a US (Mineka & Cook 1984).

There is evidence that CS which are natural precursors to US result in rapid and more durable associations (Domjan 2005). From this functional perspective, using biologically meaningful stimuli such as predator calls that precede painful stimuli may result in rapid and stronger associations. For example, Leigh and Chamberlain (2008) used barking dogs as a conditioned stimulus preceding rubber buckshot US on crop-raiding bears, which yielded stronger responses than non-conditioned individuals. A conditioned stimulus that precedes the US must be discriminable, salient and consistent. Additionally, the reinforcement schedule (how often to pair CS/US versus CS alone) can be modified depending on the nature of the conflict.

### 3.2.3 Theme 3. Behavioral indicators

Our final section is brief: there may be a variety of behavioral indicators that can be used to reflect an animal's past experiences, and knowledge of past experiences may be useful to wildlife management. The brevity of this section should not undermine its potential importance, and future research should focus on identifying other situations and indicators that can be used to inform management.

#### 3.2.3.1 Flight Initiation Distance

As discussed above, in order to understand the behavioral imprint of humans, flight initiation distance can be used as a behavioral indicator of disturbance (see Chapter 11 for more details). Assuming that all else is equal between sites (e.g. Gill *et al.* 2001), the difference in FID between two sites can provide a measure of the degree to which humans have modified risk assessment. When measured longitudinally, FID can also be used as a proxy for habituation (Ikuta & Blumstein 2003, Rodríguez-Prieto *et al.* 2010b).

#### 3.2.3.2 Socially learned traits

Socially learned traits can diffuse through populations. After reintroductions or translocations, social transmission can be used to track the spread of behaviors through groups. This may indicate how well reintroduced individuals are being incorporated or adapting behaviorally to life in the wild. In a well-documented case of reintroduction, captive-bred Arabian oryx (*Oryx leucoryx*) foraging behavior was suggested to have been influenced by interactions with conspecifics (Tear *et al.* 1997). Social learning is believed to have enhanced foraging behaviors of reintroduced individuals during periods of low food availability (Tear *et al.* 1997). This study suggests that after reintroduction/translocation, managers can probe individuals in a group to assess whether behaviors have spread indirectly through social transmission.

In several species, social learning underlies stable inter-population behavioral variation. Apes, songbirds and cetaceans are believed to exhibit long-term, socially learned traditions or cultures (Whiten et al. 1999, Rendell & Whitehead 2001, Laiolo & Tella 2007). Since these socially learned behaviors are often functional (i.e. they are foraging tactics or social signals with fitness consequences) these traits could be used to indicate population viability (Laiolo & Tella 2007, Whitehead 2010). Laiolo and Tella (2005, 2007) were able to use bird song (a socially learned trait) diversity to show that fragmentation has eroded both cultural and population diversity. These studies suggest that cultural diversity can be used as a proxy for population viability as well as a tool for targeting subpopulations likely to be threatened (Whitehead 2010).

## 3.3 SUMMARY

We believe that the fundamental mechanisms involved in animal learning are of practical importance to conservation/management practitioners and central to integrating behavioral ecology with conservation and wildlife management. The necessity of incorporating learning into conservation is further discussed in subsequent chapters. In Chapter 6, for instance, Fernández-Juricic describes how species-specific input channels and sensory systems influence the stimuli that will be learned, which can be applied to repelling or attracting animals. The role of learning in behavioral modification is further discussed by Shier in Chapter 10, including case studies involving reintroduction/translocations. From a broader perspective, learning is a mechanism of phenotypic plasticity, and the range and limits to plasticity in endangered and threatened species can be used to predict and manage species responses to anthropogenic change (Chapter 5).

## ACKNOWLEDGEMENTS

Z.S. was supported by an NSF predoctoral fellowship and by a grant from the LaKretz Center for California Conservation Science. D.T.B. was supported by NSF during manuscript preparation. We thank the editors and two anonymous reviewers for astute and very constructive suggestions that helped us improve this chapter.

## REFERENCES

Akins, C.K. and Zentall, T.R. 1996. Imitative learning in male Japanese quail (*Coturnix japonica*) using the two-action method. *Journal of Comparative Psychology*, 110: 316–320.

Akins, C.K. and Zentall, T.R. 1998. Imitation in Japanese quail: The role of reinforcement of demonstrator responding. *Psychonomic Bulletin & Review*, 5: 694–697.

Amiel, J.J., Tingley, R. and Shine, R. 2011. Smart moves: effects of relative brain size on establishment success of invasive amphibians and reptiles. *PLoS ONE*, 6: e18277.

Avarguès-Weber, A., Dawson, E.H. and Chittka, L. 2013. Mechanisms of social learning across species boundaries. *Journal of Zoology*, 290: 1–11

Bateson, P. 1978. Sexual imprinting and optimal outbreeding. *Nature*, 273: 659–660.

Berger, J., Swenson J.E. and Persson, I. L. 2001. Recolonizing carnivores and naïve prey: conservation lessons from Pleistocene extinctions. *Science*, 291: 1036–1039.

Berger-Tal, O., Polak, T., Oron, A. *et al.* 2011. Integrating animal behavior and conservation biology: a conceptual framework. *Behavioral Ecology*, 22: 236–239.

Blanchard, C.D. 2008. Defensive behaviors, fear, and anxiety. In Blanchard, C.D., Blanchard, R.J., Griebel, G. and Nutt, D.J. (eds.), *Handbook of Anxiety and Fear*. pp. 63–79. Oxford: Elsevier.

Blanchard, C.D., Griebel, G., Pobbe, R. and Blanchard, R.J. 2011. Risk assessment as an evolved threat detection and analysis process. *Neuroscience and Biobehavioral Reviews*, 35: 991–998.

Blumstein, D.T. and Fernández-Juricic, E. (2010). *A Primer of Conservation Behavior*. Sunderland: Sinauer Associates.

Brown, C., Davidson, T. and Laland, K.N. 2003. Environmental enrichment and prior experience of live prey improve foraging behaviour in hatchery-reared Atlantic salmon. *Journal of Fish Biology*, 63: 187–196.

Brown, C. and Laland, K.N. 2001. Social learning and life skills training for hatchery-reared fish. *Journal of Fish Biology*, 59: 471–493.

Carlier, P. and Lefebvre, L. 1997. Ecological differences in social learning between adjacent, mixing, populations of Zenaida doves. *Ethology*, 103: 772–784.

Carrete, M. and Tella, J.L. 2010. Individual consistency in flight initiation distances in burrowing owls: a new hypothesis on disturbance-induced habitat selection. *Biology Letters*, 6: 167–170.

Chiyo, P.I., Moss, C.J. and Alberts, S.C. 2012. The influence of life history milestones and association networks on crop-raiding behavior in male African elephants. *PLoS ONE*, 7: e31382.

Coussi-Korbel, S. and Fragaszy, D.M. 1995. On the relation between social dynamics and social learning. *Animal Behaviour*, 50: 1441–1453.

Davis, J.M. and Stamps, J.A. 2004. The effect of natal experience on habitat preferences. *Trends in Ecology & Evolution*, 19: 411–416.

Dawkins, R. and Krebs, J.R. 1979. Arms races between and within species. *Proceedings of the Royal Society of London. Series B, Biological Sciences*, 205: 489–511.

Deecke, V.B., Slater, P.J.B. and Ford, J.K.B. 2002. Selective habituation shapes acoustic predator recognition in harbour seals. *Nature*, 420: 171–173.

DeRuiter, S.L. and Doukara, K.L. 2012. Loggerhead turtles dive in response to airgun sound exposure. *Endangered Species Research*, 16: 55–63.

Dickinson, A. 1980. *Contemporary Animal Learning Theory*. Cambridge: Cambridge University Press.

Doligez, B., Danchin, E. and Clobert, J. 2002. Public information and breeding habitat selection in a wild bird population. *Science*, 297: 1168–1170.

Domjan, M. 2005. Pavlovian conditioning: a functional perspective. *Annual Review of Psychology*, 56: 179–206.

Domjan, M. and Burkhard, B. 1986. *The Principles of Learning & Behavior*. Monterey: Brooks/Cole Publication Company.

Dugatkin, L.A. and Alfieri, M.S. 2003. Boldness, behavioral inhibition and learning. *Ethology Ecology & Evolution*, 15: 43–49.

Dukas R. 2009. Learning: mechanisms, ecology, and evolution. In Dukas, R. and Ratcliffe, J.M. (eds.), *Cognitive Ecology II*. pp. 7–26. Chicago: University of Chicago Press.

Dukas, R. 2008. Life history of learning: performance curves of honeybees in settings that minimize the role of learning. *Animal Behaviour*, 75: 1125–1130.

Ellenberg, U., Mattern, T. and Seddon, P.J. 2009. Habituation potential of yellow-eyed penguins depends on sex, character and previous experience with humans. *Animal Behaviour*, 77: 289–296.

Estes, J.A., Tinker, M.T., Williams, T.M. and Doak, D.F. 1998. Killer whale predation on sea otters: linking oceanic and nearshore ecosystems. *Science*, 282: 473–476.

Fanselow, M.S. and Ponnusamy, R. 2008. The use of conditioning tasks to model fear and anxiety. In Blanchard, C.D., Blanchard, R.J., Griebel, G. and Nutt, D.J. (eds.), *Handbook of Anxiety and Fear*. pp. 29–48. Oxford: Elsevier.

Fanselow, M.S. 1984. What is conditioned fear? *Trends in Neurosciences*, 7: 460–462.

Franz, M. and Nunn, C.L. 2009. Network-based diffusion analysis: a new method for detecting social learning. *Proceedings of the Royal Society B: Biological Sciences*, 276: 1829–1836.

Frid, A. and Dill, L. 2002. Human-caused disturbance stimuli as a form of predation risk. *Conservation Ecology*, 6: 11–26.

Galef, B. and Laland, K.N. 2005. Social learning in animals: empirical studies and theoretical models. *BioScience*, 55: 489–499.

Garcia, J. and Koelling, R.A. 1966. Relation of cue to consequence in avoidance learning. *Psychonomic Science*, 4: 123–124.

Gill, J.A., Norris, K. and Sutherland, W.J. 2001. Why behavioural responses may not reflect the population consequences of human disturbance. *Biological Conservation*, 97: 265–268.

Gotz, T. and Janik, V.M. 2011. Repeated elicitation of the acoustic startle reflex leads to sensitisation in subsequent avoidance behaviour and induces fear conditioning. *BMC Neuroscience*, 12: 1–12

Greenberg, R. 2003. The role of neophobia and neophilia in the development of innovative behaviour of birds. In Reader, S. and Laland, K. (eds.), *Animal Innovation*, pp. 175–196. New York: Oxford University Press.

Griffin, A.S. and Evans, C.S. 2003. The role of differential reinforcement in predator avoidance learning. *Behavioural Processes*, 61: 87–94.

Griffin, A.S., Evans, C.S. and Blumstein, D.T. 2001. Learning specificity in acquired predator recognition. *Animal Behaviour*, 62: 577–589.

Grillon, C. 2008. Models and mechanisms of anxiety: evidence from startle studies. *Psychopharmacology*, 199: 421–437.

Groves, P M. and Thompson, R.F. 1970. Habituation a dual process theory. *Psychological Review*, 77: 419–450.

Guillette, L.M., Reddon, A.R., Hurd, P.L. and Sturdy, C.B. 2009. Exploration of a novel space is associated with individual differences in learning speed in black-capped chickadees, *Poecile atricapillus*. *Behavioural Processes*, 82: 265–270.

Hawkins, L.A., Magurran, A.E. and Armstrong, J.D. 2008. Ontogenetic learning of predator recognition in hatchery-reared Atlantic salmon, *Salmo salar*. *Animal Behaviour*, 75: 1663–1671.

Healy, S.D. and Rowe, C. 2007. A critique of comparative studies of brain size. *Proceedings of the Royal Society B-Biological Sciences*, 274: 453–464.

Hemmi, J.M. and Merkle, T. 2009. High stimulus specificity characterizes anti-predator habituation under natural conditions. *Proceedings of the Royal Society B-Biological Sciences*, 276: 4381–4388.

Heyes, C. 2012. What's social about social learning? *Journal of Comparative Psychology*, 126: 193–202.

Heyes, C.M. 1994. Social-learning in animals – categories and mechanisms. *Biological Reviews of the Cambridge Philosophical Society*, 69: 207–231.

Heyes, C.M., Ray, E.D., Mitchell, C.J. and Nokes, T. 2000. Stimulus enhancement: controls for social facilitation and local enhancement. *Learning and Motivation*, 31: 83–98.

Hill, P.S., Laake, J.L. and Mitchell, E.D. 1999. Results of a pilot program to document interactions between sperm whales and longline vessels in Alaskan waters. US Department of Commerce, Report No. NOAA TM-NMFS-AFSC-108.

Hogan, J.A. and Bolhuis, J.J. 2005. The development of behaviour: trends since Tinbergen (1963). *Animal Biology*, 55: 371–398.

Hollis, K. 1982. Pavlovian conditioning of signal-centered action pattern and autonomic behavior: a biological analysis of function. *Advances in the Study of Behavior*, 12: 1.

Hoppitt, W. and Laland, K.N. 2008. Social processes influencing learning in animals: a review of the evidence. *Advances in the Study of Behavior*, 38: 105–165.

Hoppitt, W., Boogert, N. J. and Laland, K. N. 2010. Detecting social transmission in networks. *Journal of Theoretical Biology*, 263, 544–555.

Ikuta, L.A. and Blumstein, D.T. 2003. Do fences protect birds from human disturbance? *Biological Conservation*, 112: 447–452.

Jefferson, T.A. and Curry, B.E. 1996. Acoustic methods of reducing or eliminating marine mammal-fishery interactions: do they work? *Ocean Coastal Management*, 31: 41–70.

Johnston, T.D. 1982. Selective costs and benefits in the evolution of learning. *Advances in the Study of Behavior*, 12: 65–106.

Laiolo, P. and Tella, J.L. 2005. Habitat fragmentation affects culture transmission: patterns of song matching in Dupont's lark. *Journal of Applied Ecology*, 42: 1183–1193.

Laiolo, P. and Tella, J.L. 2007. Erosion of animal cultures in fragmented landscapes. *Frontiers in Ecology and the Environment*, 5: 68–72.

Laland, K.N. 2004. Social learning strategies. *Learning & Behavior*, 32: 4–14.

LaRowe, S.D., Patrick, C.J., Curtin, J J. and Kline, J.P. 2006. Personality correlates of startle habituation. *Biological Psychology*, 72: 257–264.

Lefebvre, L. 1995. Culturally-transmitted feeding-behavior in primates – evidence for accelerating learning rates. *Primates*, 36: 227–239.

Leigh, J. and Chamberlain, M. 2008. Effects of aversive conditioning on behavior of nuisance Louisiana black bears. *Human–Wildlife Interactions*, 51: 175–182.

Li, C., Monclús, R., Maul, T.L., Jiang, Z. and Blumstein, D.T. 2011. Quantifying human disturbance on antipredator behavior and flush initiation distance in yellow-bellied marmots. *Applied Animal Behaviour Science*, **129**: 146–152.

Lima, S.L. and Dill, L.M. 1990. Behavioral decisions made under the risk of predation – a review and prospectus. *Canadian Journal of Zoology*, **68**: 619J.

Lorenz, K. 1970. Companions as factors in the bird's environment. In *Studies in Human and Animal Behaviour*. pp. 101 – 258. Cambridge: Harvard University Press.

Lowry, H., Lill, A. and Wong, B.B.M. 2011. Tolerance of auditory disturbance by an avian urban adapter, the noisy miner. *Ethology*, **117**: 490–497.

Mackintosh, N. J. 1974. *The Psychology of Animal Learning*. Oxford: Academic Press.

Macphail, E.M. and Barlow, H.B. 1985. Vertebrate intelligence: the null hypothesis [and discussion]. *Philosophical Transactions of the Royal Society of London. B, Biological Sciences*, **308**: 37–51.

Mazur, R. and Seher, V. 2008. Socially learned foraging behaviour in wild black bears, *Ursus americanus. Animal Behaviour*, **75**: 1503–1508.

Mineka, S., Davidson, M., Cook, M. and Keir, R. 1984. Observational conditioning of snake fear in rhesus-monkeys. *Journal of Abnormal Psychology*, **93**: 355–372.

Mineka, S. and Ohman, A. 2002. Phobias and preparedness: the selective, automatic, and encapsulated nature of fear. *Biological Psychiatry*, **52**: 927–937.

Nunn, C.L., Thrall, P.H., Bartz, K., Dasgupta, T. and Boesch, C. 2009. Do transmission mechanisms or social systems drive cultural dynamics in socially structured populations? *Animal Behaviour*, **77**: 1515–1524.

Ohman, A. and Mineka, S. 2002. Fears, phobias, and preparedness: toward an evolved module of fear and fear learning. *Psychological Review*, **108**: 483–522.

Raderschall, C.A., Magrath, R.D. and Hemmi, J.M. 2011. Habituation under natural conditions: model predators are distinguished by approach direction. *The Journal of Experimental Biology*, **214**: 4209–4216.

Ramp, D., Foale, C.G., Roger, E. and Croft, D.B. 2011. Suitability of acoustics as non-lethal deterrents for macropodids: the influence of origin, delivery and antipredator behaviour. *Wildlife Research*, **38**: 408–418.

Rankin, C.H., Abrams, T., Barry, R.J. *et al.* 2009. Habituation revisited: an updated and revised description of the behavioral characteristics of habituation. *Neurobiology of Learning and Memory*, **92**: 135–138.

Rau, V. and Fanselow, M.S. 2009. Exposure to a stressor produces a long lasting enhancement of fear learning in rats. *Stress – the International Journal on the Biology of Stress*, **12**: 125–133.

Rendell, L., Boyd, R., Cownden, D. *et al.* 2010. Why copy others? Insights from the social learning strategies tournament. *Science*, **328**: 208–213.

Rendell, L. and Whitehead, H. 2001. Culture in whales and dolphins. *Behavioral and Brain Sciences*, **24**: 309

Rescorla, R.A. 1968. Probability of shock in the presence and absence of CS in fear conditioning. *Journal of Comparative and Physiological Psychology*, **66**: 1–5.

Rescorla, R.A. and Wagner, A.R. 1972. A theory of Pavlovian conditioning: variations in the effectiveness of reinforcement and nonreinforcement. *Classical Conditioning II: Current Research and Theory*, pp. 64–99.

Rescorla, R.A. 1988. Behavioral studies of Pavlovian conditioning. *Annual Review of Neuroscience*, **11**: 329–352.

Rodríguez-Prieto, I., Martín, J. and Fernández-Juricic, E. 2010a. Individual variation in behavioural plasticity: direct and indirect effects of boldness, exploration and sociability on habituation to predators in lizards. *Proceedings of the Royal Society B: Biological Sciences,* **278**: 266–273

Rodríguez-Prieto, I., Martín, J. and Fernández-Juricic, E. 2010b. Habituation to low-risk predators improves body condition in lizards. *Behavioral Ecology and Sociobiology,* **64**: 1937–1945.

Schlaepfer, M.A., Runge, M.C. and Sherman, P.W. 2002. Ecological and evolutionary traps. *Trends in Ecology & Evolution,* **17**: 474–480.

Seferta, A., Guay, P.J., Marzinotto, E. and Lefebvre, L. 2001. Learning differences between feral pigeons and Zenaida doves: the role of neophobia and human proximity. *Ethology,* **107**: 281–293.

Schakner, Z.A. and Blumstein, D.T. 2013. Behavioral biology of marine mammal deterrents: a review and prospectus. *Biological Conservation,* **167**: 380–389.

Schakner, Z.A., Lunsford, C., Straley, J., Eguchi, T. and Mesnick, S.L. 2014. Using models of social transmission to examine the spread of longline depredation behavior among sperm whales in the gulf of Alaska. *PLoS ONE,* **9**: e109079.

Shalter, M. 1984. Predator–prey behavior and habituation. In *Habituation, Sensitization, and Behavior,* pp. 349–391. Orlando: Academic Press Inc.

Shapiro, K.L., Jacobs, W.J. and Lolordo, V.M. 1980. Stimulus-reinforcer interactions in Pavlovian conditioning of pigeons – implications for selective associations. *Animal Learning & Behavior,* **8**: 586–594.

Shettleworth, S.J. 1975. Reinforcement and the organization of behavior in golden hamsters: hunger, environment, and food reinforcement. *Journal of Experimental Psychology: Animal Behavior Processes,* **1**: 56–87.

Shettleworth, S.J. 2010. *Cognition, Evolution, and Behavior.* Second edition. New York: Oxford University Press.

Sih, A., Ferrari, M.C.O. and Harris, D.J. 2011. Evolution and behavioural responses to human-induced rapid environmental change. *Evolutionary Applications,* **4**: 367–387.

Sih, A. and Giudice, M.D. 2012. Linking behavioural syndromes and cognition: a behavioural ecology perspective. *Philosophical Transactions of the Royal Society of London B,* **367**: 2762–2772.

Sneddon, LU. 2003. The bold and the shy: individual differences in rainbow trout. *Journal of Fish Biology,* **62**: 971–975.

Sol, D. 2003. Behavioral innovation: a neglected issue in the ecological and evolutionary literature. In Reader, S. and Laland, K. (eds.), *Animal Innovation,* pp. 63–82. New York: Oxford University Press.

Sol, D., Bacher, S., Reader, S.M. and Lefebvre, L. 2008. Brain size predicts the success of mammal species introduced into novel environments. *The American Naturalist,* **172**: S63–S71.

Sol, D., Duncan, R.P., Blackburn, T.M., Cassey, P. and Lefebvre, L. 2005. Big brains, enhanced cognition, and response of birds to novel environments. *Proceedings of the National Academy of Sciences of the United States of America,* **102**: 5460–5465.

Sol, D., Timmermans, S. and Lefebvre, L. 2002. Behavioural flexibility and invasion success in birds. *Animal Behaviour,* **63**: 495–502.

Stamps, J.A. and Swaisgood, R.R. 2007. Someplace like home: experience, habitat selection and conservation biology. *Applied Animal Behaviour Science*, 102: 392–409.

Swaney, W., Kendal, J., Capon, H., Brown, C. and Laland, K.N. 2001. Familiarity facilitates social learning of foraging behaviour in the guppy. *Animal Behaviour*, 62: 591–598.

Tear, T.H., Mosley, J.C. and Ables, E.D. 1997. Landscape-scale foraging decisions by reintroduced Arabian oryx. *The Journal of Wildlife Management*, 61: 1142–1154.

Thode, A., Straley J., Tiemann, C.O., Folkert, K. and O'Connell, V. 2007. Observations of potential acoustic cues that attract sperm whales to longline fishing in the Gulf of Alaska. *Journal of the Acoustical Society of America*, 122: 1265–1277.

Thorpe, W.H. 1956. *Learning and Instinct in Animals*. Cambridge: Harvard University Press.

Thorndike, E.L. and Bruce. D. 1911. *Animal Intelligence: Experimental Studies*. Lewiston: Macmillan Press.

Trimmer, P.C., McNamara, J.M., Houston, A.I. and Marshall, J.A.R. 2012. Does natural selection favour the Rescorla–Wagner rule? *Journal of Theoretical Biology*, 302: 39–52.

Whitehead, H. 2010. Conserving and managing animals that learn socially and share cultures. *Learning and Behavior*, 38: 329–336.

Whiten, A., Goodall J., McGrew, W.C. *et al.* 1999. Cultures in chimpanzees. *Nature*, 399: 682–685.

Yeomans, J.S., Li, L., Scott, B.W. and Frankland, P.W. 2002. Tactile, acoustic and vestibular systems sum to elicit the startle reflex. *Neuroscience and Biobehavioral Reviews*, 26: 1–11.

# Part II

# Anthropogenic impacts on animal behavior and their implications for conservation and management

The behavior of animals enables them to better confront a constantly changing environment. Specifically, the behavior of animals modifies the environmental conditions an animal experiences in a manner that is expected to improve their fitness. This can be achieved in two main ways: (a) By shifting (moving) from a poorer to a better environment, such as moving to improve foraging efficiency, safety or thermoregulation. (b) By modifying the present environment, for example – attracting mates using various signaling techniques or increasing vigilance if perceived risk increases. Most animals possess a rich portfolio of behavioral responses, which may range in their flexibility from being entirely fixed (i.e. the same behavior will be displayed regardless of the environmental conditions) to being completely flexible (i.e. the behavioral response will change at the same rate as the environment). The level of behavioral flexibility will be dictated by the animal's evolutionary history, past experience, and genetic, physical and physiological constraints.

Human-Induced Rapid Environmental Change (HIREC) is expected to elicit a behavioral response in the animals experiencing this change. However, because these changes may be novel and rapid in evolutionary terms and were not previously experienced by the animal, it may either: fail to recognize the change, fail to respond, respond inappropriately or respond in a manner that initially or seemingly is beneficial but might have long-term negative consequences. The two chapters in this section address conservation concerns stemming from the behavioral responses of animals as their environment is rapidly modified by anthropogenic activity. Chapter 4 addresses problems stemming from behavioral rigidity resulting in an inappropriate response to novel stimuli, and gives conservation practitioners the means to identify the source of the behavioral rigidity and manage it accordingly. Chapter 5 addresses plastic responses to anthropogenic changes, their benefits to wildlife, their usefulness as a management tool and their possible long-term negative consequences.

# Behavioral rigidity in the face of rapid anthropogenic changes

## ODED BERGER-TAL AND DAVID SALTZ

## 4.1 WHAT IS RIGID BEHAVIOR?

For well over a century, biologists and psychologists have been arguing about the origins of behavior. Is behavior fixed and innate? Is it only determined by the genetic composition of the individual or is it flexible and shaped by the individual's environment? This heated argument, also known as the "nature versus nurture" debate, is yet unsettled (Ridley 2003), although the common consensus (at least among biologists, Bolhuis 2013) is that the dichotomy between innate and learned behaviors is false and that the development of behavior is a complex process involving continual interactions between the characteristics of an individual and its environment (Lehrman 1953).

The relevant aspect of this debate for our purposes is that some behaviors are mostly fixed, and do not change, regardless of the environment the individual is in, compared to other behaviors that are much more plastic. Furthermore, even plastic behaviors are constrained within limits, and these limits may vary depending on the behavior and the environment. We term the display of fixed behaviors in the face of a changing environment as "behavioral rigidity." There are three main causes for behavioral rigidity: fixed or "instinctive" behaviors for which the individual displays no learning or that cannot be changed due to physical or physiological constraints, imprinted behaviors that are plastic only during the early period of an organism's life and afterwards become fixed, and behaviors that are flexible, but this flexibility is too slow to keep up with environmental change.

### 4.1.1 Fixed behaviors

Darwin was one of the first to note that behavioral and personality traits are inherited, using the hereditary nature of behavior in domesticated animals as a compelling example (Darwin 1871). What Darwin

*Conservation Behavior: Applying Behavioral Ecology to Wildlife Conservation and Management*, eds. O. Berger-Tal and D. Saltz. Published by Cambridge University Press. © Cambridge University Press 2016.

was implying, had he possessed the terminology we now have, is that many behaviors are strongly influenced and constrained by genetic factors. It was later shown that roughly 40% of the variation in personality in humans is genetic in nature (Bouchard 1994), and that animal personalities have a similarly strong genetic basis as well (van Oers *et al.* 2005). One of the more famous examples of such fixed behavior is the cuckoo. Despite never seeing a parent or sibling of the same species, a cuckoo individual is able to sing species-specific songs and attract fellow cuckoos as mates. Such a fixed behavior was termed *instinctive* behavior by Darwin, a term later adopted by the "fathers" of ethology Tinbergen and Lorentz, as well as by many others. However, there is a considerable dispute in the scientific literature regarding the term "instinct," with some scientists even refusing to use the term altogether (Ridley 2003), and thus for our purposes we will not use the term instinctive but rather define any behavior or behavioral process that remains unchanged in the face of a changing environment as a fixed behavior. Indeed, such behavior does not have to be innate.

Animals may fail to recognize the detrimental nature of novel disturbances that are not part of the animal's evolutionary history (Sih *et al.* 2011), and therefore may keep exhibiting a fixed behavior even when the genetic composition of the animal allows for flexibility in the relevant trait. Moreover, in addition to cognitive skills that may limit an animal's ability to learn, the flexibility of a behavior may also be constrained by physical or physiological limitations, and therefore, learning to recognize a novel and potentially detrimental disturbance does not necessarily mean that an animal will be able to respond to it (Sih *et al.* 2010, 2011). For example, animals relying on camouflage as an anti-predatory strategy (such as the white plumage of several sub species of ptarmigans, *Lagopus Muta*, or the white coat of snowshoe hares, *Lepus americanus*, during the winter) may be behaviorally adapted to moving less and staying inconspicuous (Steen *et al.* 1992). Climate change may create a mismatch between the background color and the animal's seasonal coat (Imperio *et al.* 2013, Mills *et al.* 2013), making fast movement a better strategy to avoid predators. However, animals adapted to crypsis behavior may be physiologically unable to outrun their predators, increasing their vulnerability to predation in the absence of adequate camouflage (Imperio *et al.* 2013).

The perceptual abilities of animals are another type of physiological constraint that might keep a behavior rigid. Birds colliding with glass windows are an excellent example (Klem 1989). Birds that survive a collision with a glass window may not be able to avoid hitting glass windows again if they simply can't detect them. Thus, even though the flight pattern of the bird is very flexible, it will not be able to change it in order to adapt to the glass windows (of course, a bird can learn to avoid a certain location in which it hit a window, especially if there is another cue that is linked to the window location, but this will not prevent it from hitting other windows in other locations). In a similar way, leatherback turtles, *Dermochelys coriacea*, prey on jellyfish and are therefore looking for floating transparent objects as cues for food items. Ever since humans began dumping trash into the oceans, the same cues cause the turtles to consume plastic bags, which can lead to a series of health problems and eventually death (Bjorndal *et al.* 1994). If the turtles cannot tell between jellyfish and plastic bags, then this behavior will remain rigid despite its dire consequences, and may only change by natural selection on an evolutionary time scale.

### 4.1.2 Imprinted behaviors

Imprinting is defined as a genetically canalized learning process characterized by a relatively short critical period occurring early in development, which has long-lasting effects (Lorenz 1937). Most importantly, under the right conditions and stimuli, the process is almost irreversible (Bolhuis 1991). Thus, behavior that is shaped by imprinting is fixed. The most well-known form of imprinting is filial imprinting, where young animals learn parental characteristics (e.g. ducklings that learn to recognize their mother and follow it around; Lorentz 1937, Bateson 2003). However, there are other forms of imprinting such as sexual imprinting – in which exposure to a stimulus affects future behavior toward potential mates (Rantala & Marcinkowska 2011), and various "ecological" imprinting processes such as a learned preference for certain diets or habitats (Immelmann 1975).

The concept of habitat imprinting and the related and more general concept of natal habitat preference induction (NHPI) are central to certain aspects of conservation, such as conservation translocations (Stamps & Swaisgood 2007, Davis 2010). NHPI occurs when an animal's experience in its natal habitat affects its preference for post-dispersal habitats (Davis & Stamps 2004). Positive NHPI is the more common form of this behavioral phenomenon, where animals prefer to settle in habitats that contain the same cues as their natal habitats, but there are also examples of negative

NHPI where dispersers with poor experiences in their natal habitat reject habitats with similar cues (Swaisgood 2010).

Whatever the imprinting process is, the concept of critical periods has important conservation implications. Virtually all organisms (including humans) have to be exposed to certain stimuli at critical periods of their growth in order to develop physiologically and behaviorally so they will have an adequate expected fitness. For example, young chaffinches must be exposed to the singing of adults before they mature, or they will never properly learn to sing (Nottebohm 1970). This critical period represents a crucial window of opportunities for the management of species that does not exist in fixed behaviors.

The best opportunities to manage or manipulate imprinted behaviors are usually (but not strictly) found during captive breeding and reintroductions. This is done by using the appropriate cues while taking the relevant critical period into account, e.g. rearing young chicks using puppets that mimic adults of their species (Wallace *et al.* 2007), or planting cues from an intended release site in the natal habitat (captive or wild) of a source population targeted for translocation (Stamps & Swaisgood 2007). Shier (Chapter 10) describes in detail the various behavioral manipulations that can be used to manage captive-bred animals.

### 4.1.3 Slow plastic response

A behavior does not necessarily have to be innately fixed to produce rigid responses to anthropogenic changes. As is detailed in Chapter 3 on learning, there are various internal and external constraints on the ability of animals to learn and adapt to novel conditions and these constraints may determine whether a behavioral change in response to anthropogenic alterations is fast enough to be effective. If the response of an animal is too slow, it can be considered as rigid since the consequences for the animal will be the same as for animals displaying fixed behaviors (at least from a management point of view). But, in this case, in contrast to innate and imprinted behaviors, it may be possible to speed up the learning process and mitigate the threat.

The anti-predator responses of captive animals serve as a good example. Anti-predator behaviors are often very costly. For this reason, individuals in populations that are isolated from predation often lose their formerly adaptive anti-predator behaviors, either over evolutionary time, between generations or even during the course of the animal's life (Berger 1998, Griffin et al. 2000). When faced with predators again, these animals can usually re-learn anti-predator behaviors. However, the process of re-learning is long, and without the proper anti-predatory skills, an animal is unlikely to survive its first encounter with a predator. This naïveté is a serious problem when reintroducing captive-bred animals that have been living in a predator-free environment for generations (Short *et al.* 1992). The anti-predator behavior in this case is not fixed, and these animals can learn to behave adaptively, but the learning process is not fast enough to ensure their survival. The solution in this case is to train captive animals prior to their release back to the wild in order to enhance the expression of preexisting anti-predator behaviors (Griffin *et al.* 2000, chapter 10). A slow habituation process to a non-threatening disturbance, such as the presence of tourists, is another example of how a slow behavioral response can lead to a reduction in fitness (see section on ecological scarecrows).

The rapidity and lethality of anthropogenic disturbances and the consistency and reliability of their signals will also determine the rate of the behavioral response to these disturbances. An animal can only learn to change its behavior if the rate of the learning is faster than the rate of environmental change (Chapter 3). Since many anthropogenic changes are very rapid, any behavioral response to these changes is bound to be relatively slow. In such cases, there will be strong selection against those animals with an inadequate behavioral response. Very lethal disturbances may also elicit strong selection pressures, since there cannot be any learning taking place when all individuals with the relevant experience die. Lastly, the consistency and reliability of the disturbance's cues will determine whether the animal is able to effectively couple a stimulus and a response to induce an efficient learning behavior (Chapter 3). Cues that are not consistent or reliable will slow the learning process and, consequently, any change in behavior will be slow relative to the anthropogenic changes driving it.

## 4.2 CONSEQUENCES AND MECHANISMS OF BEHAVIORAL RIGIDITY

The inability of animals to change their behavior and adapt to anthropogenic changes is one of the main causes for the rapid decline in wildlife populations worldwide. The most direct consequence of behavioral rigidity is a reduction in individual fitness because of reduced survival (higher mortality) or because of lower reproductive success. Some of the most common causes of wildlife mortality, such as collisions with vehicles, are, in many cases, the result of behavioral rigidity. Millions of animals get run over and die each year in the United States alone (Coffin 2007). Most of these animals failed to learn that roads pose a grave risk for them and did not change their movement behavior accordingly (avoiding roads altogether or crossing at "safe" hours or locations). In many cases animals do not perceive cars as a threat, as they are not "chased" by them (i.e. the cars never leave the road in the direction of the animal), and the speed of vehicles is probably something they cannot comprehend. It is interesting to note that in areas where vehicles go off road and "chase" animals, e.g. 4x4s or snowmobiles, animals do consider the vehicles as a threat (Manor & Saltz 2005). Mitigating the extremely complex issue of animal–vehicle collision must also take animal behavior into account, considering the need to make animals learn that roads (or specific accident-prone sections of roads) are dangerous so they could change their behavior accordingly. Behavioral rigidity can also lower the fitness of animals by reducing their reproductive success. For example, anthropogenic noise, such as traffic noise, can "drown out" the male mating songs in bird and amphibian species, therefore interfering with acoustic assessment of male quality, or disturbing parents–offspring communications (Halfwerk et al. 2011). Individuals that do not alter their breeding behavior to accommodate for the background noise or alternatively move away to a quieter area will suffer reproductive consequences. Similarly, chemical pollutants can also drown out communication in aquatic species and disturb various aspects of reproductive behavior such as mate choice (Fisher et al. 2006, Rosenthal & Stuart-Fox 2012).

Not all consequences of behavioral rigidity are so easily observed, and behavioral rigidity may have indirect consequences by facilitating behavioral and non-behavioral processes that can have tremendous impacts on both the demography and the evolution of most wildlife species that may not be noticeable over the short term. In the following sections we discuss the main mechanisms through which behavioral rigidity directly influences individual fitness or indirectly impacts animal populations. These mechanisms may be internal (e.g. stress) or external (e.g. Allee effects).

### 4.2.1 Stress

Stress responses appear when an animal experiences a reduction in the familiarity, predictability or controllability of its environment (Wiepkema & Koolhaas 1993). This is predicted to occur in novel situations or when the environment changes abruptly. Short-term stress responses are considered beneficial as they assist individuals in coping successfully with adverse conditions by enhancing physical readiness and learning (Joëls et al. 2006) and increasing short-term survival rates (Sapolsky et al. 2000).

However, if the stress is prolonged, it causes chronically elevated levels of "stress hormones" (such as glucocorticoid steroid hormones or corticosterones), which can have negative impacts on the animal's physiology by increasing susceptibility to disease, reducing fertility and lowering survival (Spolsky et al. 2000). When an animal cannot alter its behavior to avoid or adapt to anthropogenic changes, it may suffer from chronic stress, which will reduce its fitness and may even lead to population decline. For example, capercaillie, Tetrao urogallus, which are suffering strong population declines throughout central Europe, have been shown to have significantly higher corticosterone metabolites in their droppings (indicative of higher stress levels) in areas that are used for ski tourism (Thiel et al. 2008), and endangered yellow-eyed penguins, Megadyptes antipodes, show higher stress responses, lower breeding success and lower fledgling weights at sites with unregulated tourism (Ellenberg et al. 2007). Furthermore, stress can also impair the cognitive abilities of animals, such as their learning and memory capabilities (Teixeira et al. 2007). Reduced learning capabilities can induce maladaptive decision-making and can also increase the animal's overall behavioral rigidity, further decreasing its fitness.

### 4.2.2 Evolutionary and ecological traps

Most species rely on environmental cues to make behavioral and life-history decisions. An evolutionary trap occurs when a rapid environmental change (usually due to anthropogenic activities) causes a previously reliable cue to no longer result in an adaptive decision, leading to reduced survival or reproduction (Schlaepfer et al. 2002, 2005). While this is relevant to a wide variety of decisions, such as what to feed upon, who to mate with, and when to disperse, the most well-studied behavior in this context is habitat selection. An ecological trap is a specific type of evolutionary trap that concerns habitat selection. It happens when environmental changes result in a discrepancy between the cues that individuals use to assess habitat quality and the true quality of the habitat, causing individuals to preferentially settle in habitats of lower quality and therefore experience lower survival or reproductive success relative to other available habitats (Robertson & Hutto 2006, Madliger 2012).

A species' past environment provides the evolutionary history that shapes the sensory and cognitive processes controlling behavior. When the environment changes, a mismatch is created between the cues that individuals use to make decisions and the state of the environment (Sih et al. 2011). In many cases individuals are able to learn to change their preferences and use new cues or, alternatively, evolution will favor individuals within the population that react to the new cues. However, when the response rate of species is too slow (because of no learning, slow learning or morphological and sensory constraints, previously mentioned in this chapter), an evolutionary trap is created. Hence, in most cases, evolutionary traps can be considered at their core the result of behavioral rigidity. Examples of evolutionary traps abound. Schlaepfer et al. (2002) and Robertson et al. (2013) list many of them and are recommended readings on the subject. We will only mention a few. Many animal species react to day-length cues to time their breeding and migration decisions. With global warming, these cues are becoming increasingly decoupled from their previous climatic connotations, and while many species show behavioral plasticity and begin breeding earlier in the season (Grieco et al. 2002, Gienapp et al. 2008), other species do not, and this behavioral rigidity drives population decline (Both et al. 2006, Ludwig et al. 2006). Anthropogenic changes to the environment may include the introduction of a variety of novel pathogens. Male house finches, *Carpodacus mexicanus*, were found to preferentially feed near conspecifics infected with the directly transmitted pathogen *Mycoplasma gallisepticum* (MG), probably because infected individuals show reduced aggression. This behavior is likely to contribute to MG prevalence in this species (Bouwman & Hawley 2010). A very well-known example for an ecological trap is that of mayflies (*Ephemeroptera*) looking for a water surface to lay their eggs on. Asphalt roads have spectral qualities that actually attract the mayflies more than water surfaces and the mayflies end up ovipositing on roads, essentially reducing their reproductive success to zero, despite the availability of nearby suitable ponds (Krista et al. 1998, Horvath et al. 2010). Many species of butterflies will readily lay eggs on introduced plant species, even though their larval offspring may be unable to develop on it, severely reducing recruitment to the butterfly population (Keeler & Chew 2008). Lastly, NPHI, the preference for (or aversion from) habitats that contain similar cues to the ones that were found in an animal's natal habitat (see earlier in this chapter), can lead animals away from high-quality and into poor-quality habitats when the natal cues become unreliable due to anthropogenic changes to the environment (Gilroy & Sutherland 2007).

It is not necessary for individuals to *prefer* the lesser-quality habitat in order for an ecological trap to be formed. Sometimes it is enough that individuals do not distinguish between high-and poor-quality habitats (an "equal-preference trap," Robertson & Hutto 2006). Desert habitats in which trees were planted became sink habitats for the critically endangered lizard, *Acanthodactylus beershebensis*, since trees provided perching spots for avian predators (Hawlena et al. 2010). The lizards did not distinguish between the habitats and kept dispersing into the forested habitat that was relatively vacant because of the high predation rates (density-dependent dispersal behavior), generating ecological trap dynamics. This process caused local extirpation of the lizards from both the altered and the natural habitats (Gundersen et al. 2001, Hawlena et al. 2010). Perceptual traps are another variant of ecological traps in which animals avoid high-quality habitats because they are less attractive to them. Such traps can sometimes be even more detrimental to population persistence than ecological traps (Patten & Kelly 2010). It is interesting to note that evolutionary traps may also lead to Allee effects (see next section). When population size decreases because individuals within the population prefer a harmful habitat, the smaller population size would mean that there is less competition over the more attractive resource, which in this case is access to the harmful habitat. This allows more individuals to choose this habitat, reducing the population size even further, and so forth (Kokko & Sutherland 2001).

Since evolutionary and ecological traps are the result of a mismatch between the current environment and the cues that individuals respond to, they can arise for one of three reasons (sensu Robertson & Hutto 2006): (1) The environment changes for the worse while the cues that elicit a behavioral response remain unaltered. For example, grasslands that attract reptiles may be turned into managed pastures (Rotem et al. 2013). The reptiles receive the same cues and may prefer to occupy these habitats only to face increased mortality when the pasture is mechanically harvested for hay and then plowed. (2) Anthropogenic alterations to the environment cause the creation or the intensification of cues that individuals respond to, making the habitat appear more attractive, even though in reality this habitat is actually maladaptive. For example, water is channeled to the sides of asphalt roads in desert environments, which usually results in elevated vegetation growth adjacent to roads. This may act as a cue of a good habitat and many species (such as insectivorous birds) may prefer to settle just next to a road, which can lead to a substantially decreased survival due to a higher rate of collisions with vehicles (Ben-Aharon 2011). (3) A combination of the previous two mechanisms – an environment may change for the worse and the change itself may make it more attractive. For example, tractor tracks in cereal fields act as an attractive cue for skylarks, *Alauda arvensis*, looking for openings in the crop canopy to build their nests in. However, these tracks also serve as a "predator highway," causing a high rate of nesting failure (Donald et al. 2002).

Knowing the mechanism behind the evolutionary trap is essential if the problem is to be successfully mitigated (Gilroy & Sutherland 2007). If the problem is a decrease in the quality of the environment, then we should focus on improving the environment. Alternatively, if the problem is an intensified or wrong cue, then we should aim to remove or modify the cue, since once removed or modified, the environment will stop acting as a trap, even if its quality remains the same (Roberson 2012). Gilroy and Sutherland (2007) suggested an additional course of action: Creating or enhancing cues for an undervalued beneficial habitat or resource. By attracting species back to beneficial habitats (e.g. through the use of conspecific cues such as the songs of conspecific birds; Virzi *et al.* 2012) we can release them from their evolutionary traps.

Whether we wish to reduce cues that attract species to bad habitats or manipulate cues to enhance the attractiveness of good habitats, we must have extensive knowledge on the way animals perceive their environment (Madliger 2012, Robertson *et al.* 2013). Sensory ecology is becoming an indispensable conservation and management tool that can help us interpret the ways animals behave and allow us to manipulate these behavioral mechanisms in order to improve their management. Chapter 6 of this volume is devoted to the use of sensory ecology in conservation behavior.

### 4.2.3 Allee effects

An Allee effect is a positive relationship between any component of individual fitness and either numbers or density of conspecifics (Allee 1931, Stephens *et al.* 1999). In other words it means that individuals fare better when there are more of them.

We can distinguish between two kinds of Allee effects – weak and strong (Wang & Kot 2001). When the per capita growth rate of a population is reduced at low numbers or densities, but remains positive, it is termed a weak Allee effect. When, at some population size, the growth rate is reduced so much that it actually becomes negative, it is termed a strong Allee effect. The population size below which the growth rate becomes negative is known as the Allee threshold (Courchamp *et al.* 2008) because below it a positive feedback generates a snowballing effect that will bring small populations to extinction. Imagine a species that relies on group size for anti-predatory protection. A small group may not be able to efficiently protect itself against predators, which may reduce its survival, making the group even smaller and further decreasing the efficiency of their anti-predatory behavior, reducing their survival even further and so forth until the population is extinct.

Courchamp *et al.* (2008) distinguish between two types of rare species: "naturally rare" and "anthropogenically rare." Species that are naturally rare normally occur in small or sparse populations, and have been so for a very long time. Hence, these species have adapted over their evolutionary history to being rare, taking advantage of the benefits and reducing to a minimum the costs of a small population size. In contrast, anthropogenically rare species are species that were not previously rare, but have become so due to anthropogenic activities. Therefore, these species have not undergone adaptations to small population size and have a higher probability of exhibiting an Allee effect. In other words, these species display behaviors that are suitable for large groups, and once their populations decrease in size due to anthropogenic activities they may not be able to change their behaviors (at least not fast enough) to better suit the smaller group size. The cause for these species' behavioral rigidity in the face of a reduction in group size can be any of the reasons that we have provided in the beginning of the chapter, but whatever the reason may be, the result is an Allee effect, reducing the per capita growth rate of the population and increasing the risk of extinction.

There are numerous behaviors that rely on a critical amount or a minimum density of individuals and therefore may be rendered ineffective by a smaller group size. These include (but are not limited to) cooperative hunting, cooperative breeding, group vigilance and group thermal regulation (e.g. huddling together for warmth). A very famous example is the endangered African wild dog, *Lycaon pictus*. African wild dogs cooperate in hunting, as well as breed cooperatively – a single breeding pair in each pack raises their young with help from other adults in the pack (Courchamp & Macdonald 2001). African wild dog populations are declining due to fragmentation, conflicts with livestock and game farmers, disease and other anthropogenic causes (IUCN 2013). As their groups become smaller, their hunting success decreases, and they are more likely to lose their prey to kleptoparasites. In addition, small group size means fewer helpers to raise the young, which reduces offspring survival. This has been found to be a strong Allee effect with a negative per capita growth rate occurring when the group size drops below five adults (Courchamp *et al.* 2002). The wild dogs cannot change their hunting strategies or their breeding behavior to ones that are suitable for smaller groups, and due to this behavioral rigidity their numbers further decline. Flamingos serve as another well-known example. Flamingos will reproduce better when exposed to reproductive displays from other flock members, and their reproductive success will increase with group size (Stevens & Pickett 1994). Hence, this also represents a behavior that is clearly maladaptive in small populations but remains rigid nevertheless.

Behavioral adaptations to living in groups are in many cases very complex and deeply ingrained in the life-history of social species. It should therefore come as no surprise that these behaviors will be very hard to change on an ecological timescale. Consequently, we can use this knowledge to predict which species are expected to exhibit Allee effects: cooperative breeders, species with complex social behaviors or life-histories, and any species with task differentiation between individuals (see also Courchamp et al. 2008). This knowledge may also be used to combat invasive or pest species and drive them to local extinction before they establish themselves (Tobin et al. 2011).

### 4.2.4 Ecological scarecrows

We define ecological scarecrows as anthropogenic disturbances that are essentially harmless (i.e. they do not cause any direct damage), but that an animal's inability to ignore or habituate to them cause it to suffer a reduction in fitness. The main causes of ecological scarecrows are benign cues that are misinterpreted as a risk to which the animal does not habituate.

Schakner and Blumstein (Chapter 3) have raised the important question of why do some species habituate to anthropogenic stimuli while others do not, and listed several possible explanations. Regardless of the reason, it is known that many individuals do not habituate to certain disturbances. If these disturbances are harmful in nature, then all individuals will suffer immediate consequences that may reduce their fitness and in some cases even result in mortality. However, even if these disturbances are harmless, the non-habituated individuals may still suffer from a reduction in fitness because of their own, unnecessary yet costly, response to the disturbance. There are three main mechanisms through which ecological scarecrows reduce fitness: distraction, stress and the display of redundant behaviors.

The capacity of animals to attend to multiple stimuli simultaneously is limited (Dukas 2004), and therefore when irrelevant but conspicuous anthropogenic stimuli are added to the animal's environment, attention will be diverted away from important stimuli that are relevant to the survival or reproductive success of the animal toward the novel stimuli, hampering the animal's decision-making and increasing the chances of errors (Chan & Blumstein 2011). For example, the foraging efficiency of three-spined sticklebacks, *Gasterosteus aculeatus*, decreases in the presence of otherwise harmless acoustic noise, causing increased food-handling errors and reduced discrimination between food and non-food items (Purser & Radford 2011). Similarly, Caribbean hermit crabs, *Coenobita clypeatus*, allowed a simulated predator to approach closer before they hid when in the presence of boat motor playbacks (Chan *et al.* 2010). This has been shown to not be a result of masking of the predator's approaching sounds, but rather a result of the crab being distracted by the noise (and later on also by lights) produced by passing boats, which led Chan *et al.* (2010) to propose "the distracted prey hypothesis," suggesting that anthropogenic sounds may distract prey and make them more vulnerable to predation. This hypothesis was later expanded by Chan and Blumstein (2011) to include other attention-shifting stimuli and different behaviors such as foraging and mate choice.

Noise and other anthropogenic activities such as ecotourism can also lead to heightened stress levels in animals. As already stated, prolonged exposure to stress can have negative impacts on the animal's physiology by increasing susceptibility to disease, reducing fertility and lowering survival. Individuals that cannot habituate to a disturbance and keep exhibiting increased stress levels will therefore suffer fitness consequences, even when the disturbances themselves cause no direct harm. Exposure to anthropogenic noise has been shown to induce physiological stress responses in many species (Wright *et al.* 2007), as well as exposure to nature-based recreation activities such as hiking or cycling (Steven *et al.* 2011). For example, the mere presence of tourists caused elevated anxiety levels in wild macaque monkeys, even when the tourists were not directly interacting with the monkeys (Marechal *et al.* 2011).

Lastly, anthropogenic disturbances may elicit behavioral responses, usually anti-predatory behaviors, despite the fact that these behaviors are costly and, in these cases, redundant. You may ask, if the animal behaviorally reacts to the disturbance, isn't this a case of behavioral plasticity? The answer is no. The animal's response is rigid, and it treats the disturbance just as it would treat any novel disturbance (i.e. by fleeing, increasing vigilance, etc.). While this behavior is essentially an adaptive response, if the animal persists in this behavior and does not habituate to a repeated harmless disturbance, the constant expression of the behavioral response can have negative fitness consequences. Examples include: hoatzins, *Opisthocomus hoazin*, that fly or climb away from their nests every time a tourist canoe approaches, which may increase levels of nest predation (Karp & Root 2009); grey kangaroos, *Macropus fulignosus*, that stop foraging, take flight and even move out of the local area in response to an artificial whip-crack sound (Biedenweg *et al.* 2011); and Asian rhinoceros, *Rhinoceros unicornis*, that increase the time they devote to vigilance and decrease feeding when tourists are present (Lott & McCoy 1995). The hoatzins from the first example also show no habituation to the disturbance if the approach of the canoe is accompanied by the sounds of conversation (at any level). Similarly, the grey kangaroos do not habituate to the sound of a whip even when the signal is played at a rate of up to 20 times per minute.

It is important to note that habituation, or lack of it, can be context dependent. For example, Valentinuzzi and Ferrari (1997) demonstrated that the same pigeons habituate to an acoustic stimulation during the day, but do not show habituation to the same stimulus during the night. This was suggested to be the result of the pigeons being diurnal animals, and that at night nonvisual sensory systems could exhibit enhanced sensitivity to compensate for the nocturnal visual deficit. Regardless, this should alert us to the fact that just as lack of habituation to certain stimuli can be context dependent, so can ecological scarecrows be, and understanding the context may enable managers to reduce their deleterious effects.

Ecological scarecrows can also be employed by managers in order to reduce animal–human conflicts. A classic example is the response of precocial birds to "hawk" versus "goose" silhouettes (Mueller & Parker 1980). Specifically, hawks have long tails and short necks, while geese have short tails and long necks, thus depending on the direction the silhouette moves birds will respond to it either as a threat (the short end leads) or a benign occurrence (long end leads). This response has been utilized to manage bird collisions with large glass windows on office buildings by placing hawk silhouettes on the glass; having the "hawks" literally act as scarecrows.

## 4.3 RIGID BEHAVIORS AND THEIR CONSERVATION CONSEQUENCES – CASE STUDIES

In order to illustrate how behavioral rigidity affects animal populations, and how its consequences come into play in the conservation and management of species, we highlight three case studies that cover the three behavioral domains specified in the conservation behavior framework. In each of these studies we look for the causes of the rigid behavior, its consequences and the possible management solutions.

### 4.3.1 Movement and space use
#### Misoriented sea turtles
**What is the problem?** All seven species of marine turtles display the same reproductive strategy. The females come ashore on sandy beaches at night, dig burrows in the sand, lay their eggs inside the burrows and return to sea. When the eggs hatch about 50 days later, the hatchlings will dig their way almost to the surface, where they will wait for the sand to cool (a sign that it's

night time). They will then emerge *en masse* and make their way toward the sea (Salmon 2003). The hatchlings locate the sea using visual signals. They will crawl away from tall or dark objects, usually characterized by vegetated dunes behind the beach, and toward the lower, flatter horizon, which typically reflects and emits more light from the stars and moon – the sea (Salmon 2005). Thus, the hatchlings use both light intensity and the presence of elevated dark object cues to find their way. In the last few decades, human development along seashores, as well as inland, resulted in elevated levels of photopollution (the existence of detrimental artificial light in the environment, Verheijen 1985). The artificial light emitted "drowns" the natural light and hides cues such as the dark silhouette of the dunes. When the artificial lights are weak, or when natural cues are strong (e.g. during full moon), this may result in the hatchling being disoriented – crawling in circuitous paths with no apparent direction. If the natural cues are weak or the artificial light intensity strong enough, the hatchlings will be misoriented – crawling away from the sea landward, apparently attracted to the artificial lights (Tuxbury & Salmon 2005). Both of these behaviors result in very high hatchling mortality. Mis- or disoriented hatchlings die of exhaustion and dehydration in the dunes, get captured by predators, get run over by vehicles when attracted to roads and even if they do somehow find their way into the sea, they are too exhausted to swim away from the shore and are usually depredated (Witherington 1997). Hundreds of thousands of hatchlings die each year in Florida alone as a result of photopollution (Witherington 1997).

**Why is the behavior rigid?** This is a classic case of a genetically fixed behavior. The turtles emerge with no parent or adult turtle present – so there is no parental guidance, and they must find their way to the sea on their own within a limited amount of time or they will not survive – so there is little time for learning (if at all). Finally, until recently (<100 years) the cues these hatchlings used were constant and extremely reliable. Consequently, a genetically fixed response was evolutionarily advantageous.

**What is the solution?** Since the behavior that is causing conservation concerns in the turtles cannot be altered, management steps must be taken that cater to the turtles' innate behavior. There are two possible approaches: Translocate the turtles or manage the lighting (Salmon 2005). Translocating the hatchlings into a dark area may prevent them from responding to illumination. This can be done by relocating the entire nest to an artificial hatchery on a dark beach, or by caging the nest, preventing hatchlings from moving away from it, and then moving them from the nest's vicinity to a dark beach where they can find the sea without further assistance. The two main problems of this approach are that it requires a substantial and constant managing effort, and that predatory fish and squids can learn the locations of artificial hatcheries, where turtles are found in large densities, and therefore await the turtles offshore, greatly reducing their survival once they find the sea (Salmon 2005). The second approach, managing the use of artificial light, has the advantage of solving problems on a larger scale and promoting habitat restoration. While it is very difficult to demand a complete light shutdown, even just in the critical period of the turtle's hatching, there are several other options. These include the turning off of unnecessary lights to reduce the existing lights' intensity, redirecting light sources away from the sea, altering the spectral properties of the lights to reduce the turtles' attraction to them, reducing the duration of the lights (i.e. create pulsing lights which do not attract the turtles) and creating additional orientation cues for the turtles, such as restoring and vegetating dunes between the beaches and the land (Witherington 1997, Salmon 2005). One advantage of this approach is that most of the proposed actions also benefit humans by saving a substantial amount of wasted energy. The main caveat is that even if we manage the lighting regime on the seafront, the turtles can still be attracted by the glow created by all the artificial lights inland.

### 4.3.2 Foraging and vigilance
Trash-eating condors

**What is the problem?** A common behavior in the Andean condor, *Vultur gryphus*, and the California condor, *Gymnogyps californianus*, as well as in all eight species of griffon vultures, is the swallowing of indigestible, non-food items. These include fragments of glass and plastic, metal objects such as bottle tops, small rocks, sticks and fur (Houston *et al.* 2007). There are several hypotheses for the source of this behavior. The birds may be looking for small bone chips to eat as a source of calcium because of their low-calcium diet. Alternatively, as a consequence of feeding on soft tissue only, the condors and vultures may seek indigestible objects to swallow in order to enable them to create pellets and eject other indigestible materials such as hair and hoofs from their digestive tract, or they may simply be looking for novel food items to feed on when carcass availability is low (Houston *et al.* 2007). Whatever the reason may be, this behavior seems to have had a clear adaptive value in the past. However, with the anthropogenic increase in the amount and availability of trash items, as well as the existence of sharp or toxic items, this behavior has clearly become maladaptive. In some cases, this has become an issue of serious conservation concern, such as in the case of the critically endangered California condor. The condors feed their nestlings with trash items, and trash ingestion is currently the main mortality factor in California condors' nestlings, and the primary cause of nest failure in Southern California (Mee *et al.* 2007, Rideout *et al.* 2012). In this area, the survival of nestlings is mostly due to continuous monitoring of nests for trash and medically treating chicks that have ingested trash, and is likely to be reduced to zero without these intensive management efforts (Walters *et al.* 2010).

**Why is the behavior rigid?** There is a large temporal gap separating the maladaptive behavior (feeding trash to nestlings) and its deleterious consequences (death of the nestling). This hampers the learning process (see Chapter 3). Furthermore, because condors are long lived, and their populations are currently very small, it is very unlikely that they will be able to evolve an alternative behavioral strategy before going extinct.

**What is the solution?** We can choose to target the disturbance or the behavior. To reduce the disturbance, the condors' home ranges need to be cleaned of trash. Indeed, efforts have been made by the US Forest Service to clean up sites frequented by condors; however, given the extremely large home ranges of the condors (a condor can cover hundreds of km² in one day), and the vast quantities of trash out there, these efforts seem almost futile (Mee *et al.* 2007, Walters *et al.* 2010). It is also possible to provide bone fragments as a more accessible source of calcium at feeding sites. Such action had successfully reduced nestling mortality in Griffon vultures in Israel (O. Hatzofe, pers. comm.) but had no noticeable effect on the quantity of trash delivered to the nest by condors or to their nestlings survival in Southern California (Mee *et al.* 2007). As mentioned earlier, an alternative approach is to target the maladaptive behavior of the animals. Since our assumption is that the condors exhibit behavioral rigidity because of a temporal gap between the behavior and its consequences, efforts can be made to reduce this temporal gap. This can be done in the form of aversive training prior to reintroduction. Aversive training has already successfully reduced power-line-related mortality in the condors (Mee & Snyder 2007). While it is so far uncertain whether training can modify or even eliminate trash-ingestion behavior in condors, this presents a promising and important line of investigation.

### 4.3.3 Social organization and reproductive behavior
Disappearing wheatears

**What is the problem?** The mourning wheatear (*Oenanthe lugens*) is a territorial insectivore and its habitat is in high rocky mountainsides, rocky slopes and dry streambeds in desert environments. It nests in cavities, mainly under rocks. Because it is an insectivore it is attracted to desert habitats that are relatively productive (such as dry riverbeds) and are abundant with insects. Within these habitats it searches for cavities to nest in and utilizes elevated perching locations (such as large rocks) to scan its territory. In deserts, roadside habitats are typically rich with vegetation because water runoff accumulates along the roadside ditches. This makes the roadside habitat a preferred area for establishing wheatear territories. In addition, this habitat provides excellent perches in the form of signposts. However, the upper end of these posts is open, forming a narrow vertical cavity that wheatears enter, presumably because they appear as potential sites for a nesting cavity. Once inside, the birds cannot spread their wings and are literally trapped inside the post.

The survival of roadside territory holders is far lower than individuals holding territories away from the road (Ben Aharon 2011). In fact, survival is so low that reproductive success and recruitment, however high, cannot compensate for it and the roadside wheatear population has a negative growth rate. Most mortality of the roadside territory holders is due to entrapment in the sign posts, but car collisions and predation evidently play a role too. The roadside population continues to exist because the roadside territories appear superior to the wheatears and vacated territories are immediately colonized by individuals that previously held territories away from the road. Thus the road acts as an ecological trap and if the road infrastructure is extensive enough it can bring about the extirpation of complete populations.

**Why is the behavior rigid?** The search for a cavity to establish a nest is a very basic innate tendency. The optimal nest would be a cavity in a rich habitat that also provides nearby perches, so that the hunting and protecting of the territory will not require the parents to distance themselves from the nests. The posts along the road, therefore, may appear to the wheatears as optimal nesting sites. Small vertical cavities with smooth walls (such as the inside of a road post) do not exist in nature, so there would be no evolutionary history of avoiding such structures. There is nothing about the post that can be conceived as threatening. Once inside the post, there is no escape so there can be no learning process.

**What is the solution?** The solution in this case was simple and inexpensive and has, as a result of this case study (Ben Aharon 2011), been implemented in many areas. Cone-shaped metal tops are fitted to the posts so birds cannot enter the post and birds of prey cannot use it for perching. This serves to improve the quality of the road habitat for the wheatears to the point that it no longer acts as an ecological trap, although other road-related sources of mortality still remain a problem.

## 4.4 MANAGEMENT IMPLICATIONS

In this section we have distilled the information given in the chapter into short practical advice for wildlife managers. This is of course general advice and exceptions to the rules can always be found, but we strongly believe that following this advice can greatly improve wildlife conservation and management. By understanding the basis for animals' behavior rigidity, many of the problems caused by it can be foreseen and dealt with in an *a priori* manner.

(1) When individuals within a population display a behavior that is maladaptive and reduces their fitness (and therefore their population size), the source of the behavioral rigidity should determine management solutions (see also Table 4.1):

(a) If the rigid behavior originates from a fixed genetically wired "impulse," or from physiological constraints, then management should focus on eliminating or mitigating the *disturbance*.

(b) If the rigid behavior originates from an imprinted behavior, acquired at a young age, then the *behavior* can be modified, either by breeding the animals in captivity, training the new generation, and releasing them back to nature, or by manipulating the behavior *in situ*.

(c) If the rigid behavior originates from a slow learning response, then management should either improve the animals' *learning conditions* (i.e. make the connection between the disturbance and its maladaptive consequences easier to learn), and/or slow down the effects of the *disturbance*, in order to allow the animals time to adapt to it.

(2) When managing small populations, it is important to distinguish between naturally rare and anthropogenically rare populations. The latter group is prone to suffer from Allee effects, and should be carefully monitored (preferably using behavioral indicators) even when they don't show signs of demographic decline.

(3) If animals in a population are attracted to a detrimental habitat or preferentially make maladaptive decisions, it is important to understand the cause for this behavior. If the problem stems from a decrease in the quality of a once-favorite environment or behavior, then management

**Table 4.1.** *Examples of rigid behaviors from all three behavioral domains. By knowing a priori the behavioral domain that may be affected, one can determine the type of rigidity and therefore plan for the appropriate management solution.*

| Source of behavioral rigidity | Management solutions | Examples | | |
|---|---|---|---|---|
| | | Movement and space use | Foraging and vigilance | Social structure and reproduction |
| Fixed behavior | Elimination and/or mitigation of disturbances | Problem: light pollution causes hatchling turtles to navigate away from the sea (Salmon 2005)  Solution: Reducing artificial lights on beaches with turtle nests | Problem: African wild dogs suffer from Allee effects due to reduced efficiency of hunting in small group sizes (Courchamp & Macdonald 2001)  Solution: Solving the problems causing African wild dogs numbers to decline in the first place such as fragmented habitats, conflicts with game farmers or disease | Problem: Wheatears looking for cavities to build their nests in get trapped in road posts (Ben-Aharon 2011)  Solution: Covering all road posts with cone-shaped caps |
| Imprinted behavior | Modifying behavior or the environment at early life stages | Problem: Young geese and swans that have no parents to teach them migration routes do not migrate but undertake sporadic movements instead  Solution: Using Ultralight aircraft to teach imprinted young geese and swans migration routes (Sladen et al. 2002) | Problem: Foxes relying on human waste for food die shortly after sanitation is implemented  Solution: Gradual reduction of anthropogenic food sources to enable inter-generational learning on how to utilize natural food sources (Bino et al. 2010) | Problem: Bird chicks raised in captivity and released to the wild are later attracted to humans (Wallace et al. 2007)  Solution: Feeding chicks using puppets mimicking adults of their species |
| Slow plastic response | Improve learning conditions or slow down disturbance | Problem: Animals may not readily use crossings constructed over or under roads to reduce animal–vehicle collisions  Solution: Enticing animals into using crossings using dung or urine of conspecifics (B. Rosenberg, Israel Nature and Parks Authority, unpublished data) | Problem: Animals in captivity lose their anti-predatory behavior and get preyed upon when released to the wild (Griffin et al. 2000)  Solution: Pre-release training to teach anti-predatory behavior before the release (Chapter 10) | Problem: Reintroduction of highly social animals using individuals from different groups increases the probability of failure  Solution: Form groups in captivity early on and carry out reintroduction after social structure stabilizes (Gusset et al. 2008) |

should focus on improving the *environment*, or on "luring" the animals to a better one. Alternatively, if the problem stems from the animals responding to a misleading cue, then management should focus on removing the *cue*.

## 4.5 CONCLUSIONS

Behavioral rigidity in the face of anthropogenic changes is a major cause of concern in conservation biology. The inability of animals to adapt their behavior to a changing environment can reduce both survival and reproductive success of individuals, directly affecting entire populations. Behavioral rigidity is also a major determinant of evolutionary traps, Allee effects and ecological scarecrows, all of which can have severe impacts on wildlife populations, and may even cause extinctions. Understanding behavioral rigidity will allow us to predict and manage anthropogenic effects on wildlife populations more effectively. The main question wildlife managers have to ask themselves when dealing with animals that display behavioral rigidity is "what is the source for the observed behavioral rigidity?" The answer will direct managers toward the most effective management scheme – whether it involves changing the environment, its cues or the behavior of the species in question.

## REFERENCES

Allee, W.C. 1931. *Animal Aggregations, a Study in General Sociology*. Chicago: University of Chicago Press.

Bateson, P. 2003. The promise of behavioural biology. *Animal Behaviour*, 65: 11–17.

Ben Aharon, N. 2011. *Are Roads Ecological Traps for Mourning Wheatear (Oenanthe lugens)?* M. Sc. Thesis, Ben Gurion University, Israel 55pp.

Berger, J. 1998. Future prey: some consequences of the loss and restoration of large carnivores. In Caro T.M. (ed.), *Behavioral Ecology and Conservation Biology*. pp. 80–100. New York: Oxford University Press.

Biedenweg, T.A., Parsons, M.H., Fleming, P.A. and Blumstein, D.T. 2011. Sounds scary? Lack of habituation following the presentation of novel sounds. *PLoS ONE*, 6: e14549.

Bino, G., Dolev, A., Yosha, D. *et al.* 2010. Abrupt spatial and numerical responses of overabundant foxes to a reduction in anthropogenic resources. *Journal of Applied Ecology*, 47: 1262–1271.

Bjorndal, K.A., Bolten, A.B. and Lagueux, C.J. 1994. Ingestion of marine debris by juvenile sea turtles in coastal Florida habitats. *Marine Pollution Bulletin*, 28: 154–158.

Bolhuis, J.J. 1991. Mechanisms of avian imprinting: a review. *Biological Reviews*, 66: 303–345.

Bolhuis, J.J. 2013. Minding the gap. *Science*, 339: 143.

Both, C., Bouwhuis, S., Lessells, C.M. and Visser, M.E. 2006. Climate change and population declines in a long-distance migratory bird. *Nature*, **441**: 81–83.

Bouchard, T.J. 1994. Genes, environment, and personality. *Science*, **264**: 1700–1701.

Bouwman, K.M and Hawley, D.M. 2010. Sickness behaviour acting as an evolutionary trap? Male house finches preferentially feed near diseased conspecifics. *Biology Letters*, **6**: 462–465.

Chan, A.A.Y., Giraldo-Perez, P., Smith, S. and Blumstein, D.T. 2010. Anthropogenic noise affects risk assessment and attention: the distracted prey hypothesis. *Biology Letters*, **6**: 458–461.

Chan, A.A.Y. and Blumstein, D.T. 2011. Attention, noise, and implications for wildlife conservation and management. *Applied Animal Behaviour Science*, **131**: 1–7.

Coffin, A.W. 2007. From roadkill to road ecology: a review of the ecological effects of roads. *Journal of Transport Geography*, **15**: 396–406.

Courchamp, F. and Macdonald, D.W. 2001. Crucial importance of pack size in the African wild dog *Lycaon pictus*. *Animal Conservation*, **4**: 169–174.

Courchamp, F., Rasmussen, G. and Macdonald, D.W. 2002. Small pack size imposes a trade-off between hunting and pup-guarding in the painted hunting dog *Lycaon pictus*. *Behavioral Ecology*, **13**: 20–27.

Courchamp, F., Berec, L. and Gascoigne, J. 2008. *Allee Effects in Ecology and Conservation*. Oxford: Oxford University Press.

Darwin, C. 1871. *The Descent of Man and Selection in Relation to Sex*. London: John Murray.

Davis, J.M and Stamps, J.A. 2004. The effect of natal experience on habitat preferences. *Trends in Ecology & Evolution*, **19**: 411–416.

Davis, J.M. 2010. Habitat imprinting. In Michael, D.B. and Janice, M. (eds.), *Encyclopedia of Animal Behavior*. pp. 33–37. Oxford: Academic Press.

Donald, P.F., Evans, A.D., Muirhead, L.B. *et al.* 2002. Survival rates, causes of failure and productivity of skylark *Alauda arvensis* nests on lowland farmland. *Ibis*, **144**: 652–664.

Dukas, R. 2004. Causes and consequences of limited attention. *Brain, Behaviour and Evolution*, **63**: 197–210.

Ellenberg, U., Setiawan, A.N., Cree, A., Houston, D.M. and Seddon, P.J. 2007. Elevated hormonal stress response and reduced reproductive output in yellow-eyed penguins exposed to unregulated tourism. *General and Comparative Endocrinology*, **152**: 54–63.

Fisher, H.S., Wong, B.B.M. and Rosenthal, G.G. 2006. Alteration of the chemical environment disrupts communication in freshwater fish. *Proceedings of the Royal Society of London Series B*, **273**: 1187–1193.

Gienapp, P., Teplitsky, C., Alho, J.S., Mills, J.A. and Merila, J. 2008. Climate change and evolution: disentangling environmental and genetic responses. *Molecular Ecology*, **17**: 167–178.

Gilroy, J.J. and Sutherland, W.J. 2007. Beyond ecological traps: perceptual errors and undervalued resources. *Trends in Ecology and Evolution*, **22**: 351–356.

Grieco, F., van Noordwijk, A.J. and Visser, M.E. 2002. Evidence for the effect of learning on timing of reproduction in blue tits. *Science*, **296**: 136–138.

Griffin, A.S., Blumstein, D.T. and Evans, C.S. 2000. Training captive-bred or translocated animals to avoid predators. *Conservation Biology*, **14**: 1317–1326.

Gundersen, G.E., Johannesen, E., Andreassen, H.P. and Ims, R.A. 2001. Source–sink dynamics: how sinks affect demography of sources. *Ecology Letters*, 4: 14–21.

Gusset, M., Ryan, S.J., Hofmeyr, M. *et al.* 2008. Efforts going to the dogs? Evaluating attempts to re-introduce endangered wild dogs in South Africa. *Journal of Applied Ecology*, 45: 100–108.

Halfwerk, W., Holleman, L.J.M., Lessells, C.M. and Slabbekoorn, H. 2011. Negative impact of traffic noise on avian reproductive success. *Journal of Applied Ecology*, 48: 210–219.

Hawlena, D., Saltz, D., Abramsky, Z. and Bouskila, A. 2010. Ecological trap for desert lizards caused by anthropogenic changes in habitat structure that favor predator activity. *Conservation Biology*, 24: 803–809.

Horvath, G., Blaho, M., Egri, A. *et al.* 2010. Reducing the maladaptive attractiveness of solar panels to polarotactic insects. *Conservation Biology*, 24: 1644–1653.

Houston, D.C., Mee, A. and McGrady, M. 2007. Why do condors and vultures eat junk? The implications for conservation. *Journal of Raptor Research*, 41: 235–238.

Immelmann, K. 1975. Ecological significance of imprinting and early learning. *Annual Review of Ecology, Evolution, and Systematics*, 6: 15–37.

Imperio, S., Bionda, R., Viterbi, R. and Provenzale, A. 2013. Climate change and human disturbance can lead to local extinction of Alpine rock ptarmigan: new insights from the western Italian Alps. *PloS ONE*, 8: e81598.

IUCN 2013. *The IUCN Red List of Threatened Species v. 2013.1.* www.iucnredlist.org [Last accessed August 13, 2014].

Joëls, M., Pu, Z., Wiegert, O., Oitzl, M.S. and Krugers, H.J. 2006. Learning under stress: how does it work? *Trends in Cognitive Sciences*, 10: 152–158.

Karp, D.S. and Root, T.L. 2009. Sound the stressor: how hoatzins (*Opisthocomus hoazin*) react to ecotourist conversation. *Biodiversity and Conservation*, 18: 3733–3742.

Keeler, M.S. and Chew, F.S. 2008. Escaping an evolutionary trap: preference and performance of a native insect on an exotic invasive host. *Oecologia*, 156: 559–568.

Klem, D. 1989. Bird: window collisions. *The Wilson Bulletin*, 101: 606–620.

Kokko, H. and Sutherland, W.J. 2001. Ecological traps in changing environments: ecological and evolutionary consequences of a behaviourally mediated Allee effect. *Evolutionary Ecology Research*, 3: 537–551.

Krista, G., Horvath, G. and Andrikovics, S. 1998. Why do mayflies lay their eggs en masse on dry asphalt roads? Water-imitating polarized light reflected from asphalt attracts ephemeroptera. *Journal of Experimental Biology*, 201: 2273–2286.

Lehrman, D.S. 1953. A critique of Konard Lorenz's theory of instinctive behavior. *The Quarterly Review of Biology*. 28: 337–363.

Lorenz, K.Z. 1937. The companion in the bird's world. *The Auk*, 54: 245–273.

Lott, D.F. and McCoy, M. 1995. Asian rhinos *Rhinoveros unicornis* on the run? Impact of tourist visits on one population. *Biological Conservation*, 73: 23–26.

Ludwig, G.X., Alatalo, R.V., Helle, P. *et al.* 2006. Short- and long-term population dynamical consequences of asymmetric climate change in black grouse. *Proceedings of the Royal Society of London Series B*, 273: 2009–2016.

Madliger, C.L. 2012. Toward improved conservation management: a consideration of sensory ecology. *Biodiversity and Conservation*, 21: 3277–3286.

Manor, R. and Saltz, D. 2005. Effects of human disturbance on use of space and flight distance of mountain gazelles. *Journal of Wildlife Management,* **69**: 1683–1690.

Marechal, L., Semple, S., Majolo, B. *et al.* 2011. Impacts of tourism on anxiety and physiological stress levels in wild male Barbary macaques. *Biological Conservation,* **144**: 2188–2193.

Mee, A. and Snyder, N.F.R. 2007. California condors in the 21st century – Conservation problems and solutions. In Mee, A. and Hall, L.S. (eds.), *California Condors in the 21st Century.* pp. 243–279. Series in Ornithology, no. 2. Washington, DC: American Ornithologists' Union and Nuttall Ornithological Club.

Mee, A., Rideout, B.A., Hamber, J.A. *et al.* 2007. Junk ingestion and nestling mortality in a reintroduced population of California condors *Gymnogyps californianus. Bird Conservation International,* **17**: 119–130.

Mills, L.S., Zimova, M., Oyler, J. *et al.* 2013. Camouflage mismatch in seasonal coat color due to decreased snow duration. *Proceedings of the National Academy of Science, USA,* **110**: 7360–7365.

Mueller, H.C. and Parker, P.G. 1980. Naïve ducklings show different cardiac response to hawk than to goose models. *Behaviour,* **74**: 101–113.

Nottebohm, F. 1970. Ontogeny of bird song. *Science,* **167**: 950–956.

Patten, M.A. and Kelly, J.F. 2010. Habitat selection and the perceptual trap. *Ecological Applications,* **20**: 2148–2156.

Purser, J. and Radford, A.N. 2011. Acoustic noise induces attention shifts and reduces foraging performance in three-spined sticklebacks (Gasterosteus aculeatus). *PLoS ONE,* **6**: e17478.

Rantala, M.J. and Marcinkowska, U.M. 2011. The role of sexual imprinting and the Westermarck effect in mate choice in humans. *Behavioral Ecology and Sociobiology,* **65**: 859–873.

Rideout, B.A., Stalis, I., Papendick, R. *et al.* 2012. Patterns of mortality in free-ranging California condors (*Gymnogyps californianus*). *Journal of Wildlife Disease,* **48**: 95–112.

Ridley, M. 2003. *Nature via Nurture: Genes, Experience, and What Makes us Human.* New York: Harper Collins Publishers.

Robertson, B.A. and Hutto, R.L. 2006. A framework for understanding ecological traps and an evaluation of existing evidence. *Ecology,* **87**: 1075–1085.

Robertson, B.A. 2012. Investigating targets of avian habitat management to eliminate an ecological trap. *Avian Conservation and Ecology,* **7**: 2.

Robertson, B.A., Rehage, J.S. and Sih, A. 2013. Ecological novelty and the emergence of evolutionary traps. *Trends in Ecology and Evolution,* **28**: 552–560.

Rotem, G., Ziv, Y., Giladi, I. and Bouskila A. 2013. Wheat fields as an ecological trap for reptiles in a semiarid agroecosystem. *Biological Conservation* **167**: 349–353.

Rosenthal, G.G. and Stuart-Fox, D.M. 2012, Environmental disturbance and animal communication. In Wong B.B.M. and Candolin U. (eds.), *Behavioural Responses to a Changing World: Mechanisms and Consequences.* pp. 16–31. Oxford: Oxford University Press.

Salmon, M. 2003. Artificial night lighting and turtles. *Biologist,* **50**: 163–168.

Salmon, M. 2005. Protecting sea turtles from artificial night lighting at Florida's oceanic beaches. In Rich, C. and Longcore, T. (eds.), *Ecological Consequences of Artificial Night Lighting*. pp. 141–168. Washington DC: Island Press.

Sapolsky, R.M., Romero, L.M. and Minck, A.U. 2000. How do glucocorticoids influence stress responses? Integrating permissive, suppressive, stimulatory, and preparative actions. *Endocrine Reviews*, 21: 55–89.

Schlaepfer, M.A., Runge, M.C. and Sherman, P.W. 2002. Ecological and evolutionary traps. *Trends in Ecology and Evolution*, 17: 474–480.

Schlaepfer, M.A., Sherman, P.W., Blossey, B. and Runge, M.C. 2005. Introduced species as evolutionary traps. *Ecology Letters*, 8: 241–246.

Short, J., Bradshaw, S.D., Giles, J., Prince, R.I.T. and Wilson, G.R. 1992. Reintroduction of macropods (*Marsupialia: Macropodoidea*) in Australia: a review. *Biological Conservation*, 62: 189–204.

Sih, A., Bolnick, D.I., Luttbeg, B. *et al.* 2010. Predator-prey naïveté, antipredator behavior, and the ecology of predator invasions. *Oikos*, 119: 610–621.

Sih, A., Ferrari, M.C.O. and Harris, D.J. 2011. Evolution and behavioural responses to human-induced rapid environmental change. *Evolutionary Applications*, 4: 367–387.

Sladen, W.J.L., Lishman, W.A., Ellis, D.H., Shire, G.G. and Rininger, D.L. 2002. Teaching migration routes to Canada geese and trumpeter swans using Ultralight aircraft, 1990–2001. *Waterbirds*, 25: 132–137.

Stamps, J.A. and Swaisgood, R.R. 2007. Someplace like home: experience, habitat selection and conservation biology. *Applied Animal Behaviour Science*, 102: 392–409.

Steen, J.B., Erikstad, K.E. and Hoidal, K. 1992. Cryptic behaviour in moulting hen willow ptarmigan *Lagopus l.* lagopus during snow melt. *Ornis Scandinavica*, 23: 101–104.

Stephens, P.A., Sutherland, W.J. and Freckleton, R.P. 1999. What is the Allee effect? *Oikos*, 87: 185–190.

Steven, R., Pickering, C. and Castley, J.G. 2011. A review of the impacts of nature based recreation on birds. *Journal of Environmental Management*, 92: 2287–2294.

Stevens, E. and Pickett, C. 1994. Managing the social environments of flamingoes for reproductive success. *Zoo Biology*, 13: 501–507.

Swaisgood, R.R. 2010. The conservation–welfare nexus in reintroduction programmes: a role for sensory ecology. *Animal Welfare*, 19: 125–137.

Teixeira, C.P., De Azevedo, C.S., Mendel, M., Cipreste, C.F. and Young, R.J. 2007. Revisiting translocation and reintroduction programmes: the importance of considering stress. *Animal Behaviour*, 73: 1–13.

Thiel, D., Jenni-Eiermann, S., Braunisch, V., Palme, R. and Jenni, L. 2008. Ski tourism affects habitat use and evokes a physiological stress response in capercaillie *Tetrao urogallus*: a new methodological approach. *Journal of Applied Ecology*, 45: 845–853.

Tobin, P.C., Berec, L. and Liebhold, A.M. 2011. Exploiting Allee effects for managing biological invasions. *Ecology Letters*, 14: 615–624.

Tuxbury, S.M. and Salmon, M. 2005. Competitive interactions between artificial lighting and natural cues during seafinding by hatchling marine turtles. *Biological Conservation*, 121: 311–316.

Valentinuzzi, V.S. and Ferrari, E.A.M. 1997. Habituation to sound during morning and night sessions in pigeons (*Columba livia*). *Physiology and Behavior*, 62: 1203–1209.

van Oers, K., de Jong, G., van Noordwijk, A.J., Kempenaers, B. and Drent, P.J. 2005. Contribution of genetics to the study of animal personalities: a review of case studies. *Behaviour*, 142: 1191–1212.

Verheijen, F.J. 1985. Photopollution: artificial light optic spatial control systems fail to cope with. Incidents, causations, remedies. *Experimental Biology*, 44: 1–18.

Virzi, T., Boulton, R.L., Davis, M.J., Gilroy, J.J. and Lockwood, J.L. 2012. Effectiveness of artificial song playback on influencing the settlement decisions of an endangered resident grassland passerine. *The Condor*, 114: 846–855.

Wallace, M.P., Clark, M., Vargas, J. and Porras, M.C. 2007. Release of puppet-reared California condors in Baja California: evaluation of a modified rearing technique. In Mee, A. and Hall, L.S. (eds.), *California Condors in the 21st Century*. pp. 227–242. Series in Ornithology, no. 2. Washington DC: American Ornithologists' Union and Nuttall Ornithological Club.

Walters, J.R., Derrickson, S.R., Fry, D.M. *et al.* 2010. Status of the California condor (*Gymnogyps californianus*) and efforts to achieve its recovery. *Auk* 127: 969–1001.

Wang, M.H. and Kot, M. 2001. Speeds of invasion in a model with strong or weak Allee effects. *Mathematical Biosciences*, 171: 83–97.

Wiepkema, P.R. and Koolhaas, J.M. 1993. Stress and animal welfare. *Animal Welfare*, 2: 195–218.

Witherington, B.E. 1997. The problem of photopollution for sea turtles and other nocturnal animals. In Clemmons J.R. and Buchholz, R. (eds.), *Behavioral Approaches to Conservation in the Wild*. pp. 303–328. Cambridge: Cambridge University Press.

Wright, A.J., Soto, N.A., Baldwin, A.L. *et al.* 2007. Anthropogenic noise as a stressor in animals: a multidisciplinary perspective. *International Journal of Comparative Psychology*, 20: 250–273.

# Anthropogenic impacts on behavior: the pros and cons of plasticity

DANIEL I. RUBENSTEIN

Many threatened and endangered species inhabit environments where people have altered the landscape and have created novel and powerful selective pressures. These often disrupt "bottom-up" factors associated with resource acquisition, change "top-down" factors involving predation or disease transmission or alter "side-ways" factors involving competitive or mutualistic interactions happening at the same trophic level. Since one of the basic precepts of behavioral ecology is that environmental conditions shape behavior, such changes should lead to changing responses by the species experiencing them. But this presupposes that there is enough plasticity in a species' behavioral repertoire to cope with the environmental change. If there is, then contingent responses displayed in the past may help species cope with a changing present. If the degree of environmental change is so great or the species has a limited ability to adjust its behavior, then the species could be pushed beyond its limits to adapt, with the stress it experiences lowering its fecundity or survival, leading to its demise (Chapter 4). Alternatively, the species could be pushed into new behavioral space, revealing novel responses. These might be sufficient to cope with the environmental changes, creating new evolutionary potential, or they could be pathological, creating unintended negative consequences that may at first appear benign, but in the long run may impact the species' viability or involve cascading effects on other species.

Recent reviews by Rubenstein (2010) and Caro and Sherman (2011, 2013) have provided insights into many dimensions of this problem. Caro and Sherman (2011) argue that species exhibit remarkable degrees of plasticity and that behavioral diversity per se should be

*Conservation Behavior: Applying Behavioral Ecology to Wildlife Conservation and Management*, eds. O. Berger-Tal and D. Saltz. Published by Cambridge University Press. © Cambridge University Press 2016.

added to the traditional list of attributes to be conserved – genes, species and ecosystems. If behaviors are not conserved, then many behavioral variants will remain latent and unexpressed, and if given enough time, will vanish from a species' repertoire forever. This could potentially change a species' evolutionary trajectory by reducing a species' ability to escape from new stressors or by eliminating the possibility of conservationists seeding threatened populations with corrective behavioral variants. Such a doom and gloom scenario need not be cast in stone since dramatic changes in the past have often disrupted communities and altered selective pressures. Adaptations in the past have led to successful coping strategies and they can do so again going forward. Much will depend on the degree of variation populations manifest or can create and the magnitude and rate of environmental change. Rubenstein (2010) has shown that populations may exhibit many small-scale behavioral variations in relation to habitat change, changes in population size and composition, as well as in acquiring resources or mates, or in avoiding predators. But species show limits in the plasticity of their behavior. Yet, only by identifying where, when and why these limits are reached will it be possible to predict which species will adapt of their own accord and where, and in what manner, species may need human assistance in order to adapt and retreat from the edge of extinction.

## 5.1  WHAT IS BEHAVIORAL PLASTICITY, FLEXIBILITY AND BEHAVIORAL DIVERSITY?

Behavior is typically the means by which an animal first reacts to its environment. While some behavior is fixed in form and intensity (Morris 1957), it need not be rigidly controlled by the genome. Imprinting during sensitive periods, for example, can tune behavior to local environmental circumstances. As noted in Chapter 2 of this volume, behavior of this type is considered *plastic*. Such behavior emerges when one genotype generates different phenotypes in different environments. Usually this occurs when genotypes are reared in different environments. Reaction norm graphs (e.g. Figure 2.1) nicely illustrate the environmental control of behavioral expression during development. However, when individuals change behavior as they move among habitats, behavior is considered *flexible*. The distinction between plasticity and flexibility is important because the time course over which change occurs is different, but both demonstrate that change is contingent on context. Both are also features of individuals – or the genotypes they are made of – and when taken together generate population-wide levels of behavioral diversity.

## 5.2 WHY DOES BEHAVIORAL DIVERSITY MATTER?

As Caro andSherman note (2011), behavioral diversity is the raw material that allows populations to cope with environmental change. The more diversity a population or a species has, the more likely there will be a variant that can cope with altered conditions. In a series of clever experiments with artificial feeders in Wytham Woods, Oxford, Morand-Ferron and Quinn (2011) showed that in mixed flocks of blue and great tits the ability to open feeders increased in both efficiency and frequency; large groups comprised a larger pool of competence then smaller groups. Although, from the perspective of an omnipotent conservation biologist having a wide range of variants to manage or introduce into small captive populations, this benefit is group-selectionist in nature. From an individual's evolutionary perspective how it – perhaps along with its close kin – is able to cope with a changing world is what matters most. Ultimately, this means understanding what creates and maintains individual patterns of plasticity and flexibility. In terms of plasticity, behavior is likely to change with changes in age and status. In many amphibians, for example, the same individual might be favored to behave as a territorial male when fully mature, but as a sneak or a satellite as a sub-adult (Rubenstein 1980). But changing social circumstances could also select for plastic behavior. At high density young may disperse, whereas at low density they might be retained in the natal territory, as is the case with foxes (MacDonald & Johnson 2015).

Behavioral flexibility will also enable adaptive responses, but at a faster timescale. Optimality models (Davies *et al.* 2012) universally assume that individuals can assess costs and benefits of behaviors in different environments. Game theoretic models of behavior do the same, but instead the assessments are based on frequency-dependent payoffs associated with the actions of others (Rubenstein 1980, Maynard Smith 1982). Examples of switching diets or mating systems are legion.

An individual's ability to cope with changing environments will be shaped by its decision-making process and how it is controlled. At one end of a continuum (Figure 5.1), behavior mostly under the control of genes will be relatively rigid and will only vary as each new batch of young pass through selective filters associated with critical developmental stages. Novel variants of genetically controlled behaviors will arise slowly via

**Behavioral Contingencies: Control and Consequences**

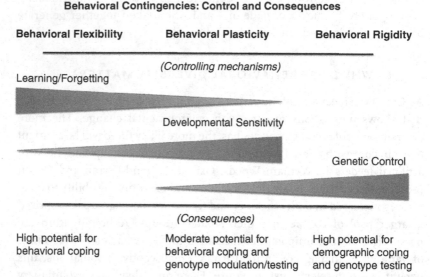

Behavioral Flexibility        Behavioral Plasticity        Behavioral Rigidity

*(Controlling mechanisms)*

Learning/Forgetting

Developmental Sensitivity

Genetic Control

*(Consequences)*

High potential for          Moderate potential for          High potential for
behavioral coping          behavioral coping and          demographic coping
                           genotype modulation/testing          and genotype testing

**Figure 5.1:** Three behavioral contingencies operate in different ways. The width of the wedges depicts the strength of rapid environmental feedbacks through learning and forgetting, developmental sensitivities to different environments, or genetics in controlling each contingency shown below the top line. Consequences of behavior or demography for coping with environmental change and leading to subsequent evolutionary change are below the bottom line.

mutation, but they could be lost quickly, either by bad luck through genetic drift if populations are eliminated or habitats are fragmented, or by selection winnowing out individuals that can't adapt to the change. Chapter 4 of this volume provides a plethora of examples in which lack of plasticity can increase mortality or reduce reproductive success, typically because individuals respond to appropriate cues in inappropriate contexts in ways that lead them to inferior habitats or miss good ones as they fall into "ecological traps" (Gilroy & Sutherland 2011).

At the other end of the continuum, behavior controlled mostly by instantaneous assessment of the environment will be the most flexible. But such assessments are likely to be the most transient since cues change as environments change, with behavior changing as well. The degree of flexibility that individuals show will often be shaped by learning and past experience. Most involve some form of associative learning (Chapter 3) and often increase robustness (the ability to succeed under a wide variety of circumstances; Dukas 2013). Fruit fly males, for example, learn with whom to mate by negative reinforcement from heterospecific females and positive reinforcement by conspecific females. What begins as random mating attempts by males becomes operantly tuned by feedback from females to become highly selective (Dukas 2004, Dukas & Dukas 2012). In the wild, blue tits (*Parus caeruleus*) have been shown to adjust the time of egg-laying based on the relative timing of the peak abundance of insects and an early warning cue – bud break. Via artificially increasing the availability of food a week later than normal, Grieco and colleagues (2002) showed that blue tits could push back egg-laying by a week in the next season. Since matching the emergence of offspring to the peak availability of food will enhance survival, learning to use and optimize attention to cues that foreshadow and predict essential foraging behaviors will be under intense selection. Similarly, it is also well known that for many bird species, those that successfully rear a brood in a particular area are very likely to return and nest in that area in subsequent years (Greenwood & Harvey 1982). In this case an association between a location and breeding success leads to a contingent "rule of thumb" – succeed–stay, fail–switch. Typically, the most basic learning rules are likely to profoundly affect fitness. And lastly, it is often through "trial and error" learning that individuals learn to assess costs and benefits associated with optimality decision-making.

While associative learning is largely individually based, individuals can also learn from the experiences of others. Birds can increase diet breadth by adopting foraging techniques of other species (Rubenstein *et al.* 1977) or by associating with, or even parasitizing, conspecifics using different foraging styles (Barnard & Sibley 1981, Giraldeau & Lefebvre 1987). Again using mixed flocks of blue and great tits (*Parus major*) in Wytham Wood, Oxford, Farine and collaborators (2014) showed that social networks of conspecifics increased the ability to discover novel feeding sites 22-fold and that even networks of heterospecifics increased the discovery rate 12-fold. Clearly, social learning increases the spread of ideas and behavioral rules within populations. The degree to which individuals and species vary in their reliance on public information is likely to have dramatic consequences for coping with environmental change.

Not all learning need be associative. Repeated presentation of a single stimulus to individuals is perhaps the most basic form of learning and can either produce diminished (habituation) or heightened (sensitization) responses to a cue over time. Because habituation is tied to a specific cue it is not the by-product of sensory or motor fatigue. This suggests that diminishing responsiveness is adaptive. Individuals reacting quickly and often to repeated stimuli that are non-lethal and not particularly risky will over-react and generate high "opportunity costs." Natural selection thus should down-regulate responses to such cues. But over-reacting to a threat that is not real is likely to be less costly then not reacting to a threat that is real – too much caution may be costly, but too little could be deadly. Animals typically balance committing errors of these types by identifying and associating particular cues with particular types of risk. Mouflon sheep, for example, adjust their behavior to different levels of human disturbances, habituating at times to some but not others (Marchand et al. 2014). While some mouflon in southern France live in protected areas where contact with people is minimal, others share their habitats with hikers and hunters. High levels of human activity generally shift mouflon activity from day to night and lead to straighter movement trajectories. But in the area of the highest hunting and tourist activity, the response of mouflon to people was highly context specific. During hunting season the typical shifts in habitat use and behavior described above occurred, but not during the height of the tourist season. Mouflon did not alter their normal behavior and behaved as if humans were absent, showing that mouflon can distinguish among types of risk and adjust their habituation levels accordingly. Human disturbance in natural areas is increasing and not surprisingly animal populations living in such areas are experiencing stress. But whether most species living in these areas can display such context-specific responses is unknown. In urban environments, however, it is clear that urbanized birds habituate rapidly to humans along with their beneficial as well as detrimental actions (Sol et al. 2013). For many bird species flight initiation distances in response to human approaches are reduced in urban environments. Given that urban birds often depend on humans for food supplements this is not too surprising. Eurasian magpies even reduce their attacks on particular humans once they learn to recognize those that are not threats to their nests (Lee et al. 2011).

In between these two extremes are behavioral variants that are under epigenetic control. Changing DNA will clearly alter gene expression. But inactivating genes by changing levels of methylation or histone modification can do the same (Ledon-Rettig 2013). Usually, greater methylation reduces gene expression (Szyf 2011). In laboratory rats, type of mothering not only affects the behavior of young later as adults, it is also associated with differential methylation of the offspring's glucocorticoid receptor (GR) promoter in the hippocampus. Rats receiving sufficient maternal care grew to be bold and neophilic, whereas those that were stressed as young were anxious, neophobic and shy when they matured. In this case, greater methylation of the GR promoter reduced GR expression and increased anxiety. Most examples of behavioral epigenetic inheritance involve "soma-to-soma" transmission.

In many mammals bearing large litters, neighboring siblings can experience very different in utero environments. Females surrounded by two males tend to experience high circulating levels of testosterone and as adults tend to be more aggressive than sisters surrounded by two females. Testosterone levels are known to increase methylation and histone acetylation of developing brains, especially in the estrogen receptor a, thus providing a possible mechanism for the development of different adult behaviors (Ledon-Rettig 2013).

A recent study of honeybees reveals that epigenetics can control behavioral plasticity. Although comparisons of methylation levels of the brains of queen and worker castes showed no difference immediately after hatching, differences among nurse and forager castes were large and reversible. By experimentally altering the relative abundance of nurses in hives, some workers reverted to nurses, and when they did their methylation patterns returned to those typical of nurses (Greenberg *et al.* 2012). Clearly, the ability for gene expression to respond to changing environmental circumstances offers the possibility of an individual drawing on the diversity of its genome and tuning it to cope with changing environments via a form of "soft Lamarckism." Given that epialleles can operate in many genomes simultaneously, epigenetics can potentiate rapid adaptation to environmental change. It even offers the possibility of generating new phenotypes better matched to coming changes than any existing forms. Some of this plasticity may be eroded through time if it is not locked into the germ-line. If the behavioral changes wrought by environmental change stabilize the response to environmental change, however, then epigenetic control could persist across generations until the environment changes again. Thus, unlike when genes directly control behavior, epigenetic control enables reversibility and the ability to erase phenotypic matches that become mismatches within a generation. While much more work is needed to understand the extent of, and mechanism responsible for, persistence of epialleles, epigenetic control of behavior appears to enable moderate levels of plasticity. While unable to match the extreme responsiveness of learning, it does not suffer from its "forgetfulness." By anchoring the expression of the trait to modifications of the genome, the underlying genetic template remains unchanged. But novel patterns of methylation could produce novel variants and these could be passed on from generation to generation if the control is in the germ-line. If not, then degree of permanence will depend on feedbacks between the trait and its ability to recapitulate the initial response to the environment that produced it.

## 5.3 WHAT ARE THE DOWNSIDES OF PLASTICITY?

Behavioral variations that are plastic or flexible offer individuals the potential to cope with environmental change. Although speed and permanence differ for plastic and flexible responses, either is likely to enhance an individual's options. However, increasing potential inevitably comes with drawbacks. The ability to perceive and then react to change requires large numbers of sensory and motor neurons and usually a large brain as well. And a large brain does matter. Larger-brained birds typically survive better in the wild (Sol *et al.* 2007) and larger-brained mammals do better in novel habitats (Sol *et al.* 2008). Apart from investment of energetic and material resources into growing new neurons, subsequent creation of connections and learning, even by trial and error, all take time. Opportunities lost and the costs incurred will be associated with naïveté (Dukas 1998) or trading off exploitation for exploration (Berger-Tal *et al.* 2014). This means that even when selection favors exploration and sampling among contingencies and making decisions, the degree of plasticity or flexibility that emerges will be constrained by costs. When plasticity or flexibility have high costs, would birds with or without these behavioral contingencies be favored by selection, facing the onslaught of urbanization and the noise pollution that accompanies it? Would birds able to raise the frequency of their calls to avoid the low frequencies of road noise do better than those that avoid the noise and leave the area? Would birds expressing contingent behavior do better than those who might be pre-adapted to cope with low-frequency background noise because they live in habitats with abundant babbling brooks? It is likely that the luck of having a pre-adapted fixed strategy of a high-frequency call will trump a contingent strategy mediated by plasticity (Slaberkoorn 2013). But it is much harder to say that the costs of maintaining multiple sensory and motor systems are more or less than the costs of moving elsewhere, especially if there might not be an elsewhere to go. At least with respect to snails showing varying degrees of plasticity in response to different styles of predatory behavior by crayfish, those families showing the greatest plasticity were those also showing the greatest reduction in growth. Apparently, those families best able to change morphologically grew more slowly because their coupled plastic behavior response also reduced their feeding success (Dewitt *et al.* 1998).

Indirect cues that allow individuals to anticipate future states or conditions are often adaptive, and as the example with egg-laying in blue tits shows, the ability to make detection–reaction associations is likely to be under strong selection (Grieco *et al.* 2002). But if the environment changes dramatically and extremely rapidly, normal rules of thumb may no longer apply and lead individuals into "ecological traps" (Schlaepfer et al. 2002). The classic example of birds preferring to nest near forest edges because they prefer heterogeneous vegetation inadvertently leading to decreased reproductive success, since nest parasitic cow birds – who are increasing in human-disturbed landscapes – were more active along forest edges (Gates & Gysel 1978), illustrates that unless birds can adjust their behavior by one of the three mechanisms presented in Figure 5.1, fitness will be reduced.

Fitness, however, may be reduced even if a flexible response is mounted. It is quite likely that cues, such as temperature, that shape developmental responses of insects also shape the phenology and developmental responses of the plants, especially in terms of flower timing and tissue maturation that the insects need to fuel breeding and to sustain larval development. If the response rates to these cues, however, are not synchronized between plant and insect, then it is likely that plant development may become faster or slower than that of the insect. As a result, the normal degree of plasticity and the ability to mount such a response would no longer prove beneficial to the insect, especially if plasticity comes with a cost of diminished longevity or fecundity.

Limits to the beneficial effects of flexibility in the face of anthropogenic change may also arise because of indirect effects – change that directly affects one behavior in a favorable way may also indirectly and negatively affect another. For example, Manor and Saltz (2005) showed for gazelles (*Gazella gazella*) that disturbance increased vigilance but also decreased group size. And since the per capita increase in vigilance was higher in large rather than small groups, individuals in the small groups were likely to be at more risk than their counterparts in large groups. Similarly, the normal response of territorial male wild asses (*Equus hemionus*) is to reduce territory size as population density and bachelor male intruder pressure increase (Saltz et al. 2000). But if this increase in numbers makes the wild asses a nuisance and leads to their being hunted, then flexibility will reverse the process. Territory sizes will increase as will each territorial male's reproductive success. But from a population level and conservation perspective, the flexible response of increasing territory size will lead to decreased effective population size and loss of genetic diversity. Finally, the indirect effects of behavioral change by one species may change the strength of interactions with other species. Werner and Peacor (2003) review many examples of where a trait-mediated indirect interaction reshapes the nature and stability of entire ecological communities. In simple food webs, such as one involving sunfish consuming salamanders consuming isopods, in the absence of fish, salamanders markedly reduced isopod survivorship. When fish were present, however, salamanders had little impact on the isopods; not because the fish preyed on the salamanders, but rather because they changed the foraging behavior of the salamanders (Huang & Sih 1991). Thus a flexible change in how salamanders managed risk impacted the structure of the ecological community, sometimes stabilizing it, sometimes not. But the relative importance of change in behavior, as opposed to density, on shaping the structure and stability of communities is likely to be system dependent. In fact, the balance may depend on the magnitude of environmental change itself. If the change is catastrophic, for example, it should impact both behavior and density and the density impact will likely take precedence. But if the change is substantial but more moderate, it is likely to affect only behavioral flexibility, thus making it a force to be reckoned with. Clearly, some behavioral adjustments will simply deepen the trap in which a species finds itself or that it imposes on other species.

## 5.4 WHAT CONDITIONS FAVOR THE GENERATION AND MAINTENANCE OF BEHAVIORAL VARIANTS?

Given that increases in costs will eventually limit the ability of any species to develop neural architectures and strategies to cope with novel and faster than usual changing conditions, are there particular biological traits or environmental features that pre-dispose some species to perform better than others? Much depends on the time scale over which the environment changes (Snell-Rood 2013). If change is coarse-grained and varies between generations, then plastic responses will be possible, since genotypes in the sense of reaction norms typically produce a range of variants, one for each new environment. If the range of past environments experienced by a species has been narrow, then mismatches between phenotype and environment are likely to appear and species may fall into ecological and evolutionary traps. Otherwise, coping is likely, especially if genetic expression can be modulated by epigenetic factors. If, however, change is fine-grained and occurs within generations, the ability to cope may require certain preconditions. Species that migrate and change their location in space (Rubenstein & Hack 2013) or hide in time by hibernating or entering diapause (Tauber & Tauber 1976) may be able to coarsen the grain of environmental change. By doing so they will experience conditions with which they are familiar, thus eliminating the need for developing novel behavioral responses. For all other species, however, environmental changes will be felt within generations, with behavioral flexibility being the best option for coping. I contend that behavioral flexibility will be facilitated by 5 factors: (1) long life; (2) large brains; (3) diversified personalities; (4) ease at using public as well as private information; and (5) having a long history of living in non-equilibrium ecological systems. The first four are intrinsic to a species' make-up, while the last is extrinsic and a feature of the environment it inhabits. When taken together these five factors should enable rapid adaptive coping. This does not mean that species displaying one or more of these phenotypes will always display behavioral flexibility. Nor does it mean that species exhibiting none of these phenotypes or living in non-equilibrium systems will not exhibit behavioral flexibility. Rather, it is my contention that when these conditions apply, the benefits of flexibility will generally outweigh its costs.

**Life span.** In general, K-selected species (Horn & Rubenstein 1984) or animals with "slow" life styles live a long time, are large bodied, have very few young, but invest copiously in each. Being able to buffer the effects of environmental variation helps reduce adult mortality, which in turn fosters high population densities. With few opportunities available for their young, adults iterate breeding and provide extensive parental care to boost future prospects for the few young produced in each reproductive episode. While large body size may reduce the frequency and extent of physical environmental variation, social surprises will inevitably occur over a long life under crowded conditions. And since optimal social strategies are those shaped by the action of others, the behavioral flexibility that learning provides should enable long-lived species to cope with unpredictably appearing social stressors. For long-lived species past experiences will accumulate, which should allow more robust responses to changing conditions (Dukas 2013). Such robustness is essential for survival because any massive decline in numbers will leave long-lived species vulnerable to demographic collapse.

For short-lived species the trade-off between strong behavioral responses and demographic ones tips the other way. High reproductive potential is the antidote that short-lived species possess to cope with population collapse. Armed with limited behavioral flexibility, short-lived species must rely on high turnover rates to test how well new genotypes and the phenotypes they produce through developmental plasticity can cope with environmental change.

**Brain size.** Large-bodied individuals have long lives, but they also typically have large brains. Ever since Jerison's (1973) path-breaking work on brain evolution, it has been assumed that large brains enhance cognitive ability, otherwise why would animals pay the price to develop and maintain such an energy-demanding organ. Evidence supporting this contention has been scant, in part because cognitive abilities among species are also tuned to the tasks each species needs to solve (Macphail 1985). Until recently, such contingencies have thwarted attempts to compare species via standardized tests on cognitive tasks, even though strong correlations have been shown to exist between brain size in primates and primate group (Dunbar 1992) and clique size (Kudo & Dunbar 2001), as well as between brain size, learning and tool use (Lefebvre *et al.* 2004). By comparing forty-six primate species across seventeen genera, and by using a model selection approach correcting for phylogenetic relatedness, Shultz & Dunbar (2010) showed that hippocampus size was the best predictor of overall cognitive ability based on performance in solving eight memory and detection tasks. Overall brain and neocortex size were also included in the best model. Although these studies were limited to primates, they provide mechanistic support for the large number of studies on other taxa highlighting the relationship between brain size and learning and an ability to innovate when confronting novel situations (Sol *et al.* 2005).

**Personality.** Most animal societies consist of individuals with different personalities (Sih & Bell 2008). Two of the most common traits are "boldness" and "shyness." Bold individuals tend to be active explorers, whereas shy individuals act as reticent followers. Having both variants in a population clearly benefits the population, since behavioral variation in general provides multiple ways for solving novel problems. Hence, from a conservation and management perspective, species with a range of personality types are more likely to cope with environmental change on their own than are species that are more monomorphic in makeup. But each behavioral variant in itself is likely to benefit from complimentary mutualistic or reciprocal interactions. Bold individuals, by being proactive, may sometimes benefit by being first; finding hidden or rare food items falls into this category. But at times, shy individuals, by being reactive, might be better off adopting a wait and learn strategy; foraging in areas where humans are also hunting or fishing may favor such caution. Even if bolder and more proactive individuals may occasionally get led into evolutionary traps (Sih 2013), mixing personalities may enable a collective solution to the exploration–exploitation trade-off that each would have found difficult on its own.

**Public information.** When individuals are in groups, "Wisdom of the Crowd" or the "Many Wrongs Principle" (Guttal & Couzin 2011) often result in wiser decisions for individuals than when they are on their own, because individuals in groups lead or attend to the behavior of immediate neighbors. With respect to populations of fish and fishermen, in a Mediterranean carnivorous species that foraged on food resembling the types of bait offered by recreational fishers, the frequency of shy fish tended to increase (Alós 2014). In other studies where the behavior of individuals could be monitored, a mixture of learning from personal experience and copying the actions of others contributed to this increase in timidity. When fish were being preyed upon by other fish, reciprocal interactions between predators and prey were shaped by learning. Dynamical feedbacks occurring within encounters, and the personality types of the interactants shaped the behavioral sequence (McGhee *et al.* 2013).

At even greater scale, the behavior of an entire species can be shaped by the interactions of a few individuals, especially if there are different behavioral strategies within the population. One example involves the evolution of migratory behavior (Guttal & Couzin 2011). When the cost of detecting resource gradients is high, selection favors individuals that remain stationary in one area. But as this cost declines, the species will be selected to switch from being stationary to migratory, with only a few individuals investing in gathering gradient information about resource locations and quality. All others need only invest in attending the action of the few leaders. What these examples illustrate is that for species that can use public information, the social learning that results can be a powerful way of rapidly changing the behavioral make up of a population, a species or even a community of interacting species.

**Non-equilibrium environments.** Two schools of thought dominated the ecological debate on how population sizes are regulated. In one, populations were assumed to grow exponentially until extremely harsh environmental conditions reduced numbers drastically, after which they resumed growing exponentially when favorable conditions resumed. The resulting saw-toothed curves of numbers versus time depicted a world regulated by density-independent factors (Andrewartha & Birch 1964). In the other, environmental fluctuations were assumed to be minimal, and, as a result, populations initially grew exponentially until increases in density reduced per capita survival and fecundity to the point where birth and death rates equilibrated and the population reached carrying capacity at equilibrium. The resulting S-shaped curve of numbers versus time was often described mathematically by the logistic growth equation (Gause 1932). While small-bodied species such as insects, fish and rodents were assumed to be governed by density independent factors, larger species were assumed to be regulated by density dependence. Such a worldview implied that only small-bodied species would cope with environmental change via boom and bust demographic activity rather than by employing behavioral flexibility. Given their high reproductive rate, their ability to respond and adapt demographically was high. But Ellis and Swift (1988) argued that many dryland systems are non-equilibrial as well. As a result, all species inhabiting them – both large and small – would experience unpredictable and persistent extreme environmental fluctuations. Given that large-bodied species have small clutches and large inter-birth intervals, coping would instead require behavioral plasticity, flexibility or both.

Recent comparative studies by Jetz and Rubenstein (2011), and Rubenstein and Lovette(2007) argue that environmental uncertainty in climate is a powerful selective force on behavior as well as on physiology and morphology. First using starlings in Africa and then a collection of over 95% of the world's birds, they show that cooperative breeding tends to occur in habitats where climate is highly unpredictable even after accounting for phylogenies and common ancestry. They argue that cooperation allows breeding both in good times and in bad, when the ability of helpers to pitch-in matters most. Coburn and Russel (2011) counter with the possibility that young accumulate during good years and when unable to breed in bad years, they "make the best of a bad job," helping out their parents instead. In a sense, these explanations are opposite sides of the same coin, suggesting that key decision-making switches are integral parts of the behavioral repertoire of species living in non-equilibrium systems that are characterized by large unpredictable environmental changes. Even many social insects vary the ratio of castes when environmental conditions change. While populations in equilibrium systems also experience environmental change, the behavioral responses they show will likely be quantitative and small scale rather than qualitative in scope.

These four features of a species' biology along with the environment it lives in are likely to characterize species capable of coping with environmental change by drawing on behavioral flexibility, especially for those that have limited reproductive potential. Copious reproducing species ordinarily live in changing environments and their high fecundity, often coupled to high vagility, enables them to respond to environmental change demographically and genetically (Horn & Rubenstein 1984). But for so-called K-selected, or "slow," species these features, especially when taken together, increase the likelihood that flexible responses will be deployed.

## 5.5 EQUIDS AS A MODEL TAXON

In exploring the causes and consequences of plasticity and flexibility on coping with environmental change, examples have been drawn from a range of birds and mammals. But by focusing on one taxon – the equids – a better understanding of the mechanisms by which flexibility operates and where its limits lie will emerge. In this way, this will help identify when and how conservation practitioners and resource managers should act.

Equids are members of the horse family (Rubenstein 2011). Today there are only seven extant species and they are all members of the same genus, *Equus*. As a result, they are close evolutionary kin and share similar body plans, physiology (hind gut fermentation) and behavior. In addition, some of the extant species are endangered (Grevy's and Mt. zebra as well as Przewalski's horse) and within the last 12,000 years two species of horses across two continents were driven to extinction by human activity in conjunction with climate change (Lorenzen *et al.* 2011). Recent transloca-tions of wild species and escapees of domesticated variants have restored these species, once extinct in the wild, to their traditional ranges. And given that they are long lived, have large brains, reveal personality differences within populations, readily share public information and live in environ-ments where climates are often seasonally harsh and unpredictable and where there is pressure from humans in the form of poaching and land encroachment, equids are a model taxon for exploring the ways in which plasticity and flexibility enable populations to cope with rapidly changing environments.

**Adapting to repeated human activity.** One of the most basic forms of learning that can enable individuals to respond to a series of environmental stresses involves habituation. In a study designed to simulate all the fea-tures of a typical translocation apart from the long-distance moving of the animals, Rubenstein and Constantino (unpublished data) gathered data on the activity patterns, particularly those of grazing and vigilance, of plains zebras in a population of approximately 300 inhabiting the Pyramid Conservancy in Laikipia, Kenya. They also recorded a series of critical distances: (1) detection distance (distance at which at least one zebra in a group lifted its head at the approach of a slow-moving vehicle); (2) flight distance (distance at which detection changed to flight at the continued approach of vehicle); and (3) the stop distance (the distance at which flight ceased). Baseline data were taken on randomly selected plains zebras that were each only approached once during the 10-day pre-trial period. Then four focal individuals were immobilized, fitted with VHF radio collar, released and tracked. Afterwards, time budget and distance data were collected from the collared zebras as well as from another set of randomly chosen zebras that were approached only once during a 10-day post-trial period. Overall, both detection and flight distance declined significantly for zebras in groups with the collared zebras that were repeatedly approached compared to pre-test animals. Interestingly, smaller but still significant reductions in these distances were found for randomly chosen post-test individuals when compared to randomly chosen pre-test zebras, suggesting that information spread is rapid and occurs via social learning. Given that harem groups regularly change associations as they move in and out of herds, this spread of information among non-test individuals is not surprising, but only as long as there are benefits associated with detecting and assessing that particular vehicles are non-threatening. And such benefits do exist. After a lengthy period of habituation, feeding rate of the collared zebras that were repeatedly approached, as well as randomly chosen zebras that were only approached once, increased three-fold. Habituation clearly benefits plains zebras and cues spread quickly and distantly via social learning.

Making distinctions about humans and how to respond to the risks they pose is not limited to plains zebras. Grevy's zebras also learn, but in different ways. Grevy's zebras have been impacted by people over the last 40 years. Despite outlawing shooting for their skins in the 1970s, numbers continued to decline precipitously in Kenya from 14,000 in 1980 to around 2000 by 2005, mostly because of heavy grazing and the monopolization of water by pastoral herders (Rubenstein 2010). Not surprisingly, Grevy's zebras on pastoral lands in Laikipia avoid human activities during the day. Instead, they graze and drink on neighboring commercial ranches or within group ranch conservation areas. At night, when livestock are corralled in "bomas" and when people are asleep, Grevy's zebras come to drink at watering points in the grazing area because they are cleaner and better maintained then those in the conservation areas, and also to nibble on the short but productive grass swards in the grazing area (Rubenstein 2010). Similarly, on the community conservancies in Samburu, lactating females with their young spend nights close to the manyattas, where people sleep to reduce predation risk. Once the herders awake, the zebras move away to forage and drink (Low et al. 2009). Although this form of associative learning is different from habituation, Grevy's zebras learn to discriminate among different types of human actions, avoiding those that might lead to harm, while taking advantage of those that can enhance survival and reproductive success.

**Coping with changing ecological pressures.** Human impacts need not always be direct. In the case of Grevy's zebras, pastoral herds have grown both in numbers and size, so that watering points are occupied virtually continuously from just after dawn to just before dusk. This means that Grevy's zebras can rarely drink at midday, their preferred drinking time (Rubenstein 2010). By having to wait until night, the chances of being attacked by lions increase. An unintended consequence of this delay is that group size increases, which helps reduce the per capita risk of being eaten. Since equids typically live in non-equilibrium settings, changes in herd size are common and can be deployed in novel ways in novel settings to help mitigate human-induced threats.

This flexibility likely emerges from the fact that individuals in equid societies often experience a variety of demographic conditions that typically lead to changes in social relationships. When densities are high, for example, it is common for harems of feral horses to become multi-male groups. Sharing matings would seem to incur reproductive costs, but sharing defense duties, even if unequally because of personality differences, largely offsets them, as does reducing time spent as sub-adult bachelors (Rubenstein & Nunez 2011). Thus it is not surprising that when human activities disrupt so-called normal patterns of behavior, equids have a propensity to adapt. This is clearly the case with Przewalski's horses, which went extinct in the wild in the late 1960s, but have been brought back to life by many reintroductions from zoos that have created a series of populations throughout their historical range. Since the last sighted Przewalski's horse was seen in the area of the Kalimaili Nature Reserve in Jingjang Provence, China, a population was established there. Normally, horses thrive in mesic conditions where abundant food and water are close together. When they are not, then horses have to radically adjust their daily rhythms (Zhang *et al.* 2015). Remaining near watering points throughout the day and drinking frequently means that foraging has to shift to the night when temperatures are lower, thus making travel in search of food less stressful. In this case, being large-bodied and having an evolutionary history of living in non-equilibrium systems was enough to compensate for a mismatch between mesic expectations and arid realities.

Similarly, when feral horses are treated with non-steroid contraceptives to control population growth, as is the case with horses living on a barrier island off the east coast of the US, the unexpected consequences of continued estrus cycling is increased sexual harassment by males (Madosky *et al.* 2010, Nunez *et al.* 2010). Normally, females reduce this stress by associating with high-ranking males or those rising rapidly in rank

(Rubenstein 1994). But since all males associating with continuously cycling females become overly aggressive, females instead fall back on forming alliances with as many females as they can to reduce this stress (Rubenstein *et al.* 2016), a tactic that they draw upon in other contexts to increase reproductive success (Cameron *et al.* 2009).

The ability to modify social networks in response to human-induced changing environmental circumstances is perhaps the most flexible of all equid responses. Equids generally exhibit two seemingly very different social systems. In one, typified by horses, plains and mountain zebras, females live in closed membership family groups with their young offspring and one male. In the other, typified by Grevy's zebras and the wild asses, associations among individuals are much more ephemeral, lasting for a few days or even only a few hours (Klingel 1975, Rubenstein 1986). But when the social relationships are analyzed in more detail (Rubenstein *et al.* 2007, Sundaresan *et al.* 2007) the structures of these two seemingly similar societies are very different. Asiatic wild asses reveal weaker connections among individuals than do Grevy's zebras, which show more modularity and stronger associations with fewer partners. Today asses live in habitats where humans have eliminated predators and made water widely available to support livestock ranching. Grevy's zebras also share the landscape with livestock, but water availability and predator abundance continue to vary unpredictably. When spreads of disease or memes are simulated on these different networks, those of wild asses function as facilitators as long as spread is easy to initiate, but they are hopeless in preventing losses. The opposite applies to those of Grevy's zebras, where modularity helps retain information. The correlation suggests that modular networks may be favored in an uncertain world (Rubenstein *et al.* 2015).

## 5.6 LESSONS FOR MANAGERS AND CONSERVATIONISTS

As a model taxon, the equids show a surprising ability in a range of social contexts to respond flexibly to human-induced environmental changes. The few examples explored in detail reveal how having a history of living in non-equilibrium and unpredictably changing environments when coupled with being long lived, having large bodies and big brains, multiple personalities and the ability to modulate public information sharing helps make flexible responses possible. But there are limits to how much flexibility a species can successfully exhibit, and these limits underscore the problem both wild species and their conservators will face in tempering the impacts of human-induced environmental change. Conservation behaviorists often draw upon models of reaction norms to illustrate how genotypes typically generate variants across environmental states. But reaction norms usually increase or decrease monotonically. Rarely do they have kinks, and if kinks are needed for coping it is hard to imagine, let alone predict, how they will arise. As conservation behaviorists we would like to predict which species can adapt and how they will do it. Sih (2013) proposes that sensory detection theory and an emerging theory of adaptive reaction norms will help do just this, but only if the dimensionality of the problem is well characterized. However, at the moment this is rarely the case. Nevertheless, even if it is not yet possible to develop models that can predict success and in what form, bettering understanding of the behavior and ecology of a species is essential. In the short run, the five-feature framework presented here can serve as a qualitative guide for identifying which species are most likely to have tool kits laden with flexible or plastic behavioral options or with responses that are more demographic than behavioral. As Figure 5.1 shows, these three response domains will be coupled to different underlying mechanisms, ranging from learning and forgetting to hard-wired genetics to epigenomic modulation of gene action.

Even if we can identify which species have high probabilities of adapting to change by adopting flexible responses, no species is sufficiently omnipotent to have the right response at the right level for the right stressors at the right time. As the above examples have shown, excluding Grevy's zebras from water during the day increased the sizes of the groups using the same watering points at night. While this social change helped reduce per capita predation risk, not drinking during the day most likely created other problems associated with thermoregulation, since it is common for Grevy's zebras to seek shade under acacias during the heat of the day. Thus, by working with communities and engaging pastoral herders as scouts to collect data to better understand Grevy's zebra movement and interaction patterns, we were able to show local herders that sharing water during the day would not harm their livestock, but would benefit the zebras (Low *et al.* 2009). As a result, sharing has become the norm, group sizes have been reduced, predation has declined, recruitment has increased and population numbers have stabilized and are now increasing. Had the zebras had to rely solely on their own anti-predator response to cope with the consequences of daytime exclusion from water, demographic benefits would have been much smaller.

If the benefits of diluting the risks of being eaten or increasing the number of eyes detecting predators only offer minimal gain when predation levels were at normal levels, what gains could have been expected when predation levels were much higher, as is often the case within national parks or on private conservancies where tourists are eager to see lions? In those situations, the fission–fusion social system – even with high degrees of modularity – could barely cope because the territorial male is usually apart from females while he wanders alone far and wide, searching for additional sexually receptive females. Even plains zebras that live in highly cohesive core family units and who manage predation risk by adopting a variety of different behavioral alternatives (Fishhoff *et al.* 2007) find it hard to avoid high levels of predation. In such human-constructed settings the only solution is to reduce the density of lions while making it easier for tourists to find those that remain. To carry out this strategy lions need to be translocated from high-to low-density areas. By doing so predation intensity would be lowered to levels where Grevy's zebras can again manage risk by themselves. But this requires changing policies and statutes that treat endangered species differently in different regions based on relative risk of extinction. So far, all attempts to change current practice have not been successful (Rubenstein 2010). But to put forth evidence to justify such a change requires understanding if a species is able to draw upon flexible responses, why it is pre-adapted to do so and at what levels a species' particular responses are likely to succeed or fail. Detailed knowledge of the behavioral ecology of the species both before and after the environment has changed will be essential.

## 5.7 WHAT IS A CLEVER MANAGER OR CONSERVATIONIST TO DO?

Managers and conservationists often have to make tough decisions on which species, landscape or ecosystem to support and which to let fend for themselves. In such situations, the species that are likely to cope with a minimum of support are those that exhibit behavioral flexibility, where the opportunity costs or unintended consequences of this flexibility are low. In general, large-bodied, long-lived species exhibiting a variety of behavioral syndromes or personalities, having easy access to public information and living in non-equilibrium systems are ones in which flexibility is mostly like part of its evolutionary tool kit. For these species, interventions that can limit the scope of human-induced change to what has been "seen" in a species' evolutionary past should suffice. Thus, managers and conservationists need to know what are the key features of the ecology and natural history of such species and then try to change human behavior to keep the environment within regions where normal adaptive adjustments can take place.

For other species and systems, however, all need not be lost. Fast-living species have high demographic potential and thus can respond to intense selection genetically, if not epigenetically. So finding ways of boosting reproduction will be essential. If habitat fragmentation and the ethos of enjoying nature continue to grow at their current pace, few places on Earth will not feel some form of human impact. Some argue that if species can habituate to human activity, then attempts to habituate them should be encouraged. Nesbit (2000) even suggests that water bird colonies should be managed for multiple uses, and to do so requires habituating them to humans and their activities. Such a strategy, however, could have long-term negative implications because their survival will be dependent on continued human activity, and the consequences of such a shift in a species' evolutionary trajectory are yet unknown. But it is hard to see an alternative when people and their actions are seemingly everywhere even if these actions themselves have some adverse effects (Bejder et al. 2009).

If flexibility in animals has more pros than cons, then human flexibility should also be targeted for action. Human pressure on wildlife comes when people strive to improve their livelihoods or well-being. Knowing what species can do and how they interact with other species in communities can reveal where are the pressure points that reduce success or ecosystem function that need to be managed. Once these are identified, then managers and conservationists can work with people to find common ground where limited restraint by humans is likely to have the greatest effect in letting the natural adjustments that species can make work effectively. This will buy time until the underlying additional plastic or flexible responses can arise in stressed species. Whether the emergence of novel behavioral variants merge from learning, genes or epigenes, they are more likely to do so if the most egregious levels of stress are ratcheted back by an engaged and empowered public. Managers and conservationists must show flexibility in the way they define and understand the problem and in how they work with others to tune the solution to the abilities of the species or systems they hope to conserve.

## REFERENCES

Alós, J., Palmer, M., Trías, P., Díaz-Gil, C. and Arlinghaus, R. 2014. Recreational angling intensity correlates with alteration of vulnerability to fishing in a carnivorous coastal fish species. *Canadian Journal of Fisheries and Aquatic Sciences,* 72:1–9.

Andrewartha, H.G. and Birch, L.C. 1964. *The Distribution and Abundance of Animals.* Chicago: University of Chicago Press.

Barnard, C.J. and Sibly, R.M. 1981. Producers and scroungers: a general model and its application to captive flocks of house sparrows. *Animal Behaviour*, 29:543–550.

Bejder, L, Samuels, A., Whitehead, H., Finn, H. and Allen, S. 2009. Impact assessment research: use and misuses of habituation, sensitization and tolerance in describing wildlife responses to anthropogenic stimuli. *Marine Ecology Progress Series*, 395:177–185.

Berger-Tal, O., Nathan, J., Meron, E. and Saltz, D. 2014. The exploration–exploitation dilemma: a multidisciplinary framework. *PloS one*, 9:e95693

Cameron, E.Z., Setsaas, T.H. and Linklater, W.L. 2009. Social bonds between unrelated females increase reproductive success in feral horses. *Proceedings of the National Academy of Sciences*, 106:13850–13853.

Caro, T. and Sherman, P.W. 2011. Endangered species and a threatened discipline: behavioural ecology. *Trends in Ecology &Evolution*, 26:111–118.

Caro, T. and Sherman, P.W. 2013. Eighteen reasons animal behaviourists avoid involvement in conservation. *Animal Behaviour*, 85:305–312.

Cockburn, A. and Russell, A.F. 2011. Cooperative breeding: a question of climate? *Current Biology*, 21:R195–R197.

Davies, N.B., Krebs, J.R. and West, S.A. 2012. *An Introduction to Behavioural Ecology.* West Sussex: John Wiley & Sons.

Dukas, R. 1998. Evolutionary ecology of learning. In Dukas, R. (Ed.), *Cognitive Ecology*, pp. 129–174. Chicago: University of Chicago Press.

Dukas, R. 2004. Male fruit flies learn to avoid interspecific courtship. *Behavioral Ecology*, 15:695–698.

Dukas, R. 2010. Insect social learning. In Breed, M. and Moore I. (Eds.) *Encyclopedia of Animal Behavior*, pp. 176–179, Oxford: Academic Press.

Dukas, R. 2013. Effects of learning on evolution: robustness, innovation and speciation. *Animal Behaviour*, 85:1023–1030.

Dukas, R. and Dukas, L. 2012. Learning about prospective mates in male fruit flies: effects of acceptance and rejection. *Animal Behaviour*. 84:1427–1434.

Dunbar, R.I. 1992. Neocortex size as a constraint on group size in primates. *Journal of Human Evolution*, 22:469–493.

Ellis, J.E. and Swift, D.M. 1988. Stability of African pastoral ecosystems: alternate paradigms and implications for development. *Journal of Range Management Archives*, 41:450–459.

Farine, D.R., Aplin, L.M., Sheldon, B.C. and Hoppitt, W. 2015. Interspecific social networks promote information transmission in wild songbirds. *Proceedings of the Royal Society of London B: Biological Sciences*, 282:20142804.

Franceschini, M.D., Rubenstein, D.I., Low, B. and Romero, L.M. 2008. Fecal glucocorticoid metabolite analysis as an indicator of stress during translocation and acclimation in an endangered large mammal, the Grevy's zebra. *Animal Conservation*11:263–269.

Fischhoff, I.R., Sundaresan, S.R., Cordingley, J. and Rubenstein, D.I. 2007. Habitat use and movements of plains zebra (*Equus burchelli*) in response to predation danger from lions. *Behavioral Ecology*, 18:725–729.

Gates, J.E. and Gysel, L.W. 1978. Avian nest dispersion and fledging success in field-forest ecotones. *Ecology*, 59:871–883.

Gause, G.F. 1932. Experimental studies on the struggle for existence I. Mixed population of two species of yeast. *Journal of Experimental Biology*, 9:389–402.

Gilroy, J.J., Anderson, G.Q.A., Grice, P.V. and Sutherland, W.J. 2011. Identifying mismatches between habitat selection and habitat quality in a ground-nesting farmland bird. *Animal Conservation*, 14:620–629.

Giraldeau, L.A. and Lefebvre, L. 1987. Scrounging prevents cultural transmission of food-finding behaviour in pigeons. *Animal Behaviour*, 35:387–394.

Greenberg, J.K. *et al.* 2012. Behavioral plasticity in honey bees is associated with differences in brain microRNA transcriptome. *Gene Brain and Behavior*, 11:660–670.

Grieco, F., van Noordwijk, A.J. and Visser, M.E. 2002. Evidence for the effect of learning on timing of reproduction in blue tits. *Science*, 296:136–138.

Greenwood, P.J. and Harvey, P.H. 1982. The natal and breeding dispersal of birds. *Annual Review of Ecology and Systematics*, 13:1–21.

Guttal, V. and Couzin, I.D. 2011. Leadership, collective motion and the evolution of migratory strategies. *Communicative & Integrative Biology*, 4:294–298.

Horn, H.S. and Rubenstein, D. I. 1984. Behavioural adaptations and life history. In Krebs, J.R. and Davies, N.B. (Eds.), *Behavioural Ecology*, pp. 279–300. Oxford: Blackwell Scientific Publications.

Huang, C.F. and Sih, A. 1991. Experimental studies on direct and indirect interactions in three trophic-level stream systems. *Oecologia*, 85:530–536.

Jerison, H. 2012. *Evolution of the Brain and Intelligence*. New York: Academic Press.

Jetz, W. and Rubenstein, D.R. 2011. Environmental uncertainty and the global biogeography of cooperative breeding in birds. *Current Biology*, 21:72–78.

Klingel, H. 1975. Social organization and reproduction in equids. *Journal of Reproduction and Fertility, Supplement* 23:7–11.

Kudo, H. and Dunbar, R.I.M. 2001. Neocortex size and social network size in primates. *Animal Behaviour*, 62:711–722.

Lahiri, M., Tantipathananandh, C., Warungu, R., Rubenstein, D.I. and Berger-Wolf, T.Y. 2011. Biometric animal databases from field photographs: identification of individual zebra in the wild. *Proceedings of the ACM International Conference on Multimedia Retrieval* (ICMR 2011), Trento, Italy.

Ledón-Rettig, C.C., Richards, C.L. and Martin, L.B. 2013. A place for behavior in ecological epigenetics. *Behavioral Ecology*, 24:329–330..

Lee, W.Y., Choe, J.C. and Jablonski, P.G. 2011. Wild birds recognize individual humans: experiments on magpies, *Pica pica. Animal Cognition*, 14:817–825.

Lefebvre, L., Reader, S.M., and Sol, D. 2004. Brains, innovations and evolution in birds and primates. *Brain, Behavior and Evolution*, 63:233–246.

Linklater, W.L., Cameron, E.Z., Minot, E.O. and Stafford, K.J. 1999. Stallion harassment and the mating system of horses. *Animal Behaviour*, 58:295–306.

Lorenzen, E.D., Nogués-Bravo, D., Orlando, L., Weinstock, J., Binladen, J., Marske, K.A. and Cooper, A. 2011. Species-specific responses of Late Quaternary megafauna to climate and humans. *Nature*, 479:359–364.

Low, B., Sundaresan, S.R., Fischhoff I.R. and Rubenstein, D.I. 2009. Partnering with local communities to identify conservation priorities for endangered Grevy's zebra. *Biological Conservation*, 142:1548–1555.

Macdonald, D.W. and Johnson, D.D.P. 2015. Patchwork planet: the resource dispersion hypothesis, society, and the ecology of life. *Journal of Zoology*. 295:75–107.

Macphail, E.M. and Barlow, H.B. 1985. Vertebrate intelligence: the null hypothesis [and discussion]. *Philosophical Transactions of the Royal Society B: Biological Sciences*, **308**:37–51.

Madosky, J.M., Rubenstein, D.I., Howard, J.J. and Stuska, S. 2010. The effects of immuno-contraception on harem fidelity in a feral horse (*Equus caballus*) population. *Applied Animal Behavior Science*, **128**:50–56.

Manor, R. and Saltz, D. 2005. Impact of human nuisance disturbance on vigilance and group size of a social ungulate. *Ecological Applications* 13:1830–1834.

Marchand, P., Garel, M., Bourgoin, G., Dubray, D., Maillard, D. and Loison, A. 2014. Impacts of tourism and hunting on a large herbivore's spatio-temporal behavior in and around a French protected area. *Biological Conservation*, **177**:1–11.

McGhee, K.E., Pintor, L.M. and Bell, A.M. 2013. Reciprocal behavioral plasticity and behavioral types during predator–prey interactions. *American Naturalist*, 182:704–717.

Morand-Ferron, J. and Quinn, J.L. 2011. Larger groups of passerines are more efficient problem solvers in the wild. *Proceedings of the National Academy of Sciences*, **108**:15898–15903.

Morris, D. 1957. "Typical intensity" and its relation to the problem of ritualisation. *Behaviour*, 11:1–12.

Nuñez, C.M.V., Adelman, J.S. and Rubenstein, D.I. 2010. Immunocontraception in wild horses (*Equus caballus*) extends reproductive cycling beyond the normal breeding season. *PLoS ONE*. 5(10): e13635.

Rubenstein, D.I. 1980. On the evolution of alternative mating strategies. In Stadden, J.E.R. (Ed.), *Limits to Action: The Allocation of Individual Behaviour*, pp. 65–100. New York: Academic Press.

Rubenstein, D.I. 1986. Ecology and sociality in horses and zebras. In Rubenstein D.I. and Wrangham, R.W. (Eds.), *Ecological Aspects of Social Evolution*, pp. 282–302. Princeton: Princeton University Press.

Rubenstein, D.I. 1991. The greenhouse effect and changes in animal behavior: effects on social structure and life-history strategies. In Peters, R. (Ed.), *Consequences of Global Warming for Biodiversity*, pp. 180–192. New Haven: Yale University Press.

Rubenstein, D.I. 1994. The ecology of female social behavior in horses, zebras, and asses. In Jarman, P. and Rossiter, A. (Eds.), *Animal Societies: Individuals, Interactions, and Organization*, pp. 13–28. Kyoto: Kyoto University Press.

Rubenstein, D.I. 2010. Ecology, social behavior, and conservation in zebras. In Macedo, R. (Ed.), *Advances in the Study Behavior: Behavioral Ecology of Tropical Animals*, Vol. 42, pp. 231–258. Oxford: Elsevier Press.

Rubenstein, D.I. 2011. Family equidae (Horses and relatives). In Wilson, D.E. and Mittermeier, R.A. (Eds.), *Handbook of Mammals of the World, Vol. 2, Hoofed Mammals*, pp. 106–143. Barcelona: Lynx Edicions.

Rubenstein, D.I., Barnett, R.J., Ridgely, R.S. and Klopfer, P.H. 1977. Adaptive advantages of mixed species feeding flocks in Costa Rica. *Ibis*, 119:10–21.

Rubenstein, D.I. and Hack, M. 2004. Natural and sexual selection and the evolution of multi-level societies: insights from zebras with comparisons to primates. In Kappeler, P. and van Schaik, C.P. (Eds.), *Sexual Selection in Primates: New and Comparative Perspectives*, pp. 266–279. Cambridge: Cambridge University Press.

Rubenstein, D.I., Sundaresan, S., Fischhoff, I. and Saltz, D. 2007. Social networks in wild asses: comparing patterns and processes among populations. In Stubbe, A., Kaczensky, P., Wesche, K., Samjaa, R. and Stubbe, M. (Eds.), *Exploration into the Biological Resources of Mongolia, Vol. 10,* pp.159–176. Halle: Martin-Luther-University Halle-Wittenberg.

Rubenstein, D.I.and Nuñez, C. 2009. Sociality and reproductive skew in horses and zebras. In Hager, R.and Jones, C.B. (Eds.), *Reproductive Skew in Vertebrates: Proximate and Ultimate Causes,* pp. 196–226. Cambridge: Cambridge University Press.

Rubenstein, D.I. and Hack, M.A. 2013. Migration. In Levin, S.A. (Ed.), *Encyclopedia of Biodiversity, second edition, Volume 5,* pp. 309–320. Waltham: Academic Press.

Rubenstein, D.I., Sundaresan, S.R., Fischhoff, I.R., Tantipathananandh, C. and Berger-Wolf, T.Y. 2015. Similar but different: dynamic social network analysis highlights fundamental differences between the fission-fusion societies of two equid species, the Onager and Grevy's zebra. *PLoS ONE,* 10:e0138645.

Rubenstein, D.I., Cao, Q. and Chui, J. 2016. Equids and ecological niches: behavioral and life history variations on a common theme. In Ransom, J. & Kaczensky, P. (Eds.), *Wild Equids: Ecology, Conservation, and Management.* Baltimore: Johns Hopkins Press.

Rubenstein, D. R. and Lovette, I.J. 2007. Temporal environmental variability drives the evolution of cooperative breeding in birds. *Current Biology,* 17:1414–1419.

Saltz, D., Rowen, M. and Rubenstein, D.I. 2000. The effect of space-use patterns of reintroduced Asiatic wild ass on effective population size. *Conservation Biology,* 14:1852–1861.

Saltz, D., Rubenstein, D.I. and White, G.C. 2006. The impact of increased environmental stochasticity, due to climate change on the dynamics of Asiatic wild ass. *Conservation Biology,* 20:1402–1409.

Schlaepfer, M.A., Runge, M.C. and Sherman, P.W. 2002. Ecological and evolutionary traps. *Trends in Ecology & Evolution,* 17:474–480.

Shultz, S. and Dunbar, R.I.M. 2010. Species differences in executive function correlate with hippocampus volume and neocortex ration across nonhuman primates. *Journal of Comparative Psychology,* 124:252–260.

Slabbekoorn, H. 2013. Songs of the city: noise-dependent spectral plasticity in the acoustic phenotype of urban birds. *Animal Behaviour,* 85:1089–1099.

Snell-Rood, E.C. 2013. An overview of the evolutionary causes and consequences of behavioural plasticity. *Animal Behaviour,* 85:1004–1011.

Smith, J.M. 1982. *Evolution and the Theory of Games.* Cambridge: Cambridge University Press.

Sih, A. 2013. Understanding variation in behavioural responses to human-induced rapid environmental change: a conceptual overview. *Animal Behaviour,* 85:1077–1088.

Sih, A. and Bell, A.M. 2008. Insights for behavioral ecology from behavioral syndromes. *Advances in the Study of Behavior,* 38:227–281.

Sol, D., Duncan, R.P., Blackburn, T.M., Cassey, P. and Lefebvre, L. 2005. Big brains, enhanced cognition, and response of birds to novel environments. *Proceedings of the National Academy of Sciences of the United States of America,* 102:5460–5465.

Sol, D., Székely, T., Liker, A. and Lefebvre, L. 2007. Big-brained birds survive better in nature. *Proceedings of the Royal Society B: Biological Sciences,* 274:763–769.

Sol, D., Bacher, S., Reader, S.M. and Lefebvre, L. 2008. Brain size predicts the success of mammal species introduced into novel environments. *American Naturalist*, 172:S63–S71.

Sundaresan, S.R., Fischhoff, I.R., Dushoff, J. and Rubenstein, D.I. 2007. Network metrics reveal differences in social organization between two fission-fusion species, Grevy's zebra and onager. *Oecologia*, 151:140–149.

Szyf, M. 2011. The early life social environment and DNA methylation: DNA methylation mediating the long-term impact of social environments early in life. *Epigenetics*, 6:971–978.

Zhang, Y., Cao, Q.S., Rubenstein, D.I., Zang, S., Songer, M., Leimgruber, P. et al. 2015. Water use patterns of sympatric Przewalski's horse and khulan: interspecific comparison reveals niche differences. *PLoS ONE*, 10:e0132094.

# Part III

# Behavior-based management: using behavioral knowledge to improve conservation and management efforts

In the previous section we saw how the behavioral responses of animals to anthropogenic activities, and in particular, the level of behavioral plasticity or flexibility an animal can express in response to these activities, influence the fitness of individuals, and by extension, the persistence of wildlife populations. The flip side of the coin is that understanding animal behavior may allow us to predict their responses to anthropogenic changes and to design management protocols that will minimize the risks to wild populations and maximize their fitness under the given conditions.

In one of the first contributions to the emerging field of conservation behavior, Clemmons and Buchholtz' edited volume on behavioral approaches to conservation, Steve Beissinger lists the seven "tools" that have emerged from the development of conservation biology and that can be applied to conserve biological diversity (Beissinger 1997). The tools are: (1) Reserve and landscape design. (2) Ecosystem management (i.e. management of non-protected areas). (3) Population Viability Analysis (PVA). (4) Sustainable development. (5) Field recovery of endangered species. (6) Captive breeding and reintroduction. (7) Ecosystem restoration (which is nowadays usually considered to include reintroductions of species). While the field of conservation biology has substantially developed and grown in complexity over the last two decades, these seven tools or approaches still compellingly encompass the essence of what conservation management is all about.

The chapters in this section aim to address the seven tools and expand on them from a behavioral point of view. Chapter 7 challenges the traditional spatial approach to reserve and landscape design, within and outside protected areas. Chapters 8 and 10 discuss captive breeding and reintroductions from two perspectives, respectively – the behavioral-sensitive management approach and the behavioral modification approach (see

Chapter 1 for more details on these approaches), and Chapter 9 explores the use of behavior ecology in wildlife population modeling, including PVAs and models of sustainable harvesting.

One of the most notable trends of the past few years is the growing interest and rapid increase in research on sensory ecology (Blumstein & Berger-Tal 2015). Advancing technologies and the novel insights they have led to have transformed the understanding of sensory mechanisms into a vital tool in the conservationist's toolbox, shedding light on how animals make decisions and suggesting ways of manipulating these decisions. To reflect this emerging trend, this section opens with a chapter dedicated to this promising field and its applications in conservation and management.

### REFERENCES

Beissinger, S.R. 1997. Integrating behavior into conservation biology: potentials and limitations. In Clemmons, J.R. and Buchholtz, R. (eds.), *Behavioral Approaches to Conservation in the Wild.* pp. 23–47, Cambridge: Cambridge University Press.
Blumstein D.T. and Berger-Tal, O. 2015. Understanding sensory mechanisms to develop effective conservation and management tools. *Current Opinion in Behavioral Sciences*, 6:13–18.

# The role of animal sensory perception in behavior-based management

ESTEBAN FERNÁNDEZ-JURICIC

## 6.1 INTRODUCTION

At the core of the conservation behavior framework is behavior-based management, which takes into consideration animal behavior in making conservation decisions (Chapter 1). Often, behavior-based management requires manipulation of the behavior of a species in order to accomplish specific conservation or management goals (Sutherland 1998) or avoiding actions producing stimuli that may elicit unwanted behavioral responses. Manipulating behavior may involve repelling an invasive nest parasite from a breeding site, attracting a species to a restored habitat, or sensitizing newly re-introduced individuals to predators. Obviously, the specific means of manipulating behavior will be a function of the biology of the species.

One strategy to modify the behavior of animals is to develop stimuli (visual, auditory, olfactory, etc.) intended to grab their attention and generate a specific type of response. For instance, songs of conspecifics have been used successfully to attract individuals of the endangered Cape Sable seaside sparrow (*Ammodramus maritimus mirabilis*) to suitable breeding areas in the Florida Everglades (Virzi *et al.* 2012). But, some situations can be more challenging. For example, in trying to cause aversive responses in rabbits close to agricultural fields, Wilson and McKillop (1986) tested the effectiveness of a commercially available scaring device that would broadcast sounds at high frequencies (9–15 kHz). They found that the device effect was limited to only 3 m and only while it was playing back the sounds, but most importantly animals habituated after just a few days. Despite the different characteristics of the acoustic stimuli and the different taxa, these opposite results suggest that some species may perceive our stimuli, but that perception alone does not guarantee a response. Is there any strategy

*Conservation Behavior: Applying Behavioral Ecology to Wildlife Conservation and Management*, eds. O. Berger-Tal and D. Saltz. Published by Cambridge University Press. © Cambridge University Press 2016.

to increase the success of stimuli developed for conservation or wildlife management purposes?

The first limitation we should acknowledge is that there may be a discrepancy between the perceptual world of the biologists and that of the target species (Lim *et al.* 2008, Blumstein & Fernández-Juricic 2010). To illustrate this, let us think of the following hypothetical example. Imagine that in a large exhibition tank in an aquarium, we need to attract spotted wobbegong sharks (*Orectolobus maculatus*) to one part of the tank to feed them and avoid interactions with other species. So, we decide to use yellow LED lights to attract wobbegongs to the feeding portion of the tank, and blue LED lights to discourage them from going to other parts of the tank. The expectation is that individuals would go for the yellow lights as this color matches their body coloration (i.e. reducing their saliency to potential prey and predators). After several days, we realize that wobbegongs choose at random between colors. To us, these colors are easy to discriminate, so what could be the problem? One likely reason is the way they visually perceive these patches. Color vision is associated with the presence of more than one type of visual pigment in the cone photoreceptors (Land & Nilsson 2012). Humans have three types of visual pigments in their cone photoreceptors; however, spotted wobbegongs have recently been found to have a single type of cone visual pigment (Theiss *et al.* 2012), and are essentially color blind! This means that wobbegongs see their world in a very different way from us as they only rely on achromatic visual cues to make decisions.

The point of this example is that many times we try to develop novel stimuli using a trial-and-error approach motivated by our own human sensory system. Using stimuli outside of the perceptual world of the target species can (a) reduce the chances of observing the intended behavioral responses, (b) miss the limited time and opportunities we have to steer a change at the individual, population or community levels, and (c) waste the generally limited financial/logistical resources available to conservation/management projects (Lim *et al.* 2008). The magnitude of this problem can be substantial considering the diversity of the animal sensory systems beyond our sensory reach, including ultraviolet and polarized vision, echolocation, electroreception and magnetoreception (Dusenbery 1992, Stevens 2013).

The goal of this chapter is to illustrate conceptually how to tackle the problem of designing stimuli that are tuned to an animal's sensory system and to discuss some scenarios where this approach can be applied. For the sake of space, the focus of the chapter is on vision and birds, although some examples from other sensory modalities and taxa are provided.

Considering the perceptual world of non-human species is not necessarily a new idea (Endler 1997, Lim *et al.* 2008, Fernández-Juricic *et al.* 2010, Martin 2011, Van Dyck 2012). However, there is relatively little guidance in the literature as to how to go about doing this. The key is for conservation scientists and managers to embrace sensory ecology and physiology. Such a task should be feasible given the inherent multi disciplinary nature of conservation biology (Chapter 1).

There are two elements that sensory ecologists study that are particularly relevant from an applied perspective: (1) what information animals gather from their environment, and (2) how that information is gathered. The first point defines the properties of the physical environment that a species makes use of (e.g. ultraviolet, infrasound, etc.); whereas the second establishes the configuration of the sensory organs and consequently the degree of spatial and temporal sensitivity (i.e. distance and rate at which signals can be gathered). Understanding these two components is essential to narrow down the range of sensory stimuli that can trigger changes in behavior.

## 6.2   SENSORY SYSTEMS

Sensory systems are not cheap! Devoting tissue to the peripheral sensory system as well as to the sensory centers in the brain is generally associated with solving specialized tasks under specific environmental conditions. For instance, star-nosed moles (*Condylura cristata*) have more than twenty appendages around their nostrils covered with somatosensory organs that are highly represented in the neocortex (Catania 2011). These sensory organs allow moles to detect food by touch with high precision and speed under low light conditions. Furthermore, processing sensory information in the brain is also costly due to the high energy needed to maintain neurons not only during signaling but also at rest (Niven & Laughlin 2008). For example, energetic consumption of the rat olfactory glomerulus can increase 400% in a single sniff with an increase of two orders of magnitude in odor concentration in the environment (Nawroth *et al.* 2007). Given these constraints, the null expectation should be that our study species may share some sensory capacities with us but also differ in many others depending on the ecological conditions it lives in. This precautionary approach toward the perceptual world of a species can help us consider more carefully sensory criteria in the early stages of any kind of management strategy. Sorting out the sensory *modalities* (e.g. vision, audition, olfaction) in order of relevance for a given species will allow us to gain an initial understanding of the general and specialized tasks that its sensory system can accomplish.

However, we should keep in mind that even *within* a given sensory *modality* there are different *dimensions* (i.e. components that code for different features of a signal). For instance, the avian auditory system can process changes in the frequency as well as the temporal structure of a vocalization. Morphological constraints at the basilar membrane lead to a trade-off in the ability to process these two components of a signal: individuals with high frequency resolution cannot also have high temporal resolution. House sparrows (*Passer domesticus*) have higher temporal auditory resolution than Carolina chickadees (*Poecile carolinensis*), possibly to more efficiently process different components of conspecific vocalizations (Henry *et al.* 2011). The implication is that the different dimensions within a sensory modality may have different representation in the perceptual world of a species, and the relative relevance of these dimensions is likely to vary substantially between species.

Understanding the basic configuration of the sensory organs of our study species is a crucial step in providing an indication of the spatial and temporal ranges of its sensory systems. In this chapter, the focus will be on the visual system due to space constraints, but similar arguments can be made with any other sensory modality.

For example, let's compare human with avian vision. Humans have frontally placed and relatively large eyes that provide a wide binocular field, a single almost centrally placed center of acute vision in the retina (i.e. fovea with high density of cone photoreceptors) that projects into the binocular field, a large degree of eye movement and, as mentioned earlier in this section, three types of visual pigments. Birds have some similarities, but also many differences.

Birds have, generally, laterally placed eyes. Therefore, each eye projects a monocular field toward the sides of the head (Figure 6.1). Both monocular fields encompass the visual field, which is volume around the head from which the animal can see. There is an area in front of the head where the two monocular fields overlap giving rise to the binocular field (Figure 6.1). The areas covered by the visual field of each eye excluding the binocular field are the lateral fields (Figure 6.1). Finally, the area at the rear of the head that is not covered by either monocular field is the blind area (Figure 6.1). Because the specific location of the orbits and the degree of eye movement varies considerably between bird species, so does the

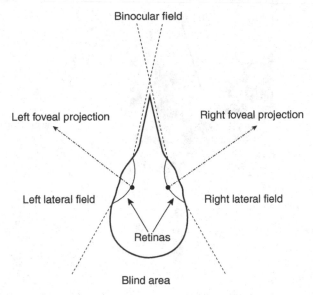

**Figure 6.1:** Schematic representation of the different components of the avian visual field, which is the projection of the margins of the retinas into visual space (not to scale). The binocular field is the overlap of the two lateral fields in front of the head, whereas the blind area at the rear of the head does not receive any visual input. Also shown is the relative position of the center of acute vision (fovea) in retina and its projection. The fovea has a high density of cone photoreceptors, which provide high visual resolution (chromatic and achromatic).

size of the binocular, lateral and blind areas (e.g. Martin 2007, O'Rourke *et al.* 2010, Fernández-Juricic *et al.* 2011, Moore *et al.* 2013), and consequently so does the amount and types of visual information available for a given bird species around its head. The configuration of the visual field can certainly affect the behavior of animals (i.e. species with wider blind areas tend to spend more time scanning head-up; Guilleman *et al.* 2002). Actually, Martin and Shaw (2010) argued that some collisions between birds and power lines may be caused by the limited visual coverage of some bird species, particularly above their heads (i.e. vertical extent of the binocular field). The goal, therefore, is to design stimuli with higher chances of detection based on the degree of visual coverage of the target species.

However, visual performance is not homogenous across the visual field. This is due to the configuration of the retina. The retina is a multi-layered tissue with different types of cells (Figure 6.2). The cells responsible for

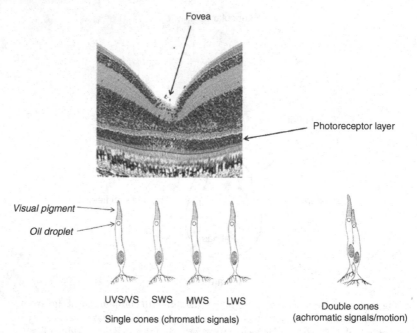

**Figure 6.2:** Cross section of the fovea (i.e. invagination of the retinal tissue) of the brown-headed cowbird showing the position of the photoreceptor layer, which has cone (involved in diurnal vision) and rod (involved in nocturnal vision) cells. There are two types of cones: single (involved in chromatic vision) and double (thought to be involved in achromatic and motion vision). Within the single cones, there are four kinds, depending on the sensitivity of the visual pigment: ultraviolet or violet sensitive (UVS/VS), short-wavelength sensitive (SWS), medium-wavelength sensitive (MWS) and long-wavelength sensitive (LWS). Each type of single cone also has an associated type of oil droplet, which is an organelle that filters light before it reaches the visual pigment.

phototransduction (i.e. conversion of light into electric signals) are rods (for low light conditions) and cones (for day light conditions). Birds (but not humans) have two types of cones that differ morphologically: double cones and single cones (Figure 6.2). Double cones appear to be involved in achromatic vision and motion detection, whereas single cones are responsible for color vision (Hart & Hunt 2007).

Birds have a single type of double cone, with a principal and an accessory member (Figure 6.2). The density of double cones tends to be higher than that of single cones (Hart & Hunt 2007). Birds have four types of single cones, each with a visual pigment sensitive to different parts of the spectrum (Hart & Hunt 2007): (1) ultraviolet- or violet-sensitive cone (UVS/VS) depending on the species, (2) short-wavelength sensitive (SWS),

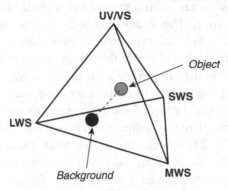

**Figure 6.3:** Schematic representation of the avian color space, which is limited by the visual pigment types present in the four types of single cones (Figure 6.2). Perceptual models estimate the relative distance in color space between the object and the visual background. The larger the distance, the higher the ability of the visual system to detect the object.

(3) medium wavelength sensitive (MWS) and (4) long-wavelength sensitive (LWS). Other retinal cells (e.g. ganglion cells) compare the levels of stimulation of the four different single cones when stimulated by light, which eventually leads to perception of color in the brain (Land & Nilsson 2012). The avian color space can be represented by a tetrahedron (Figure 6.3), which is bounded by four vertices, each corresponding to one of the four types of single cones (Goldsmith 1990, Neumeyer 1992, Cuthill 2006). To put things into perspective, the human color space is only bounded by three types of single cones, and hence it can be represented by the single triangular side at the bottom of the avian color space (Figure 6.3). Consequently, the avian color space is much wider than that of humans, which means that birds can perceive colors that humans cannot even imagine.

Furthermore, birds have within their cone photoreceptors organelles filled with carotenoids called oil droplets (Figure 6.2) that filter the light before it reaches the visual pigments. Because oil droplets act as wavelength-specific filters, they constrain the range of wavelengths that stimulate the visual pigments. This makes it easier to compare the degree of stimulation of different cones in response to light, thereby enhancing the ability to tell different colors apart (Cuthill 2006). Humans do not have oil droplets, and thus color discrimination may not be as refined.

Single and double cones are not homogeneously distributed, which means that visual performance varies across the retina. Recent studies mapping the density of both single and double cones have shown that their highest densities are around the fovea (Fernández-Juricic *et al.* 2013, Baumhardt *et al.* 2014). The implication is that the fovea is not only the center of chromatic but also achromatic/motion vision. Still, birds can detect stimuli with the periphery of their retina (i.e. outside of the foveal area). However, after detection, they align the fovea with the stimulus of interest by moving their eyes or heads in order to inspect it visually with the high acuity provided by the high density of cone photoreceptors.

Another very important visual dimension to consider is visual resolution, which has spatial and temporal components. Spatial visual resolution can be thought of as the ability of an individual to resolve two objects from a distance: the higher the spatial resolution, the farther away these two objects can be differentiated. Spatial visual resolution is estimated in cycles per degree, which generally represent the number of different objects (e.g. black bars) that can be distinguished from the background in 1° of angular distance in the retina. For instance, humans have a spatial visual resolution of about 30 cycles/degree (Hodos 2012). By knowing the spatial visual resolution of a species and the size of a stimulus, we can calculate the threshold distance at which an object can be resolved from the visual background under optimal ambient light conditions. Beyond that threshold distance, the animal would have some difficulty telling that the object is there (i.e. it would blend with the background).

Spatial visual resolution has important implications for developing targeted stimuli. For instance, Blackwell *et al.* (2009) estimated that brown-headed cowbirds (*Molothrus ater*), with a spatial visual resolution of about 5 cycles/degree, would be able to resolve an object 2 m high (e.g. large vehicle approaching) from about 1000 m. Spatial visual resolution depends upon the size of the eye and the density of cone photoreceptors (along with ganglion cells). Consequently, spatial visual resolution varies considerably between species. Some bird species, such as raptors (e.g. brown falcon *Falco beribora*, 73 cycles/degree; Reymond 1987), have higher spatial visual resolution than humans, whereas other birds have much lower resolution, such as Passeriformes (e.g. European starlings *Sturnus vulgaris*, 6.3 cycles/degree; Dolan & Fernández-Juricic 2010).

There is also a temporal component of visual resolution, which estimates how fast the retina can process visual stimuli (i.e. number of snapshots it can get from the environment per unit time). Temporal visual resolution is measured as the ability to detect flicker in a pulsing light source, which varies with the intensity of light (Hodos 2012). The maximum flicker frequency that a retina can detect is called the critical flicker frequency (CFF) and is measured in Hz. Species exposed to pulsing lights at

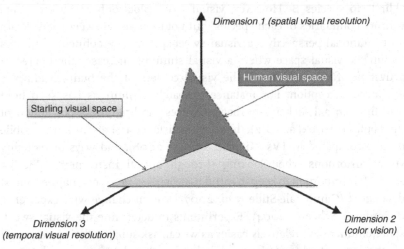

**Figure 6.4:** Schematic representation of the visual space of humans and European starlings, taking into consideration three visual dimensions: spatial visual resolution, temporal visual resolution and color vision.

frequencies higher than their critical flicker frequency would only be able to perceive it as a steady light. Imagine, for example, a ceiling fan working at high speeds. With our temporal visual resolution (~58 Hz; Hodos 2012), we would only see a blurred image of the blades rather than the individual blades rotating. However, European starlings, with much higher temporal visual resolution (~100 Hz; Greenwood *et al.* 2004), would likely see the individual blades rotating. In general, smaller-sized species that have higher metabolic rates tend to have higher temporal visual resolution (Healy *et al.* 2013). We can expect that a species with higher temporal resolution would be able to sample the approach of an object (e.g. predator) at a faster rate, perceive the looming more smoothly, and thus more accurately estimate the time to avoid danger.

Overall, there are some relevant points to make from an applied perspective (without taking into account ultimate explanations):

First, birds see their world in fundamentally different ways from humans. This is not just restricted to their ultraviolet vision, as emphasized in the literature (Cuthill 2006); but to multiple other visual dimensions (see earlier in this section).

Second, one of the implications of these taxon-specific visual traits is that if we plot the visual space (defined by different visual dimensions) of different species, there will be some degree of overlap as well as segregation (Figure 6.4). An area of sensory space occupied only by species A would be "blind" to species B. However, visual space plots only show the visual sensory boundaries of each species, but not their visual sweet-spots. From an operational perspective, a visual sweet-spot can be defined as an area within the visual space where a visual stimulus increases the degree of activation of the neurons in the visual centers of the brain, leading to enhanced perception. For instance, research in humans has shed light into the optimal flicker rate that enhances the brightness perception of light bulbs (Rieiro *et al.* 2012), the viewing distance and resolution of mobile devices compared to TVs (Knoche & Sasse 2008), and ways of reducing visual distortions when looking through optical instruments (Merlitz 2010). It is important to keep in mind that visual sweet-spots may be context dependent (Gamberale-Stille *et al.* 2007) or even change with experience (Schmidt & Schaefer 2004). Experiments to determine the visual sweet-spot in humans are relatively easier as we can ask subjects their assessment of the perceptual experience. Obviously, this becomes much more challenging in non-human systems.

Third, there is considerable between-species variation in birds in terms of the visual dimensions discussed (Martin 2007, Gaffney & Hodos 2003, Hart & Hunt 2007, Fernández-Juricic 2012). This variability suggests that different bird species may have different visual sweet-spots. One question that needs to be addressed in the future is the degree of overlap in the sweet-spots of different species. This can be particularly relevant when targeted stimuli are intended to modify the behavior of multiple species at the same time rather than a single one. Furthermore, the between-species variation in the relative position of the sweet-spot in visual sensory space is important when we want to manipulate simultaneously the behavior of one species in one direction (i.e. attract) and another species in a different direction (i.e. repel). If the sweet-spots are in the very same position, it could be more challenging than if they are in different positions.

### 6.3 HOW TO GO ABOUT DEVELOPING TARGETED STIMULI TAKING THE SENSORY APPROACH

Determining how animals perceive a stimulus based purely on their behavioral response is a challenging task. Even if the stimulus is within their sensory space, they may perceive it but it may be so far outside of their sensory sweet-spot that they may not show any behavioral response (either positive or negative). Similarly, animals may not react if the stimulus is not associated with any particular risk or motivation (e.g. food). For instance, think of a group of people having a picnic in the woods and watching how a

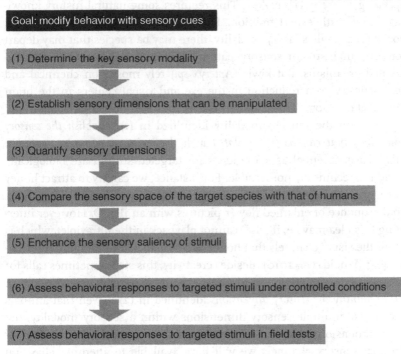

**Figure 6.5:** Suggested steps to develop targeted sensory cues to modify the behavior of animals. See text for details.

light breeze moves the leaves of trees. They can certainly detect such movements, but they are not likely to react, particularly if their attention is focused on maintaining a conversation. Given the difficulty in assessing animals' perception, what can managers and conservation biologists do?

Blackwell and Fernández-Juricic (2013) provided some suggestions, which I expand on here to formalize a seven-step approach to address this problem. The basic idea is to narrow down candidate stimuli based on the sensory configuration of the target species, possibly developing stimuli close to the sensory sweet-spot, expose animals to these stimuli and measure their behavioral responses. This approach encourages the development of sensory-based hypotheses that can make predictions about the degree of saliency, and the associated behavioral responses. The seven steps are summarized in Figure 6.5.

(1) Taking into consideration the type of behavior that we would like to modify (e.g. foraging, anti-predator, territory establishment), *determine the sensory modality that most likely could trigger the intended behavioral response*

*depending on the species biology.* This requires some natural history knowledge along with expert opinion. Even for taxa characterized as relying mostly on a single sensory modality, there may be species that may depart from the mainstream sensory pattern. For instance, birds are visually-oriented organisms, but kiwis (*Apteryx sp.*) rely mostly on chemical and tactile cues, with a reduction in the eye and visual centers in the brain (Martin *et al.* 2007). This is mostly due to the kiwis nocturnal habits.

(2) Within the sensory modality identified in (1), *establish the sensory dimensions that can be manipulated with the logistical resources available.* This step is essential as some ideas for targeted stimuli are biologically sound but technically not feasible. For instance, we can try to attract honey bees (*Apis mellifera*) to patches with nectar-rich flowers by showing them a rapid sequence of enlarged flower pictures with an iPad©. However interesting this idea may be, iPads© cannot playback in the ultraviolet, which is one of the visual channels that honey bees use to gather foraging information (e.g. Arnold *et al.* 2010). Besides creativity, this step sometimes calls for collaborations with engineers to adjust existing or develop new technology.

(3) *Quantify the sensory dimensions* identified in (2). Given that animals tend to use multiple sensory dimensions within a sensory modality, the more dimensions that can be manipulated, the higher the number of potential sensory channels we will have available to attempt behavioral manipulations. This step may benefit from establishing collaborations with sensory physiologists, obtaining the information from the literature, or estimating the sensory dimensions from life-history traits. Sensory physiologists may be willing to develop collaborations as their basic work can take an applied spin, potentially leading to new funding avenues. Nevertheless, it may be possible that many of these dimensions have already been characterized for the target species or species that are taxonomically very close. For example, with regards to the avian visual system, there are already multiple resources that provide information on the visual field configuration (Martin 2007), the type of center of acute vision (along with the density of photoreceptors and ganglion cells) (Collin 2008), the sensitivity of visual pigments and oil droplets (Hart & Hunt 2007), and so on, of several species. In cases where there is no possibility of studying a species using physiological techniques due to ethical reasons and/or availability (e.g. endangered or threatened species), we might be able to get some of this information behaviorally in the field. But, if we cannot even get the behavioral information, we can estimate certain parameters based on life-history associations found in some taxa. For instance, studies reported significant relationships between avian spatial visual resolution and body

mass (Kiltie 2000) as well as temporal visual resolution and body mass (Healy *et al.* 2013). The linear equations available can certainly be used to estimate values for these visual dimensions based on the body mass of the target species.

(4) *Compare the sensory space of the target species with that of humans to establish where the targeted stimuli will be played back.* If there are some portions of the sensory space of both species without overlap, we may need to address the question of whether humans should also perceive the stimuli or not. For instance, if the stimulus is meant to cause discomfort to repel the animal, it may be better to hide it sensorially from us (e.g. using ultraviolet, infrasound, etc.). There may also be some strict regulations. For instance, the Federal Aviation Administration does not allow white lights at airports to pulse at frequencies higher than 3 Hz (Rash 2004). In some situations, it might also be advisable to plot the sensory space of some non-target species of conservation concern that inhabit the same habitat to assess the potential indirect effects of the stimulus at the community level.

(5) Within a sensory dimension, *estimate stimuli that are more salient for the sensory system of the target species* (i.e. stimuli closer to its sensory sweet-spot). This can be done experimentally or through modeling approaches. Experimentally, it often involves using physiological approaches. For instance, in the case of the visual system, we can use electroretinograms to estimate the threshold frequency of flickering that a species can process as such at different wavelengths. Rubene *et al.* (2010) found that critical flicker frequencies of chickens (*Gallus gallus*) were higher (i.e. higher temporal visual resolution) for lights that included white plus UV components than only white, only yellow or only UV lights. Similarly, we can use auditory evoked potentials to different acoustic stimuli, which reflect the ability of the neurons in the peripheral auditory system to process sounds with different frequency, intensity and temporal characteristics, to determine the saliency of those signals. For example, Carolina chickadees (*Poecile carolinensis*) have good frequency resolution to vocal signals with relatively higher frequencies (4 kHz) compared to other bird species, whose peak frequency resolution ranges from 2 to 3 kHz (Henry & Lucas 2010). These physiological procedures yield information about the specific portions of the sensory space where targeted stimuli can be processed more finely, and thus their perceptual saliency can be enhanced.

There are some sensory dimensions in which the estimation of stimulus saliency is more complex. Modeling approaches may be necessary in these cases. One of the best examples is how to determine the saliency of a visual cue for a species where the number and the sensitivity of the visual

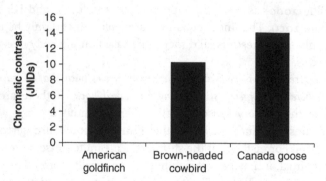

**Figure 6.6:** Chromatic contrast (in Just Noticeable Differences) of a rice panicle (object) against its leaves (visual background) from the perspective of the visual system of three species: American goldfinch, brown-headed cowbird and Canada goose. Calculations were done with the photon catch and receptor noise perceptual model (Vorobyev & Osorio 1998) using published physiological data from each species (Moore et al. 2012, Fernández-Juricic *et al.* 2013, Baumhardt *et al.* 2014).

pigments are quite different from those of humans (e.g. birds and bees). One of the solutions is to use perceptual modeling, which are mathematical algorithms that make predictions about the degree of visual contrast of a visual cue from the visual background for a specific visual system under specific ambient light conditions (Montgomerie 2006). Visual contrast can be estimated for chromatic and achromatic cues.

There are different types of perceptual models (e.g. Vorobyev & Osorio 1998, Endler & Mielke 2005, Stoddard & Prum 2008). For example, the photon catch and receptor noise perceptual model (Vorobyev & Osorio 1998) uses several visual physiological parameters of the study species, allowing us to make species-specific predictions about the saliency of visual cues. This is important because of the large between-species variability in the visual traits involved in chromatic and achromatic vision (i.e. density of cone photoreceptors, sensitivity of visual pigments and oil droplets; Hart & Hunt 2007, Moore *et al.* 2012, Fernández-Juricic *et al.* 2013, Baumhardt *et al.* 2014). For instance, using perceptual modeling, we can see the large degree of between-species variation (~130%) in chromatic contrast of the same branch of rice flowers for brown-headed cowbirds, American goldfinches, *Carduelis tristis*, and Canada geese, *Branta canadienses* (Figure 6.6). Even though we do not understand very well how these differences in modeled perception translate into behavior, perceptual models are a good starting point to establish the relative saliency of visual stimuli under different ecological conditions.

The photon catch and receptor noise model uses various parameters to calculate visual contrast (Vorobyev & Osorio 1998). Here is a summary of the three main parameters. First, we need information on some visual traits: the sensitivity of the visual pigments (and oil droplets if our target species is a bird, turtle, lizard or some fish) and the relative density of cone photoreceptors. Second, we need measurements of the reflectance of the stimulus and its background (i.e. how much light is reflected across different wavelengths). Third, we need measurements of the spectral properties of the ambient light (i.e. irradiance) upon which the animals will be modeled to be perceiving the stimulus. With all this information, the model calculates the relative distance between the stimulus and the background (i.e. degree of visual saliency) under specific ambient light conditions and within the color space of the target species (Figure 6.3), yielding values of chromatic and achromatic contrast. The units are called just noticeable differences (JNDs). At least theoretically (although this could be species-specific), JNDs < 1 indicate that the stimulus cannot be discriminated from the background, JNDs from 1 to 4 indicate that discrimination is possible but challenging, and JNDs > 4 indicate that visual discrimination is highly likely (Siddiqi et al. 2004). Based on the results presented in Figure 6.6, we can conclude that the rice panicle can be discriminated by all three species, but it would be easier to resolve from the background for Canada geese.

(6) *Expose animals to stimuli tuned to their sensory system under controlled conditions.* This step is aimed at testing different stimuli to identify those with higher chances of causing the expected behavioral responses (e.g. attraction, repulsion). This is an iterative process that may require going back to step (5) to establish the saliency of variations in the stimuli based on the results obtained in the behavioral assays. The value of using controlled experiments (lab, outdoor enclosures, etc.) is that several confounding factors can be minimized (e.g. identity effect, food availability, social interactions) or manipulated (e.g. ambient light conditions, noise levels). This will lead to a better understanding of the cause–effect relationships between the stimuli and the behavioral responses as well as the environmental conditions where the responses are enhanced (e.g. temperature, light intensity); however, it may not have a large degree of generality.

(7) *Expose animals in the wild to stimuli tuned to their sensory systems to generalize the responses.* The natural conditions may introduce factors that were not considered in the previous step (e.g. animals in a group modify their decision-making in relation to solitary conditions). Therefore, tweaking the stimuli in these experiments may require revisiting

steps (5) and (6). Additionally, if the stimulus is intended to replace an old one, the metrics to compare the performance of both would need to be established in advance.

This seven-step approach makes an important implicit assumption: stimuli close to the sensory sweet-spot would trigger an enhanced behavioral response, either in terms of attraction or repulsion. Unfortunately, there is a dearth of literature testing this key assumption and this is an area of future research that can provide much needed insights. Yet, the overall approach is still valid as we could tweak the stimulus in different directions from the sensory sweet-spot to assess at which point the behavioral response changes in type and strength.

To illustrate some of these steps let's consider a couple of hypothetical examples. First, imagine we are trying to develop stimuli to minimize the damage that European starlings cause on crops. Given how visually driven starlings are (Martin 1986), we decide to develop a visual stimulus in the form of a pulsing light, which is known to cause discomfort at high pulsing frequencies, at least to humans (Stone 1990). We choose three visual dimensions that have been characterized in starlings (Hart *et al.* 1998, Dolan & Fernández-Juricic 2010, Feinkohl & Klump 2011) to address this problem: (1) spatial visual resolution because it can provide information on the distance at which the lights would be detected, (2) temporal visual resolution because it would allow us to enhance the discomfort effect by increasing the pulsing frequency and (3) color vision because it would allow us to explore light colors that may be more noticeable to starlings. In mapping these dimensions for both starlings and humans (Figure 6.4), we find that starlings have higher temporal visual resolution and a wider color space, but their spatial visual resolution is lower than that of humans. A stimulus in the shared portions of the sensory space may also negatively affect humans. Therefore, we decide to hide the stimulus as much as possible from the human visual system by developing a light that pulses in the ultraviolet portion of the spectrum (360 nm) at high pulsing frequencies (90 Hz) for only starlings to detect. We deploy the lights in small boxes (10 x 10 x 10 cm) held by a 1-m-high dowel rod. Based on the starling visual acuity (6.3 cycles/degree), we estimate that under perfect light conditions individuals would be able to detect the light from 72 m. To increase the surprise factor (and save battery life) we install a motion detector system that would turn the lights on at 70 m from any object moving within that range. This simple system can be tweaked to enhance the behavioral response, for instance, by increasing the pulsing frequency of the lights as the starlings move closer to the lights.

A second hypothetical example involves the development of bird feeders that are more visually enticing to the real consumers, the birds themselves, as opposed to humans given the between-species differences in their visual systems. To get started, we can determine if some bird feeder colors available in the market stand out visually from the avian perspective. We can choose a feeder consisting of a semicircular dome from which three socks filled with seeds are suspended. The dome is available in three colors: red, yellow and green. We target the American goldfinch, a sexually dimorphic bird (males are brightly colored during the breeding season whereas females have a duller coloration), as this species generally visits bird feeders year round and the basic properties of its visual system have recently been characterized (Baumhardt et al. 2014). We address the basic question of which bird feeder, provided the seeds available are the same, would be more visually salient from the goldfinch perspective using perceptual modeling. This tool also allows us to explore the saliency of the feeders in different seasons as light intensity (affecting irradiance) as well as vegetation structure (affecting the reflectance of the visual background) change from the breeding to the non-breeding seasons. Additionally, goldfinches make use of both closed and open habitats, where the spectral properties of the ambient light vary substantially (Lythgoe 1979). Considering all these factors, we can model the chromatic contrast of these three feeders using the goldfinch visual traits (Figure 6.7). The results show that the yellow and red feeders are much more salient than the green feeder in open and closed habitats during the breeding season and in closed evergreen habitats during the non-breeding season. However, these differences in chromatic contrast are minimized in open and closed deciduous habitats during the non-breeding seasons. One implication is that for the bird feeders to have a similar level of visual saliency throughout the year, other color combinations would need to be explored during the non-breeding season. Another factor that could play a role is a potential seasonal difference in the visual system, as found in other taxa (e.g. Whitmore & Bowmaker 1989). Overall, these results open up interesting opportunities for novel bird feeder designs.

## 6.4 HOW TO IMPLEMENT THE APPLIED SENSORY ECOLOGY APPROACH INTO BEHAVIOR-BASED MANAGEMENT

### 6.4.1 Implementing the applied sensory ecology approach into conservation planning

Understanding the sensory system of the species in question can help improve reserve design to protect species, plan a corridor to facilitate the

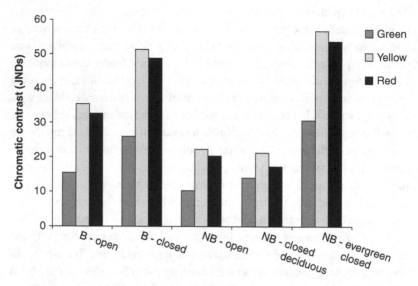

**Figure 6.7:** Chromatic contrast (in Just Noticeable Differences) of bird feeders of different colors (green, yellow, red) in the breeding (B) and non-breeding (NB) seasons and different habitat types and thus visual backgrounds (open, closed deciduous, closed evergreen), from the perspective of American goldfinches. Calculations were done with the photon catch and receptor noise perceptual model (Vorobyev & Osorio 1998) using published physiological data for goldfinches (Baumhardt *et al.* 2014).

movement of species or prepare the translocation of a species. Species interact with different components of the landscape (e.g. habitat structure), conspecifics and heterospecifics (e.g. prey, predators) through their sensory systems to make decisions (Chapter 7). Conservation measures that change the spatial and temporal distribution of these elements will likely change how the environment of a species is perceived and potentially its behavior. A couple of examples are described in this section.

Evolutionary traps occur when animals show stronger or similar preference to resources (e.g. foraging resources, shelter, breeding habitat) that provide lower fitness compared to other naturally available resources of the same type (Schlaepfer *et al.* 2002, Chapter 4). For example, water surfaces reflect polarized light, a cue that many insects use to locate areas where they lay their eggs (Schwind 1991). However, asphalt can also generate similar or even higher levels of polarized reflected sunlight, leading mayflies to lay their eggs in areas that have no reproductive value (Kriska *et al.* 1998), which can have negative population consequences (Horváth *et al.* 2009). A similar phenomenon takes place with insects and glass buildings, which urban birds take advantage of for foraging purposes due to the higher availability of food (Robertson *et al.* 2010). The aforementioned applied sensory ecology approach can be used to reduce the incidence of this sensory pollution problem. For instance, Horváth *et al.* (2010) found that the degree of polarization was smaller in solar panels with white borders and grates, leading to lower preference for egg-laying by different flying insects. A better understanding of the key components of polarized light perception in these insects could allow us to find novel strategies to reduce their levels of preference. Alternatively, the same sensory principle can be used to attract some insects to polarizing traps (Egri *et al.* 2013).

Another potential application of the sensory approach corresponds to buffer areas, which are used by conservation biologists to exclude human visitation from areas in order to increase the nesting success of some species (Blumstein & Fernández-Juricic 2010). Buffer areas are generally calculated using information on escape behavior: a human approaches an individual at a steady pace and records the distance at which the animal flushes (e.g. flight initiation distance). This distance is then taken as the radius of a circle to estimate the buffer area (reviewed in Fernández-Juricic *et al.* 2005). The rationale is that by preventing humans from encroaching into this buffer area, we could prevent a species from leaving their nesting grounds.

However, there are many sensory problems with the way buffer areas are estimated (Fernández-Juricic *et al.* 2005). First, the distance at which an individual escapes is not necessarily the same as the one it detects the approach (see earlier in this section). Actually, it is likely that individuals detect the approach very early but refrain from leaving the patch until the risk is considered too high (i.e. humans are much closer). There is evidence using heart rate telemetry, for instance, that penguins that are visited by tourists at close distances do not necessarily flee but have high pulse rate levels that are sustained as long as the exposure lasts (Ellenberg *et al.* 2013). High pulse rate levels have been associated with higher levels of stress hormones and lower reproductive output (Ellenberg *et al.* 2007). Second, the buffer area approach does not consider the temporal component of the interactions between recreationists and wildlife. In other words, a species with higher temporal visual resolution could potentially gain information about the human exposure at a quicker rate (and thus make decisions faster) than one with lower temporal visual resolution (Healy *et al.* 2013), leading to the idea of species-specific differences in the temporal perception of disturbance and, potentially, stress.

Consequently, how long and from how far away should an animal be exposed to recreationists? Knowledge on the sensory physiology of the target species can help. If we consider a visually oriented organism, we can make some recommendations based on its spatial and temporal resolving power. Spatial visual resolution can give us estimates of the distance at which the target species would not be able to resolve recreationists visually. Using these distance values to calculate buffer areas would certainly reduce stress levels. This option could particularly work with species that have lower spatial resolution than humans. Those with higher spatial resolution than humans would require having blinds whose coloration/patterning make them difficult to visually resolve from the visual background. Temporal visual resolution could provide estimates of the optimal exposure time to minimize stress levels. There is evidence in humans that the perception of time may be associated with the temporal visual resolution (Hagura et al. 2012). We can then speculate that species with higher temporal visual resolution might perceive a "short" human visit as actually a long one; while the opposite effect might occur in species with lower temporal resolution. Taking this temporal perception into account could provide new ways of managing the rate of visitation to protected areas.

### 6.4.2 Manipulating behavior

The seven-step approach to develop stimuli tuned to the sensory system of a target species (see above) can be applied to different contexts relative to the manipulation of a species' behavior. A couple of examples are described in this section.

Collisions between aircraft and birds (bird-strikes) have become a large problem despite successful management efforts to reduce the incidence of different bird species *within* airport property (Dolbeer et al. 2012). The reason is that the frequency of damaging strikes *outside* of the airport property has been increasing in the last few years (Dolbeer 2011). Commercial aircraft themselves do not currently have any specific means to minimize the chances of collisions with large species or small species flying in groups, which can damage critical mechanical elements of an aircraft (e.g. engine). This is what happened in January 2009 when a flock of migrating geese struck both engines of a flight departing from La Guardia Airport in New York City, causing the powerless aircraft to crash-land in the Hudson River (Marra et al. 2009). There have been efforts to develop lighting systems for aircraft tuned to the avian visual system that can be used to trigger early avoidance behavior on birds (Blackwell & Fernández-Juricic 2013). The rationale is simple. A recent study showed that birds involved in collisions had tried to avoid aircraft unsuccessfully, possibly due to a lack of time to respond due to high aircraft speeds (Bernhardt et al. 2010). Lights that enhance avian perception may provide slightly extra time for birds to engage in successful avoidance maneuvers (Blackwell & Fernández-Juricic 2013). For instance, Blackwell et al. (2012) estimated that the chromatic contrast of a radio-controlled (RC) aircraft with lights on was substantially higher than the same aircraft with lights off from the visual perspective of Canada geese. Additionally, they showed that geese became alert to the approaching RC aircraft about 4 s earlier with the lights on than off (Blackwell et al. 2012). At the fast speeds at which these collisions occur, this is a considerable amount of time. Interestingly, the type of light (pulsing, steady) that enhances alert behavior depends on the ambient light conditions (sunny, cloudy; Blackwell et al. 2009), suggesting the possibility of an automatic system that adjusts the light regime depending upon weather. However, the ability of birds to assess the position of approaching vehicles gets reduced at higher speeds, increasing the chances of collision (Farmer & Brooks 2012). A recent study actually found that lights tuned closer to the avian visual sweet-spot can minimize this negative speed effect, probably by improving the tracking of the object at high speeds (Doppler et al. 2015).

Attracting animals to specific spots is another problem that this sensory approach can tackle. Animals move around for multiple reasons: foraging, roosting, nesting, migrating and so on. However, the high degree of human disturbance and resulting habitat modification has made the arrangement of landscapes much more complex: (a) regularly visited habitat patches may be gone or may still be present with high levels of disturbance (e.g. recreationists); (b) remaining undisturbed habitat patches may have a higher density of conspecifics, leading to direct and indirect intra-specific competition; (c) remaining undisturbed habitat patches may still be available but the landscape matrix surrounding them may have changed, and with it the cues used to get to them; and (d) never-used but new habitat patches may be available as a result of restoration efforts. The issue is how to help individuals go from point A to point B providing cues that are tuned to their sensory systems. Two examples are worth mentioning. First, Coleen St. Clair and collaborators have championed several studies to understand the behavioral mechanisms small forest passerine birds use to cross forest gaps using (a) homing experiments (i.e. translocating birds relatively short to medium distances from their territories and measuring their ability to return to the point of capture; Bélisle & St. Clair 2001), and (b) acoustic cues (mobbing calls) to measure the degree of willingness of an individual to cross different habitat arrangements (undisturbed forest, corridor, forest gap; St. Clair *et al.* 1998). These types of studies have been relevant to understanding the differential responses of forest specialists and habitat generalists to changes due to habitat fragmentation and the value of stepping-stones in facilitating movements across different types of landscapes (e.g. Gillies & St. Clair 2010). The second example revolves around how larval fish in the pelagic phase find their way through the ocean to make it to the benthic juvenile phase. There is evidence that coral reef fish larvae can distinguish chemical cues from different habitat types (e.g. coral reef vs. open ocean odors; Atema *et al.* 2002). A recent study used an unmanned chamber in the open ocean to track the swimming behavior of larval fish and found that they swim toward the coral reef odor (Paris *et al.* 2013). These results open up the possibility of remotely cueing in larvae with chemical stimuli to enhance the possibility of a successful settlement process in areas with high levels of ocean pollution.

## 6.5 CONCLUSIONS

For decades, conservation biologists and wildlife managers have been generally developing stimuli to manipulate the behavior of target species using a trial-and-error approach, which assumes that the sensory system of the target species is the same as that of humans. If this sensory system overlaps to a large degree with the human one, the trial-and-error approach may be sufficient. But, when there are substantial differences in sensory perception between humans and the target species, the applied sensory ecology approach presented in this chapter can be a complement to open up novel strategies to manipulate the behavior of animals. A seven-step process is conceived as iterative to fine-tune the relationship between sensory input and behavioral output. This process considers sensory hypotheses (e.g., perceptual modeling) that can generate specific predictions regarding the degree of sensory saliency of the targeted stimulus to the study species. Overall, trying to perceive the world through animal senses (instead of our limited sensory experience) can improve the allocation of limited resources to management and conservation efforts.

There are many contexts to which the sensory ecology approach could be applied. Yet, one key component that future research should address is how any kind of short-term behavioral responses caused by targeted stimuli can be sustained in time. In other words, would the changes in behavior as a result of using sensory cues lead to habituation or sensitization (see Chapter 3)? Although the answer to this question is bound to be species- and context-specific, it can also provide some general understanding of the role of sensory systems in the behavioral responses of animals to human-induced environmental change.

## ACKNOWLEDGEMENTS

Gabriela Sincich kindly drew some figures and Patrice Baumhardtmade the visual contrast calculations. I thank Jeff Lucas, Amanda Ensminger, Jessica Yorzinski, Luke Tyrrell, Kelly Ronald, Diana Pita and Shannon Butler for constructive comments on an early version of the draft. I particularly thank Oded Berger-Tal and David Saltz for their support and patience during the preparation of this chapter. EFJ was funded by the National Science Foundation (IOS #1146986).

## REFERENCES

Arnold, S.E.J., Faruq, S., Savolainen, V., McOwan, P.W. and Chittka, L. 2010. FReD: the floral reflectance database – a web portal for analyses of flower colour. *PLoS ONE*, 5(12):e14287.

Atema, J., Kingsford, M.J. and Gerlach, G. 2002. Larval reef fish could use odour for detection, retention and orientation to reefs. *Marine Ecology Progress Series*, 241:151–160.

Baumhardt, P.E., Moore, B.A., Doppler, M. and Fernández-Juricic, E. 2014. Do American goldfinches see their world like passive prey foragers? A study on visual fields, retinal topography, and sensitivity of photoreceptors. *Brain, Behavior and Evolution*, 83:181–198.

Bélisle, M. and St. Clair, C.C. 2001. Cumulative effects of barriers on the movement of forest birds. *Conservation Ecology*, 5(2):9.

Bernhardt, G.E., Blackwell, B.F., DeVault, T.L. and Kutschbach-Brohl, L. 2010. Fatal injuries to birds from collisions with aircraft reveal anti-predator behaviours. *Ibis*, 152:830–834.

Blackwell, B.F, Fernández-Juricic, E., Seamans, T.W. and Dolan, T. 2009. Avian visual system configuration and behavioural response to object approach. *Animal Behaviour*, 77:673–684.

Blackwell, B.F. and Fernández-Juricic, E. 2013. Visual deterrents at airports. In DeVault, T.L., Blackwell, B.F. and Belant, J.L. (eds.), *Wildlife Management in Airport Environments*, pp. 11–22. Baltimore, MD: The Johns Hopkins University Press.

Blackwell, B.F., DeVault, T.L., Seamans, T.W., Lima, S.L., Baumhardt P. and Fernández-Juricic, E. 2012. Exploiting avian vision with aircraft lighting to reduce bird strikes. *Journal of Applied Ecology*, 49:758–766.

Catania, K.C. 2011. The sense of touch in the star-nosed mole: from mechanoreceptors to the brain. *Philosophical Transactions of the Royal Society*, 366:3016–3025.

Collin, S.P. 2008. A web-based archive for topographic maps of retinal cell distribution in vertebrates. *Australian Journal of Optometry*, 91:85–95.

Dolan, T. and Fernández-Juricic, E. 2010. Retinal ganglion cell topography of five species of ground foraging birds. *Brain, Behavior and Evolution*, 75:111–121.

Dolbeer, R.A., Wright, S.E., Weller, J. and Begier, M.J. 2012. *Wildlife strikes to civil aircraft in the United States, 1990–2011. U.S. Department of Transportation, Federal Aviation Administration, Office of Airport Safety and Standards*, Serial Report No. 18, Washington, DC, USA

Dolbeer, R.A. 2011. Increasing trend of damaging bird strikes with aircraft outside the airport boundary: implications for mitigation measures. *Human–Wildlife Interactions*, 5:235–248.

Doppler, M., Blackwell, B.F., DeVault, T.L. and Fernández-Juricic, E. 2015. Cowbird responses to aircraft with lights tuned to the avian visual system: implications for bird-aircraft collisions. *The Condor* 117:165–177.

Dusenbery, D.B. 1992. *Sensory Ecology: How Organisms Acquire and Respond to Information*. New York: W.H. Freeman.

Egri, A., Blaho, M., Szaz, D., Kriska, G., Majer, J., Herczeg, T., Gyurkovszky, M., Farkas, R. and Horvath, G. 2013. A horizontally polarizing liquid trap enhances the tabanid-capturing efficiency of the classic canopy trap. *Bulleting of Entomological Research* 103:665–674.

Ellenberg, U., Setiawan, A.N., Cree, A., Houston, D.M. and Seddon, P.J. 2007. Elevated hormonal stress response and reduced reproductive output in yellow-eyed penguins exposed to unregulated tourism. *General and Comparative Endocrinology*, 152:54–63.

Ellenberg, U., Mattern, T. and Seddon, P.J. 2013. Heart rate responses provide an objective evaluation of human disturbance stimuli in breeding birds. *Conservation Physiology*, 1: cot013.

Endler, J.A. 1997. Light, behavior and conservation of forest-dwelling organisms. In Clemmons, J.R. and Buchholz (eds.), *Behavioral Approaches to Conservation in the Wild*, pp. 330–356. Cambridge (UK): Cambridge University Press.

Endler, J.A. and Mielke, P.W. 2005. Comparing entire colour patterns as birds see them. *Biological Journal of the Linnaean Society*, 86:405–431.

Farmer, R.G. and Brooks, R.J. 2012. Integrated risk factors for vertebrate roadkill in Southern Ontario. *Journal of Wildlife Management*, 76:1215–1224.

Feinkohl, A. and G. Klump. 2011. Processing of transient signals in the visual system of the European starling (*Sturnus vulgaris*) and humans. *Vision Research*, 51:21–25.

Fernández-Juricic, E., Moore B.A., Doppler, M., Freeman, J., Blackwell, B.F., Lima, S.L. and DeVault, T.L. 2011. Testing the terrain hypothesis: Canada geese see their world laterally and obliquely. *Brain, Behavior & Evolution*, 77:147–158.

Gaffney, M.F. and Hodos, W. 2003. The visual acuity and refractive state of the American kestrel (*Falco sparverius*). *Vision Research*, 43:2053–2059.

Gamberale-Stille, G., Hall, K.S.S. and Tullberg, B.S. 2007. Signals of profitability? Food colour preferences in migrating juvenile blackcaps differ for fruits and insects. *Evolutionary Ecology*, 21:99–108.

Gillies, C.S. and St Clair, C.C. 2010. Functional responses in habitat selection by tropical birds moving through fragmented forest. *Journal of Applied Ecology*, 47:182–190.

Goldsmith, T.H. 1990. Optimization, constraint, and history in the evolution of eyes. *Quarterly Review of Biology*, 65:281–322.

Guilleman, M., Martin, G.R. and Fritz, H. 2002. Feeding methods, visual fields and vigilance in dabbling ducks (*Anatidae*). *Functional Ecology*, 16: 522–529.

Greenwood, V.J., Smith, E.L., Goldsmith, A.R., Cuthill, I.C., Crisp, L.H., Walter-Swan, M.B. and Bennett, A.T.D. 2004. Does the flicker frequency of fluorescent lighting affect the welfare of captive European starlings? *Applied Animal Behaviour Science*, 86:145–159.

Hagura, N., Kanai, R., Orgs, G. and Haggard, P. 2012. Ready steady slow: action preparation slows the subjective passage of time. *Proceedings of the Royal Society B*, 279:4399–4406.

Hart, N.S. and Hunt, D.M. 2007. Avian visual pigments: characteristics, spectral tuning, and evolution. *American Naturalist*, 169:S7–S26.

Hart, N.S., Partridge, J.C. and Cuthill, I.C. 1998. Visual pigments, oil droplets and cone photoreceptor distribution in the European starling (*Sturnus vulgaris*). *Journal of Experimental Biology*, 201:1433–1446.

Healy, K., McNally, L., Ruxton, G.D., Cooper, N. and Jackson, A.L. 2013. Metabolic rate and body size are linked with perception of temporal information. *Animal Behaviour*, 86:685–696.

Knoche, H.O. and Sasse, M.A. 2008. The sweet spot: how people trade off size and definition on mobile devices. *MM '08 Proceedings of the 16th ACM international conference on Multimedia* 21–30.

Henry, K.S. and Lucas, J.R. 2010. Auditory sensitivity and the frequency selectivity of auditory filters in the Carolina chickadee, *Poecile carolinensis*. *Animal Behaviour*, 80:497–507.

Hodos, W. 2012. What birds see and what they don't. In Lazareva, O.F., Shimizu, T. and Wasserman, E.A. (eds.), *How Animals See The World: Comparative Behavior, Biology, and Evolution of Vision*, pp. 5–24. Oxford: Oxford University Press.

Horváth, G., Blahó, M., Egri, A. Kriska, G., Seres, I. and Robertson, B.A. 2010. Reducing the maladaptive attractiveness of solar panels to insects. *Conservation Biology*, 24:1644–1653.

Horváth, G., Kriska, G., Malik, P. and Robertson, B.A. 2009. Polarized light pollution: a new kind of ecological photopollution. *Frontiers in Ecology and the Environment*, 7:317–325.

Kiltie, R.A. 2000. Scaling of *visual acuity* with body size in mammals and birds. *Functional Ecology*, 14:226–234.

Kriska, G., Horvath, G. and Andrikovics, S. 1998. Why do mayflies lay their eggs en masse on dry asphalt roads? Water-imitating polarized light reflected from asphalt attracts Ephemeroptera. *Journal of Experimental Biology*, 201:2273–2286.

Land, M.F. and Nilsson, D.-E. 2012. *Animal Eyes*. Oxford: Oxford University Press.

Marra, P.P., Dove, C.J., Dolbeer, R., Faridah Dahlan, N., Heacker, M., Whatton, J.F., Diggs, N.E., France, C. and Henkes, G.A. 2009. Migratory Canada geese cause crash of US Airways Flight 1549. *Frontiers in Ecology and the Environment*, 7:297–301.

Martin, G.R. and Shaw, J.M. 2010. Bird collisions with power lines: failing to see the way ahead? *Biological Conservation*, 143:2695–2702.

Martin, G.R. 1986. The eye of a Passeriform bird, the European starling (*Sturnus vulgaris*) – eye-movement amplitude, visual fields and schematic optics. *Journal of Comparative Physiology A*, 159:545–557.

Martin, G.R. 2007. Visual fields and their functions in birds. *Journal of Ornithology*, 148:S547–S562.

Martin, G.R. 2011. Understanding bird collisions with man-made objects: a sensory ecology approach. *Ibis*, 153:239–254.

Martin, G.R., Wilson, K.J., Wild, J.M., Parsons, S., Kubke, M.F. and Corfield, J. 2007. Kiwi forego vision in the guidance of their nocturnal activities. *PLoS ONE*, 2(2):e198.

Montgomerie, R. 2006. Analyzing colors. In Hill, G.E. and McGraw, K.J. (eds.), *Bird Coloration: Mechanisms and Measurements* (Vol 1), pp. 90–147. Cambridge: Harvard University Press.

Moore, B.A, Doppler, M., Young, J.E. and Fernández-Juricic, E. 2013. Interspecific differences in the visual system and scanning behavior of three forest passerines that form heterospecific flocks. *Journal of Comparative Physiology A*, 199:263–277.

Nawroth, J.C., Greer, C.A., Chen, W.R., Laughlin, S.B. and Shepherd, G.M. 2007. An energy budget for the olfactory glomerulus. *Journal of Neuroscience*, 27:9790–9800.

Neumeyer, C. 1992. Tetrachromatic color vision in goldfinch: evidence from color mixture experiments. *Journal of Comparative Physiology A*, 171:639–649.

Niven, J.E. and Laughlin, S.B. 2008. Energy limitation as a selective pressure on the evolution of sensory systems. *Journal of Experimental Biology*, 211:1792–1804.

O'Rourke, C.T., Hall, M.I., Pitlik, T. and Fernández-Juricic, E. 2010. Hawk eyes I: diurnal raptors differ in visual fields and degree of eye movement. *PLoS ONE*, 5(9):e12802.

Paris, C.B., Atema, J., Irisson, J.-O., Kingsford, M., Gerlach, G. and Guigand, C.M. (2013) Reef odor: a wake up call for navigation in reef fish larvae. *PLoS ONE*, 8(8):e72808.

Rash, C.E. 2004. Awareness and causes and symptoms of flicker vertigo can limit ill effects. *Human Factors & Aviation Medicine*, 51(2):1–6.

Reymond L. 1987. Spatial visual acuity of the falcon, *Falco berigora*: a behavioural, optical and anatomical investigation. *Vision Research*, 27:1859–1874.

Robertson, B.A., Kriska, G., Horváth, V. and Horváth, G. 2010. Glass buildings as bird feeders: urban birds exploit an ecological trap. *Acta Zoologica Academiae Scientiarum Hungaricae*, 56:283–293.

Rubene, D., Håstad, O., Tauson, R., Wall, H.and A. Ödeen. 2010. The presence of UV wavelengths improves the temporal resolution of the avian visual system. *Journal of Experimental Biology*, 213:3357–3363.

Schlaepfer, M.A., Runge, M.C. and Sherman, P.W. 2002. Ecological and evolutionary traps. *Trends in Ecology & Evolution*, 17:474–480.

Schmidt, V. and Schaefer, H.M. 2004. Unlearned preference for red may facilitate recognition of palatable food in young omnivorous birds. *Evolutionary Ecology*, 6:919–925.

Schwind, R. 1991. Polarization vision in water insects and insects living on a moist substrate. *Journal of Comparative Physiology A*, 169:531–540.

Siddiqi, A., Cronin, T.W., Loew, E.R., Vorobyev, M. and Summers, K. 2004. Interspecific and intraspecific views of color signals in the strawberry poison frog *Dendrobates pumilio. Journal of Experimental Biology*, 207:2471–2485.

St. Clair, C.C., Bélisle, M., Desrochers, A. and Hannon, S.J. 1998. Winter responses of forest birds to habitat corridors and gaps. *Conservation Ecology*, 2(2):13.

Stevens, M. 2013. *Sensory Ecology, Behaviour, & Evolution*. Oxford: Oxford University Press.

Stoddard, M.C. and Prum, R.O. 2008. Evolution of avian plumage color in a tetrahedral color space: a phylogenetic analysis of New World buntings. *American Naturalist*, 171:755–776.

Stone, P.T. 1990. Fluorescent lighting and health. *Lighting Research and Technology*, 24:55–61.

Sutherland, W.L. 1998. The importance of behavioural studies in conservation biology. *Animal Behaviour*, 56:801–809.

Theiss, S.M., Davies, W.I.L., Collin, S.P., Hunt, D.M. and Hart, N.S. 2012. Cone monochromacy and visual pigment spectral tuning in wobbegong sharks. *Biology Letters*, 8:1019–1022.

Van Dyck, H. 2012. Changing organisms in rapidly changing anthropogenic landscapes: the significance of the "Umwelt"-concept and functional habitat for animal conservation. *Evolutionary Applications*, 5:144–153.

Virzi, T., Boulton, R.L., Davis, M.J., Gilroy, J.J. and Lockwood, J.L. 2012. Effectiveness of artificial song playback on influencing the settlement decisions of an endangered resident grassland Passerine. *Condor*, 113:846–855.

Vorobyev, M. and Osorio, D. 1998. Receptor noise as a determinant of colour thresholds. *Proceedings of the Royal Society of London B*, 265:351–358.

Whitmore, A.V. and Bowmaker, J.K. 1989. Seasonal variation in cone sensitivity and short wavelength absorbing visual pigments in the rudd *Scardinius erythrophthalmus. Journal of Comparative Physiology A*, 166:103–115.

Wilson, C.J. and McKillop, I.G. 1986. An acoustic scaring device tested against European rabbits. *Wildlife Society Bulletin*, 14:409–411.

# Behavior-based contributions to reserve design and management

COLLEEN CASSADY ST. CLAIR, ROB FOUND,
ADITYA GANGADHARAN AND MAUREEN MURRAY

## 7.1 INTRODUCTION

All students of conservation are familiar with the quintessential model of a reserve network, in which a hostile, human-dominated matrix limits the occurrence of natural habitat and vulnerable species to scattered protected areas connected by corridors of intermediate suitability (e.g. Diamond 1975, Soule & Terborgh 1999, Bennett 2003). This conceptual model also identifies anthropogenic features, such as roads, that may create such significant barriers to animal movement that they require mitigation (reviewed by Forman et al. 2006). The resulting construct for conservation planning tends to categorize types of space as core areas, corridors, matrix and barriers while underestimating the myriad non-spatial features of both species and landscapes that exist along inconvenient and intersecting continua. Behavior is one of these factors and it contributes much to the fate of imperiled populations, but its effects have not been much synthesized in the contexts of reserve design and conservation management.

Before delving into the role of behavior in reserve design, it is worth pausing to consider some of the reasons for the traditional emphasis on spatial characteristics. First, binary and spatial constructs are readily visualized by people to facilitate common and explicit goals, such as the creation of national parks and other kinds of protected areas. Second, spatial features of reserve design are supported by foundational and extensive ecological theory, much of which emanated from Island Biogeography (MacArthur & Wilson 1967, reviewed by Lomolino & Brown 2009), to provide support for conservation

*Conservation Behavior: Applying Behavioral Ecology to Wildlife Conservation and Management*, eds. O. Berger-Tal and D. Saltz. Published by Cambridge University Press. © Cambridge University Press 2016.

predictions, management actions and enduring academic interest. A third reason that spatial attributes lead so much of reserve design is that space influences most of the physical experiences of organisms and defines most anthropogenic threats to biodiversity (Chapter 1) across a vast range of scales.

Despite the good reasons to emphasize space in reserve design, we contend that space alone does not define the experience of any individual or directly imperil populations. Space is more like a canvas on which those experiences play out. Protecting the habitat contained in space is essential to most conservation action, but that action alone cannot ensure the survival of individuals, populations, species or ecosystems. Moreover, spatial attributes are difficult to generalize as both problems and solutions in conservation (Newmark 1996, Gascon et al. 2000), which limits their proactive use in ways that could best advance conservation goals (Caughley 1994). For many species, a more direct cause of endangerment is actually the specialization that makes some species highly dependent on particular types of space with dwindling availability (Harcourt et al. 2002, Owens & Bennett 2000). Although specialization is typically viewed as a characteristic of species (e.g. Scriber 2010), a similar concept can be applied to populations, individuals or even responses to events within the life of a single individual. In each case, the behavioral capacity to use space is at least as relevant to conservation outcomes as is space itself.

We believe it will be necessary to separate more explicitly the spatial attributes of landscapes from the behavioral processes that contribute to space use, to promote the coexistence of people and other species and preserve biodiversity in the Anthropocene (sensu Steffen et al. 2011). To do that, we will need a much richer understanding of animal behavior in every conservation context, including the design and management of reserves. In this chapter, we advance a behavioral approach to this topic by reviewing and integrating existing behavioral contributions. We begin by showing how even the inherently behavioral phenomena of animal movement and habitat selection have been studied with an emphasis on spatial patterns. Then we identify four existing behavioral approaches that complement this emphasis: distinguishing pattern from process with functional connectivity; integrating sensory ecology and the concept of perceptual range; studying movement directly and deductively; and exploring individual variation. We synthesize and integrate these four approaches to suggest how more emphasis on behavioral processes could foster better conservation outcomes.

## 7.2 THE CHANGING VIEW OF SPACE AS THE CORNERSTONE OF CONSERVATION PLANNING

For thousands of years, the conservation of wildlife or ecosystems has been achieved primarily by restricting human access, either incidentally via intrinsic limits, or intentionally via prohibitions. A few species were protected from exploitation because harming or eating them invoked religious taboos (Colding & Folke 1997). Others, particularly in Europe and Asia, were protected by royal decrees that limited hunting or other exploitation (e.g. Asian elephants *Elephas maximus* in ancient India). But for most of the history of our own species, preservation of other species occurred primarily via natural limits on human population growth and mobility.

Early in the twentieth century, the unprecedented ability of humans to modify habitat, travel great distances and harvest native species prompted the spread of a preservationist ethic, championed in North America by John Muir and resulting in the modern concept of national parks (Ise 2011). Protected areas came to be viewed as island refuges surrounded by oceans of hostile, human-dominated land. This binary view of landscapes was supported by the development of the seminal theory of Island Biogeography (MacArthur & Wilson 1967), which was originally offered to explain the patterns of community assemblage in island archipelagos. Its seminal concepts – that island size drove rates of species extinction, while distance from a mainland drove rates of colonization – were easily applicable to reserve size and reserve connectivity respectively (Lomolino & Brown 2009). Early metapopulation theory (Hanski 1998) built upon and extended Island Biogeography, while retaining the core binary distinction between habitat and non-habitat. The importance of island size and position in Island Biogeography led directly to the six spatial principles of reserve design that were synthesized by Diamond (1975; reviewed by Groom *et al.* 2006). All else being equal, a single reserve was expected to be more effective at gaining and retaining species if it was large, composed of a single patch and rotund in shape. A set of reserves was expected to have more species if the patches were close in proximity, arranged in non-linear clumps and connected by linear extensions of similar habitat.

The spatial prescriptions of early reserve design were logically compelling, but the evidence for them was typically pattern-based and correlative, which invited both theoretical and empirical challenges (reviewed by Williams *et al.* 2005). The edict that reserves should be large and round was countered with empirical evidence that numerous small islands with differing abiotic conditions supported more species in the same total area

(Simberloff & Abele 1976). An ensuing debate termed SLOSS (Single Large or Several Small) raged for a decade, with support for both positions coming from studies motivated by ecological theory as well as the newer discipline of conservation biology (reviewed by Kingsland 2002). Although ultimately inconclusive, this debate helped to define the process of habitat fragmentation as a core driver of anthropogenic extinctions (Wilcox & Murphy 1985) via the familiar progression of habitat loss described by dissection, perforation, fragmentation and, finally, attrition (Forman 1995). The more recent and prevalent concept of Systematic Conservation Planning (SCP; Margules & Pressey 2000) has retained a strong emphasis on space via two primary directives: identify the core areas that harbor the most representative biodiversity of a region, and prevent its loss by choosing and managing areas to maintain that representation (Pressey & Botrill 2008).

The set of rules related to connectivity also spurred research that was focused on space, much of it involving the habitat corridors that connected Diamond's (1975) reserves. Corridors were expected to mitigate the effects of inbreeding and stochastic extinctions that characterize small, isolated populations (Wilson & Willis 1975, Hanski, 1998). However, corridor critics pointed out risks, such as facilitating movement by invasive species or disease vectors (Simberloff & Cox, 1987), and inefficiencies in the allocation of scarce conservation dollars (Hobbs 1992, Simberloff et al. 1992). Decades of debate are just now producing a consensus that corridors often increase animal movement (reviewed by Hilty et al. 2006, Gilbert-Norton et al. 2010), but they can also have detrimental effects (reviewed by Haddad et al. 2014) and they are not always the best conservation investment (Hodgson et al. 2009). There is widespread agreement that corridors are not a panacea and other habitat elements, such as stepping stones or the matrix itself, can also foster connectivity (reviewed by Bennett 2003, Castellon & Sieving 2006, Hilty et al. 2006). Importantly, any such feature must support gene flow to provide lasting conservation value (Beier & Gregory 2012). Hereafter, we use the term corridor in this broader sense of connectivity.

An important outcome of debating the importance of reserve size and corridor function was increased attention to the habitat matrix. The composition of the surrounding matrix can be extremely important to the efficacy of both reserves (e.g. Woodroffe & Ginsberg 1998) and corridors (e.g. Baum et al. 2004). Moreover, many populations persist, albeit with lower fitness, outside reserves and in landscapes that are highly degraded and fragmented (Daily et al. 2001, Vandermeer & Perfecto 2007). As evidence has accumulated of the nuance associated with the matrix, views of reserve design have moved steadily away from the relatively simplistic and categorical view of landscapes toward a more continuous one. The traditional network of core areas, buffers, matrix and connecting elements still provides a conceptual basis for conservation planning (e.g. Bennet 2003), but landscapes are increasingly depicted as continuously varying surfaces of species abundance, habitat use or fitness (Cushman *et al.* 2009).

This contemporary view of landscapes as a continuous mosaic of relative suitability for multiple functions provides both challenges and opportunities to conservation managers. On the one hand, it acknowledges the potential for a much greater diversity of approaches to advance conservation goals, which extend far beyond the traditional protection of space through legislation and agreements (e.g. Convention on Biological Diversity 2010) or private effort (e.g. The Nature Conservancy in North America). But on the other hand, viewing landscapes as continuous mosaics implicitly acknowledges much greater overlap in the distribution of people and other species. Legislation governing direct wildlife mortality through hunting or trade (e.g. Convention on International Trade in Endangered Species; reviewed by Wijnstekers 2003) and programs that foster coexistence with wildlife will be increasingly important to conservation success. We assert that meeting this challenge will require a much broader and deeper understanding of animal behavior in both the design and management of conservation actions. In particular, there needs to be a greater understanding of the specific behavioral traits that enable some species – and individuals – to persist in human-impacted areas while avoiding conflict with people. This knowledge will be essential to slowing the loss of biodiversity in the coming decades while human populations and consumption continue to increase.

## 7.3  BEHAVIORAL APPROACHES TO RESERVE DESIGN

Many of the earliest efforts to supplement the protection of species in space were based on the behavioral processes of habitat selection and movement. The formal study of habitat selection is decades old and encompasses such diverse ecological flavors as community ecology, population ecology, behavioral ecology and ecological genetics, to produce seminal theories of competitive exclusion, intermediate disturbance, ideal free distribution and many others (reviewed by Molles & Cahill 2011). The study of animal movement has equally diverse contexts including biomechanics, life history theory, navigational mechanisms and disease transmission, in addition to behavioral ecology, invasion ecology and conservation biology (reviewed by Turchin 1998, Holyoak *et al.* 2008, Nathan *et al.* 2008).

Other important behavioral aspects of reserve design include territorial behavior (Lamberson *et al.* 1994, Eads *et al.* 2014) and behaviors stemming from the social organization of species (Shumway 1999, Afonso *et al.* 2008). However, unlike habitat selection and movement, these behavioral aspects have so far received very little attention in the reserve design literature.

For the purposes of this chapter, we will review the study of habitat selection and movement behavior dating primarily from the technological transformation of the late 1990s in which the use of global positioning systems, remote sensing and geographic information systems have supported rapid development of quantitative and statistical tools, as well as the emergence of whole new branches of ecology such as Spatial Ecology and Movement Ecology. Today, researchers can take advantage of innovations such as motion-activated camera traps with satellite uplinks, automated surveillance via telemetry stations and GPS telemetry that is communicated by remote download devices, such as cell phones, to collect unprecedented volumes of information about where animals are and, by extension, where they are not. Simultaneously, they can measure attributes of the environment with continuous and categorical variables on spatial scales ranging from cm to km (reviewed by Turner *et al.* 2003, Kerr & Ostrovsky 2003).

The voluminous information about animal locations and environmental variables is often synthesized with multivariate techniques, especially logistic regression, which can be used to model both habitat selection, via resource selection functions (RSF; Boyce & McDonald 1999, Manley *et al.* 2002) and movement, via step selection functions (SSF; Boyce *et al.* 2003, Fortin *et al.* 2005) or the selection of movement trajectories (Whittington *et al.* 2004). These statistical models have been applied in hundreds of ways, but they differ mainly in whether a collection of points is homogenized over time (RSF), developed as a sequential pair of points from an individual or group (SSF), or plotted as a sequence of locations characterizing actual versus simulated trajectories. As with the relationship between still and video photography, the technical distinction between habitat use and movement is comprised only by the frame rate of our observations.

Despite their undeniable advantages, the use of complex statistical models for studying habitat selection and movement has some drawbacks. First, the demonstration of selection for a particular environmental covariate depends on its availability in the landscape to generate a functional, rather than purely numerical, response (Mysterud & Ims 1998). Even the most essential commodity, such as food, will not appear to be selected if it is abundant. Variation in availability can cause differences in selection coefficients that prevent generalization within a population. Such individual-based variation might be identified with random effects in statistical models (Mysterud & Ims 1998, Gillies *et al.* 2006), but these are notoriously difficult to interpret (Gillies & St. Clair 2010). These drawbacks fuel suggestions that models based only on spatial patterns of occupancy – however derived – are little more than statistically complex case studies with no broader application (Morrison 2012).

A second limitation of statistical models as representations of the behaviors associated with habitat selection and movement is the assumed homogeneity of purpose across all contexts (Chetkiewicz *et al.* 2006). Dispersing animals may prefer to travel through the same kinds of habitat that they select for other life-history purposes since they still have to eat, sleep, hide from predators and avoid or seek out conspecifics. But they may also have different habitat requirements for different ecological activities (reviewed by Belisle 2005) and any assumption of homogeneity of habitat selection across contexts and individuals is usually one of convenience. Understanding dispersal is especially relevant to the design of reserves (e.g. McLellan & Hovey 2001, Wiens *et al.* 2006, Di Franco *et al.* 2012) and connectivity (Chetkiewicz *et al.* 2006), but it is notoriously difficult to study (reviewed by Clobert *et al.* 2001). Worse, researchers seldom know the movement motivation or sensory capacity of an animal at any particular point in space and time.

The absence of information about behavioral context encourages researchers, especially when they are armed with piles of empirical data, to infer via induction the behavioral mechanisms underlying habitat selection and movement to create a pattern-process enigma (Chetkiewicz *et al.* 2006, Hebblewhite & Haydon 2010). The prevalence of inductive approaches to these topics clearly limits the generalizations that can emerge (Morrison 2012), but researchers also seem to lack an appropriate behavioral framework (Chapter 1) and integrated behavioral theory, especially in the context of movement ecology (Liedvogel *et al.* 2013). Regardless of the reasons for it, a mainly inductive approach will make it difficult to use studies of habitat selection and animal movement to support more generic conservation principles and actions. Advancing the integration of associated behavioral processes has the potential to provide the missing basis of more deductive approaches.

Many others have appreciated the necessity of putting more emphasis on behavioral processes and theory to inform studies of movement and habitat selection, and some have also anticipated their applications to reserve and corridor planning. We find four of these approaches to be particularly helpful and we review each in detail below. In brief, these approaches are to (a) distinguish between functional and structural connectivity in landscapes, (b) recognize the pivotal importance of sensory ecology and perception, (c) examine animal behavior as directly and deductively as possible, and (d) explore the growing understanding of individual variation and anticipate its effects. Despite substantial development of all four approaches, their associated literatures do not overlap much and we could find no study that emphasized them all. By reviewing and integrating these behavioral approaches, we hope to complement the foundational importance of spatial attributes to support an underlying theory that integrates movement behavior, habitat selection, and reserve design as contributors to the ultimate goal of coexistence.

### 7.3.1 Functional connectivity

Taylor *et al.* (1993) defined landscape connectivity as "the degree to which the landscape facilitates or impedes movement among resource patches," thereby emphasizing the interaction between structural aspects of a landscape (such as vegetation type) and functional aspects of movement behavior (Taylor *et al.* 2006). Others viewed functional connectivity as the ease of movement into a patch (Moilanen & Hanski 2001), across a habitat edge (Bakker & Van Vuren 2004) or through a putative barrier (McRae et al. 2012). In each case, functional connectivity is assumed to describe the ease with which animals can traverse some landscape type or feature and to be inversely related to the movement resistance (e.g. Ricketts 2001).

The simplest and most direct way to measure functional connectivity is to observe whether individuals move between two points across an intervening habitat feature or type (Beier & Noss 1998). This approach is achievable over small spatial scales and suits the measurement of specific features, such as crossing structures over highways (e.g. Clevenger & Waltho 2005). When it is not possible to measure functional connectivity directly, it is sometimes inferred from the geometry of defined landscape attributes, such as patch size and inter-patch distance, that are related to movement metrics such as immigration rates (reviewed by Calabrese & Fagan 2004). These metrics assume that structural characteristics correlate with the movement of individuals, but their realism is limited when habitat varies along a continuum (Bender & Fahrig 2005) or occurs in complex landscapes (Kadoya 2009).

In the early days of landscape ecology and metapopulation biology, much emphasis was placed on these structural metrics (e.g. Hanski & Gilpin 1997, Pulliam 1988), but often for small species in hypothetical contexts.

The need to apply metrics of functional connectivity to the planning of actual reserve networks fostered a tendency to infer functional connectivity using statistical models of habitat selection and movement with a three-step process (reviewed by Calabrese & Fagan 2004; Beier *et al.* 2011). The first step identifies broad areas with variation in the probability of occurrence or abundance for a target species or group of species. These areas might be dichotomized as good and bad habitat, sources and sinks, reserves and matrix, core areas and corridors, nodes and edges (Hanski & Gilpin 1997, Pulliam 1988, Bennett 2003, Minor & Urban 2008) or even as continuously distributed values above and below a threshold of estimated suitability (Chetkiewicz & Boyce 2009). The second step for promoting functional connectivity involves the estimation of a resistance surface, which depicts the impedance that a moving animal is expected to experience as it passes through each location (typically a cell in a Geographic Information System) in the study area. Because direct quantification of resistance is problematic for the rare species that attract conservation attention, early uses of it were based mainly on expert opinion (e.g. Adriaensen *et al.* 2003). More recently, estimates of resistance are often obtained by inverting statistical descriptions of habitat selection obtained via resource selection functions (above) or occupancy models (Mackenzie *et al.* 2006). As a third step, the resistance surface is used to predict functional connectivity, while assuming that: (a) animals move in a series of binary or categorical choices to choose which cells to occupy, (b) "good" habitat, as measured by an average of animal locations, is easier to move through than "bad" habitat, and (c) animals prefer to move in the same habitat they prefer for other activities (e.g., Chetkiewicz *et al.* 2006).

These three steps are assumed to be a reliable means of modeling the functional connectivity that might be achieved with corridors, but the approach glosses over much behavioral detail. For example, most algorithms that automate the three-step process target an optimal combination of lesser distance and greater habitat suitability. But there is seldom a single best route through hostile habitat. Buffering the optimal paths to identify broader linkage zones or combining the optimal paths from other species can increase the generality of corridor designs (Beier *et al.* 2008). A second way that models of functional connectivity can be misleading is their reliance on iterative and sequential steps through the matrix, which are often based on variants of random walks that may specify the degree of directional persistence, knowledge of or dedication to a particular destination, and movement efficiency (reviewed by Turchin 1998). But in most conservation contexts, these mechanistic details are assumed or inferred from best-fitting models of data; the actual behavioral mechanisms are seldom known by researchers.

Some of these limitations in models of functional connectivity have been addressed recently with extensions of graph theory (reviewed by Urban & Keitt 2001), especially circuit theory, which has the advantage of identifying multiple pathways between source patches that occur below some threshold of cumulative resistance for random walkers (McRae et al. 2008). This leaves more room for both errors of perception based on anthropocentric measurements and differences in the underlying motivation of individuals. Centrality metrics of resistance release investigators from the identification of source patches so that the connectivity of each cell to every other cell can be modeled without assuming knowledge of an animal's origin or intended destination (Estrada & Bodin 2008). Neural networks make it possible to model temporal variables, such as animal state, as if-then statements to accommodate some of this individual variation (Tracey et al. 2013). Further, resistance can be estimated using measurement of gene flow, which directly reflects the evolutionary consequences of animal movement (Cushman et al. 2009). The mechanistic detail with which animal movement can be measured and modeled has become a vast sub-literature of its own (reviewed by Nathan et al. 2008), but it is still mostly inferential.

One of the most fruitful approaches for integrating movement data with behavioral mechanisms employs individual and agent-based models to estimate functional connectivity and predict conservation outcomes. These approaches model animal movement by simulating the paths of agents, which can represent animals ranging from insects (e.g. Arrignon et al. 2007) to tigers Panthera tigris (e.g. Ahearn 2001), moving in discrete steps across a virtual landscape while following a variety of movement rules. Agent-based models have been used since the 1960s (e.g. Siniff & Jessen 1969), but modern applications have increased the emphasis on behavior by including individual variation, perceptual range, predator–prey interactions, conspecific attraction, movement states, behavioral plasticity, internal physiological state and even spatial memory as parameters governing the agent's paths (reviewed by DeAngelis & Mooij 2005, Tang & Bennett 2010).

Exploring the nature of animal behavior with agent-based models is likely to increase the specificity of conservation predictions, ultimately enhancing the benefit of conservation actions. For example, an agent-based model demonstrated that the importance of patch size for population persistence was dependent on animal state and its interaction with conspecific attraction (Fletcher 2006). Managers might use this information to manipulate conspecific distribution or density, rather than focusing on the spatial attributes of reserves. These models have tremendous potential to help researchers identify the behavioral processes underlying the patterns of movement and habitat selection contained in the mountains of observational data that are being amassed by the technological advances described above. A generalized understanding of relevant behavioral processes could vastly increase the efficiency of conservation planning. The mechanistic emphasis of agent-based models underscores the need to understand how animals perceive their environments, which taps the rich domain of sensory ecology.

### 7.3.2 Sensory systems and perceptual ranges

Encouraging the movement of animals through functional connectivity is a central goal of designing reserves and corridors to promote population persistence. Predicting the difference between functional and structural connectivity requires detailed knowledge of sensory systems, perceptual ranges and their effects on movement decisions. In general, the movement decisions of individuals are not random or uninformed because sensory systems gather cues about environment quality that are integrated in the brains of even primitive animals to elicit the motor responses we see as movement behavior. In fragmented habitats, animals must make ongoing assessments of landscape features to avoid unsuitable habitat and minimize movement risk (Schmidt *et al.* 2010). Animals undoubtedly make choices with those movements, but many are guided by anciently wired behavioral rules that may not involve conscious thought even in highly advanced organisms. Perceptions occur through many sensory modalities, all of which are influenced by space. This fact was the pivotal insight by Lima and Zollner (1996) when they coined the term perceptual range and defined it as the distance within which an animal can detect environmental cues using its sensory system.

The sensory systems of today's species have been tuned over evolutionary time to emphasize the sensory modalities that made the most efficient and reliable use of prevailing environmental cues. For example, most birds navigate a complex three-dimensional habitat primarily during the day and have high visual acuity and color vision (Martin 2012), whereas mammals, whose common ancestor was most likely nocturnal (Gerkema *et al.* 2013), tend to rely more on olfaction through chemical signals (Campbell-Palmer & Rosell 2011). Similarly, aquatic species tend to rely more on chemical cues because they are easily transmitted in water where low light limits the use of vision (Wisenden 2000). Sensory modalities can be highly different from those used by people, such as the use of sonar by bats to identify both conspecifics and prey (Ruczyñski & Bartoñ 2012). Many species rely on multiple sensory cues, which can be tuned to different purposes. For example, birds communicate with each other primarily with auditory cues, but use vision to assess vegetation type and quantity, olfaction to detect nest predators (Forsman et al. 2012), and magnetoreception to migrate in the dark (Wiltschko & Wiltschko 1996). Many insects use a combination of olfactory and visual cues to identify host plants (Patt & Sétamou 2007, Wenninger et al. 2009). For complex movements, such as catching food and navigation, animals appear to rely on simultaneous information in their perceptual landscapes from multiple sensory systems.

Apart from direct assessments of habitat quality, animals may also opt to occupy and travel through habitat patches based on indirect cues such as spatial gradients and the presence of conspecifics and predators. For some species, perception of habitat suitability is increased by the presence of conspecifics (Stamps 2012, Farrell et al. 2012) or even heterospecifics (Fletcher 2007), but for many territorial species, especially carnivores, the presence of conspecific competitors reduces that attractiveness of a patch (Moorcroft et al. 2006). Many visual, olfactory or auditory cues can alert individuals to the presence of predators and competitors to produce a holistic assessment of patch quality (Forsman et al. 2012). Perception of habitat quality can also be biased by spatial gradients in topography and soil moisture (Rothermel 2004), and the presence of edges (Ovaskainen & Cornell 2003), wherein the direction of movement may be biased by the directionality of these gradients (Desrochers & Fortin 2003).

As a consequence of different sensory modalities, environments and purposes, the interaction between environmental cues and sensory systems is adapted to both spatial and temporal scales. Of these, spatial attributes and effects have received most of the attention related to perceptual range to landscape ecology. Because animals appear to have perceptual ranges that approximate the size and spacing of relevant features in natural landscapes (Lima & Zollner 1996), efficient movement through fragmented landscapes by any organism may require a comparable distribution of resource patches (Baguette & Van Dyck 2007). When necessary resources occur in smaller sizes or at greater distances, animals may not be able to detect their presence in landscapes or they may spend more time and energy reaching them. In addition to higher movement costs, longer dispersal journeys increase the risk of starvation or predation, which combine to cause the higher mortality rates of dispersing individuals across vast taxonomic groups even in relatively pristine habitat (reviewed by Clobert et al. 2001).

Perceptual range has been viewed mainly as a spatial construct, but it is actually a combination of time and space. It is the iterative, cumulative nature of sensory cues that make time a critical component in movement decisions. A forest-dwelling mouse might make a dash across an agricultural gap to another habitat patch only on a moonless night, but it might know of the existence of the gap because it saw it from the edge of its natal patch on a night with a full moon or because it sensed its presence from scent cues on a rainy night (Zollner & Lima 1999). The ability for organisms to integrate sensory information in time and space is essential for increasing movement between patches and connectivity through fragmented landscapes (e.g. Pe'er & Kramer-Schadt 2008).

The integration of space and time is not restricted to the sensory perceptions of animals; it is also a dynamic attribute of all landscapes. This concept is easily visualized when considering vegetation growth. For small animals, perceptual range tends to increase when there is less obstruction by plants (Flaherty *et al.* 2008, Prevedello *et al.* 2010), but this relationship is changing constantly in a matrix comprising a growing crop or an old field undergoing succession. Such changes in land cover over time could cross thresholds at which functional connectivity is suddenly lost or gained for target species even if all other factors remain constant. The concept of thresholds in the effects of habitat degradation has been an intense topic of study in landscape ecology (e.g. Fahrig 2003; Bayne *et al.* 2005), but it has not been explicitly integrated with the spatio-temporal and primarily behavioral paradigm of perceptual ranges.

A mismatch between the sensory perceptions of individuals and the provisions of human-dominated landscapes can thwart conservation planning and implementation. More awareness of this issue could make it possible to identify and overcome the features that impede movement. For example, managers could take advantage of perceptual biases by simulating the essential components of an environmental cue. Duck hunters do this when they use wooden decoys to imitate landed conspecifics, which imply habitat security. The same trick was implemented with painted sticks to invite extirpated Atlantic puffins (*Fratercula arctica*; Kress 1977) to colonize a former nesting island. Similarly, pre-placed dollops of poop, collected from captive animals around the world, reassured reintroduced black rhinos (*Diceros bicornis*) that it was safe to reside in their new habitat (Stamps & Swaisgood 2007). But on the other hand, perceptual systems can fail to detect risks that did not exist in the developmental or evolutionary history of individuals to create ecological traps, for which sensory information might be equally relevant to mitigation (Chapter 4). Either direction of application will require that sensory perceptions are understood, which will first require that they be measured.

### 7.3.3 Measuring movement directly and deductively

The integration of sensory ecology, particularly perceptual range, could add important mechanistic understanding to measures of functional connectivity – if animal perceptions could be measured in reliable and repeatable ways. Ideally, these measurements should explicitly link animal perceptions with movement decisions to support deductive conclusions. Often, but not always, the purpose of measuring movement is to estimate the relative permeability or resistance of different habitat types or landscape features. Habitat resistance can be measured in many ways (reviewed by Zeller *et al.* 2012), but there is little consensus on the best way to do it (Spear *et al.* 2010). Emphasizing the familiar scientific paradigm of deduction, as it is applied to most of behavioral theory, is likely to increase the rate at which generalizations emerge.

Deductive conclusions can be supported by experiments that are manipulative or mensurative, the latter meaning that descriptive data are collected to test specific predictions based on explicit hypotheses (*sensu* Hurlbert 1984, Underwood 2009). Either approach could develop hypotheses about functional connectivity that incorporate sensory ecology. These approaches differ from descriptive studies by applying deductive logic to alternative *a priori* hypotheses about the data, rather than inferring those mechanisms from patterns in the data. Manipulative experiments typically offer greater potential to reduce the role of confounding variables that induce unwanted and unmeasured sources of variation. In the context of movement behavior relevant to reserve design, such manipulations typically seek to standardize the motivation to move, which can vary with both intrinsic factors (e.g. hunger) and extrinsic ones (e.g. competition; Getz & Saltz 2008, Nathan *et al.* 2008).

Several early studies supported a deductive approach to measuring animal perceptions of habitat by focusing on small animals in experimental microcosms of semi-natural habitat (e.g. Ims & Stenseth 1989, reviewed by Wiens *et al.* 1993). Although the relevance of this approach to conservation has been questioned (e.g. Carpenter 1996), it is highly applicable to some questions and species. For example, experimental arenas have been used to elucidate the mechanisms of navigation and cue use by migratory species (Avens & Lohmann 2004) and explore the importance of predator–prey dynamics in patch use (Hammond *et al.* 2007). Modern arena experiments can make use of video tracking systems and motion tracking software (e.g. Valente *et al.* 2007) to support both the development and testing of hypotheses relevant to movement behavior and reserve design.

A second manipulative approach to measuring habitat permeability is to manipulate whole landscapes to contain patches that differ in their size and configuration and then measure the movement of individuals among patches. This approach demonstrated the use by butterflies, birds and other animals of cleared corridors in forested landscapes (Haddad 1999, Haddad *et al.* 2003, Machtans *et al.* 2006). Such landscape-level experiments can be used to assess the effects of anthropogenic activities on community composition and species persistence (e.g. Schmiegelow *et al.* 1997, Damschen *et al.* 2006, Ferraz *et al.* 2007). But despite their spatial relevance, it is usually expensive and logistically challenging to manipulate whole landscapes.

A third manipulative approach integrates the first two by conducting experiments on individuals at small scales using the features contained in real landscapes. Clever uses of arenas demonstrated the effects of barriers on gene flow in rodents (Krebs *et al.* 1969) and the permeability to frogs of different habitat types (Stevens *et al.* 2005). Similarly, the use of audio playbacks to lure birds over particular habitat configurations (e.g. Desrochers & Hannon 1997) has revealed how they find and use corridors (St. Clair *et al.* 1998, Sieving *et al.* 2000), whether they prefer to detour rather than cross gaps in forest cover (Bélisle & Desrochers 2002) and the permeability of different kinds of barriers (St. Clair 2003, Tremblay & St. Clair 2009). This approach is also used in translocations of territorial individuals to measure the speed and success with which animals move between points of capture and release (Bélisle *et al.* 2001), to demonstrate the distance at which a small mammal can detect a forest edge (e.g. Zollner 2000), the distance over which insects can detect host plants (e.g. Schooley & Wiens 2003), the utility of fencerows for bird movement (St. Clair *et al.* 1998, Gillies & St. Clair 2010), the permeability of different matrix types (Bakker 2006, Tremblay & St. Clair 2011) and the cumulative effects of multiple barriers (Bélisle & St. Clair 2002).

Although manipulative experiments have produced clear results with conceptual relevance to reserve design, some authors have debated their ability to offer specific insights for conservation action (e.g. Beier & Noss 1998 vs. Haddad *et al.* 2000). It is usually difficult to scale the implications of playback and translocation experiments on individuals to identify population-level consequences in landscapes (e.g. Desrochers *et al.* 1999), a limitation that has also applied to translocations performed for the purposes of reintroductions

(Ottewell *et al.* 2014). In most cases, the experiments occur at spatial and temporal scales that are smaller than the ones that determine conservation outcomes for free-living animals.

Mensurative experiments (*sensu* Underwood 2009) have the potential to overcome the problems of scale and relevance that trouble manipulative experiments because they can employ more natural behaviors in real-world contexts. They may also occur over much broader spatial and temporal scales. To maximize their conservation insights, these experiments should use *a priori* and explicit hypotheses, identify measurements that can test specific predictions, measure those variables in landscapes that possess relevant variation and distinguish among multiple hypotheses. Mensurative experiments are especially well-suited to observations of animal responses in real time, with which management restrictions or reserve boundaries might be identified. For example, mensurative experiments have been used to quantify the distances at which human approaches cause alert responses (e.g. Fernández-Juricic 2001, Hererro *et al.* 2005) or flight reactions (e.g. Laursen *et al.* 2005, Ronconi & St. Clair 2001). As advantages, these metrics are usually easy to measure, apply to a wide range of species, and can be translated to management action. But a core disadvantage is that they tend to exhibit high variability across species and contexts, which limits their generality (Laursen *et al.* 2005, Stankowich 2008).

In addition to direct observations of behavior, many other kinds of data could be gathered with mensurative experiments to understand animal movement in relation to reserve design and management. Most descriptive studies of habitat selection and movement based on collaring, tracking or photographic data could increase their utility by managers if they applied the principles of mensurative experiments. An example of this application is to identify the context dependency of movement decisions, which are especially likely to characterize carnivore species, and make those contexts explicit in study design. For example, wolves (*Canis lupus*) were hypothesized to exhibit greater avoidance of both roads and trails when human use of those features was high (Whittington *et al.* 2004), and cougars (*Puma concolor*) were expected to show greater avoidance of roads and human dwellings when those features were relatively rare (Knopff *et al.* 2014). Mensurative experiments are also amenable to the use of capture-mark-recapture techniques. Ovaskainen (2004) combined mark recapture methods with diffusion models of butterfly movement to estimate rates of movement through different kinds of habitat, including corridors. Useful generalizations about the effect of sensory systems on functional connectivity are more likely to emerge from studies that emphasize direct measurements and deductive logic because they will create a balance of evidence (*sensu* Sutherland *et al.* 2004). However, this combination could be even more informative if it embraced individual variation.

### 7.3.4 The importance of individual variation

Despite the increasing emphasis on functional connectivity, the gradual integration of sensory ecology, and the increasing prevalence of studies that measure movement and habitat selection both directly and deductively, there is still little attention paid to the variation among individuals in most conservation contexts, including reserve design and management. Instead, such variation is usually treated as inconvenient noise that obscures the population signal that could otherwise be used to advance conservation goals. This tendency is certainly reasonable in the crisis-driven and resource-limited world of conservation practitioners. However, that population emphasis necessarily ignores the process of natural selection over evolutionary time as a contributor to current expressions of variation (Chapter 2). The same variation in behavioral expression, which is always related somehow to heritable variation, still influences the survival of individuals in human-dominated landscapes. An alternative view of this variation would see it as a critical contributor to conservation outcomes because of the intensity of pressure now being exerted by artificial selection over much shorter timescales.

The truism that individual variation is highly relevant to the context of reserve design and management is made clear by the surprisingly small amount of genetic exchange needed to protect populations from inbreeding or rescue sub-populations from stochastic local extinctions (Mills & Allendorf 1996). This exchange can be achieved with just a few wide-ranging individuals, which are typically unknown to conservation managers, but hold the key to the persistence of populations. Such critical individuals are typically dispersers that may exhibit very different habitat preferences than resident individuals (e.g. Selonen & Hanski 2006, Debeffe *et al.* 2012), which is likely to be underestimated if they are under-detected in abundance or occupancy surveys (Mackenzie *et al.* 2006). Lesser detectability is especially likely to characterize transient, low-ranking animals in territorial species (Foster & Harmsen 2011) and the omniscience of individuals is likely to differ in most species (Panzacchi *et al.* 2015). Even birds, with their bird's-eye view of landscapes, differ tremendously within species and guilds in their apparent ability to see and use habitat features such as corridors (Gillies & St. Clair 2008).

Nathan *et al.* (2008) appreciated the importance of individuals to the study of movement ecology when they offered a conceptual framework that distinguishes among three basic components of movement within a logical temporal sequence: internal state, navigation capacity and motion capacity. In other words, an animal must first experience a need to move before it will use its sensory systems to evaluate the environment and decide where to move, and only then does it execute its species-specific motor abilities. Unfortunately, the use of movement data in conservation contexts has often proceeded in the opposite direction by designating patches of high and low quality, predicting movement probabilistically based on patterns of occupancy or sequential positions of individuals, and then using these samples to infer the motivational context, sensory experience and movement trajectories of individuals.

Even if one agrees with the relevance of individual variation to conservation problems and their solutions, a core limitation for interpreting behavioral variation results from the difficulty of measuring it. In comparison to individual variation in morphological traits, behavior expression appears to be ephemeral and infinitely variable. Nonetheless, every behaviorist knows that behavioral variation actually stems from a combination of genes, environment and gene–environment interactions, which include the massive domain of learning. An enormous literature has blossomed recently to describe some of this variation with predictable and consistent gradients such as bold versus shy behavioral types (e.g. Wolf & Weissing 2012), temperament (e.g. Visser *et al.* 2008) and proactive versus reactive coping styles (Koolhaas *et al.* 2010). Few conservation managers are aware of this rapid development in behavioral theory or the increasing frequency with which it is being applied to wild populations.

The variants described by these gradients are conventionally described as behavioral syndromes wherein individuals exhibit some consistency in behavioral expression through time and across contexts (Sih *et al.* 2004, Reale *et al.* 2007). The resulting suites of correlated behaviors characterize gradients of identifiable behavioral or personality "types." Identifying animals by type makes their behavioral expression far more predictable and thus more useful to managers than is the seemingly random variation in behavioral expression that results from environmental or experiential stochasticity. As expected from evolutionary theory, the nature of current behavioral expression can sometimes identify the history of selective pressures. For example, the personality traits of captive-born three-spined sticklebacks (*Gasterosteus aculeatus*) could be successfully predicted from variation in presumed selective pressures for neophilia and activity levels among twelve different source habitats (Dingemanse *et al.* 2007).

Until recently, the study of behavioral syndromes was almost completely limited to animals in captivity, with modest exploration limited to capturing ungulates (Reale *et al.* 2009) and observations of free-living primates (reviewed by Itoh 2002). However, an explosion in interest by behaviorists has applied assessments of personality to wild populations of fish (e.g. Adriaenssens & Johnsson 2013), birds (e.g. Krajl-Fiser *et al.* 2010) and mammals (Gartner & Powell 2012, Carter *et al.* 2012) with a host of metrics that could be applied in management contexts. As these studies accumulate, there may emerge some reliable morphological proxies for personality that could make it more tractable to identify behavioral types in the wild. For example, in some dog breeds, aggressiveness is correlated with coat color (Konno *et al.* 2011). Because behavioral syndromes are, by definition, composed of correlated suites of behaviors, the study and measurement of one personality trait will typically increases a manager's ability to predict correlated traits. Revealing these correlations with appropriate assays will increase the efficiency of both measurement and interpretation of wildlife personality as a management tool.

One potential assay of behavioral types is contained by hormones, which mediate much of the variation among individuals in personality as both a cause and consequence of behavioral expression. This relationship makes it plausible to use hormones for assessing and predicting the range of effects on individuals of anthropogenic stressors pertinent to reserve design. For example, socially dominant individuals exhibit higher levels of the stress hormone cortisol in both wild dogs (*Lycaon pictus*) and dwarf mongooses (*Helogale parvula*; Creel *et al.* 2001). Similarly, humans that self-identified as having dominant personality types had higher levels of testosterone (Sellers *et al.* 2007). Some personalities might mitigate the stress of anthropogenically changing environments. For example, juncos (*Junco hyemalis*) that had colonized a novel, urban area more readily explored novel objects and had significantly lower levels of cortisol, whereas non-colonizing juncos tended to be neo-phobic and non-exploratory individuals (Atwell *et al.* 2012).

Another useful assay of personality may come in the form of lateralization (Tommasi 2009), for which handedness provides a visible example that is familiar to all humans. Lateralization stems from a greater compartmentalization of the cerebral hemispheres, which is assumed to support a greater division of labor for separate brain functions to increase behavioral efficiency (Tommasi 2009). Greater lateralization, such as the tendency for an animal to turn in one direction when faced with novel stimuli and in the other direction when faced with familiar stimuli (Robins & Phillips 2009) is likely adaptive in predictable environments. A highly lateralized ungulate, for example, can simultaneously survey its landscape for predators while continuing to graze (Austin & Rogers 2012). Lateralization may correlate with the degree of aggression in some fish (e.g. cichlids, *Amatitlania nigrofasciata*; Reddon & Hurd 2008) or the presence of a fear response in some mammals (Rogers 2010), ultimately contributing to survival (e.g. for some poeciliid fishes in areas of high vs. low predation; Brown *et al.* 2007). Despite the apparent advantages of more lateralized brains, the less efficient brains of animals with weaker lateralization might make simultaneous use of more brain functions to increase their creativity and flexibility (Searleman 2013) or, in the context of conservation, adaptability to changing, human-dominated landscapes.

Even if personality is partly heritable and can be measured reliably in the field, it does not appear to be a static feature of individuals and this limits the generalizations that are possible. Experience also appears to be important to the development of personality, although its effect may be more important early in development. For example, the presence of predators stimulated brain development in sticklebacks (*Pungitius pungitius*; Gonda *et al.* 2012). The relationship between a behavioral type and the environment is similarly dynamic. Chickadees (*Poecile atricapallus*) that explored their environments more slowly also did so more thoroughly, which permitted them to benefit more in novel environments relative to faster birds that benefited in more stable and homogenous environments (Guillette *et al.* 2011). Yet despite that benefit for slow-exploring chickadees, learning by association occurred faster in bold trout (*Oncorhynchus mykiss*; Sneddon 2003). As this literature grows, it will be increasingly possible to assess generalizations with relevance to conservation managers, such as the hypothesis that bolder animals exhibit higher competitive and reproductive success, but have lower survival (Smith & Blumstein 2008).

## 7.4 INTEGRATING FOUR BEHAVIORAL APPROACHES TO CONSERVATION PLANNING

With the disparate literature we cited in the last four sections, we have shown that there has been surprisingly little overlap of these behavioral approaches as they pertain to reserve design and management. We believe that better integration of these approaches will reveal more generalizations to advance conservation goals. In particular, we advocate (a) identifying and emphasizing functional over structural connectivity, while (b) acknowledging the role of sensory systems in defining perceptual ranges in space and time to (c) measure the movement of animals in reserve networks (and their component parts) directly and deductively, with a goal of (d) interpreting the variation among individuals that could support more general rules about habitat use and movement. When all four attributes are tackled simultaneously, the result may provide much tangible guidance for the design and management of reserve networks. Our approach still employs aspects of space, but adds to that emphasis an explicit and intentional integration of four behavioral approaches that could advance the ultimate goal of fostering greater coexistence between people and wildlife.

We have no prescriptions (yet) for the integration of these four behavioral approaches in studies of movement behavior and habitat selection that seek to provide relevant information to conservation managers. Instead we attempt to identify a few of the associated opportunities and challenges of this integration by describing some of our own work and the lessons that have emerged from it. We organize these around four generic assertions chosen to foster interest and dialogue by others.

(1) *Perception of landscape elements differs among species enough to defy conventional terms like barriers and corridors.* Colleen St. Clair co-led a series of studies on the effects of roads on movement by birds and mammals and found that animal perceptions of a given feature as a barrier or a corridor varied among species, individuals and contexts. For example, many forest birds are more repelled by natural rivers than by anthropogenic highways (St. Clair 2003), but responses to roads are highly divergent among similarly sized songbirds (Tremblay & St. Clair 2009), which may stem partly from their ability to accommodate road noise (Proppe *et al.* 2012). Small mammals eschew the kind of corridors constructed across roads that are preferred by grizzly bears (*Ursus arctos*) serving as "umbrella species," in favor of greater cover provided by inexpensive culverts, especially if woody cover is added to their entrances (McDonald & St. Clair 2004). The functional connectivity of barriers and corridors can be so conflated that new terms should enter the lexicon of reserve design: barridors and corriers.

(2) *Individual state can profoundly influence perceptions of habitat suitability, associated expressions of functional connectivity and the potential for human-wildlife conflict.* Aditya Gangadharan has been studying the problem of functional connectivity for large mammals in the Shencottah Gap of the Western Ghats, India. There, plantations,

human habitation, forestry and, especially, a transportation route, inhibit movement, particularly of tigers (*Panthera tigris*) and elephants (*Elephas maximus*) between two tiger reserves. Camera traps have revealed that these animals occasionally come closer to anthropogenic features than has been assumed for decades. However, the tiger that came closest to crossing the human-impacted corridor did so only when its body condition was particularly poor and an elephant that approached to within meters of the highway was a bull that was apparently trying to access water at the height of summer. These results corroborate those from other studies showing that animals in poorer condition or lower social status are more likely to tolerate humans and their infrastructure. Such contingencies create an acute challenge for conservation managers, because these atypical individual animals may promote much-needed connectivity across human-dominated landscapes but at the potential cost of conflict with the people who live in those landscapes.

(3) *Animals that exhibit greater flexibility in habitat use may be more likely to suffer mortality via ecological traps.* Maureen Murray studied the consequences of individual variation in movement behavior for urban coyotes in Edmonton, Canada. She found that coyotes that were killed in vehicle collisions were less nocturnal and crossed roads more often during the day, especially in the winter when evening rush hour occurs after sunset. At high latitudes, the relationship between daylight and rush hour changes substantially among seasons, which may reduce the reliability of evolved sensory cues and disadvantage animals that employ more reactive coping styles. Moreover, coyotes that were visibly infested with the mite *Sarcoptes scabiei* were less nocturnal and much more willing to use developed areas than apparently-healthy coyotes. Both behaviors increase the probability that mangy animals will come into conflict with people, which can prompt lethal management. Identifying these associations makes it possible for management actions to foster co existence while minimizing conflict.

(4) *Individual variation stemming from personality may explain temporal trends related to habitat selection and movement, such as the loss of migratory behavior in ungulates.* Rob Found used a series of behavioral assays to determine the role of personality in the loss of seasonal migration behavior in elk (*Cervus canadensis*) that increasingly reside year-round near townsites in mountain parks of Canada. He showed that resident elk are disproportionately

composed of individuals with bold personality types, whereas the migrants contain more shy individuals. The greater wariness of shy individuals might increase their survival during migration, whereas bold individuals might be better able to exploit the novel foods that are available in closer proximity to people. Because of the co-evolution of personality traits, reserves that create novel conditions with high human disturbance and low natural predation may unintentionally select for individuals expressing other boldness traits, such as aggression and a proclivity to habituate to humans. Similar attention to personality may support the reversal of habituated behavior via aversive conditioning that targets bolder animals with stimuli of higher salience, intensity, consistency and duration.

In each of these examples, the functional connectivity of landscapes fragmented by roads and other forms of human use was dependent on the behavioral flexibility of individuals within both species and populations. In every case, the decision individuals made to approach and cross, or not, a particular habitat element demonstrated a context- and sometimes state-dependent perception via diverse and changing sensory information. The presence of more flexible behavioral types may be relevant to the global phenomenon of lost migratory behavior in ungulates. By examining the variation in behavior more directly and deductively, we are more likely to advance many conservation goals in the context of reserve and corridor design.

We believe that the ephemeral and adaptive flexibility of individuals is an especially important and under-appreciated aspect of individual variation that might even comprise a conservation commodity that could be targeted by managers (Chapter 1, Box 1.1). In most human-dominated landscapes, more flexible species and individuals are probably more likely to survive, and even thrive. Yet flexibility itself may sometimes be encouraged in an imperiled population. For example, managers in Banff National Park in Canada noticed that wolves lost the ability to use a particular highway crossing structure with the mortality of alpha animals. The managers responded by dragging an animal carcass through the structure, successfully overcoming the reticence of remaining pack members who were then able to access elk exploiting what was otherwise a predator refuge near the townsite (T. Hurd, personal communication). By contrast, there are other circumstances in which behavioral flexibility is a liability, such as when it results in human–wildlife conflict or when it produces high rates of mortality and leads to ecological traps. Generalist species and individuals that easily exploit humans and their infrastructure might be reined in more effectively if wildlife managers can anticipate ways to exploit their sensory systems to limit their flexibility.

## 7.5 CONCLUSIONS AND MANAGEMENT IMPLICATIONS

In this chapter, we have shown that space has been the primary commodity of reserve design in conservation biology. There are excellent pragmatic and theoretical reasons for this emphasis and it supports ongoing development in the growing field of spatial ecology, particularly of animal movement and habitat selection. Unfortunately, these methods rely mainly on statistical inference to describe patterns of animal locations and movement trajectories that are often case-specific and seem to offer few generalizations for conservation practice. Such insights are badly needed because the preservation of space alone cannot prevent the loss of biodiversity, especially in the regions of the world where high ecological productivity also hosts the highest densities of people (Luck 2007).

We suggest that a greater emphasis on the behavioral bases of habitat selection and movement behavior could complement the traditional reliance on spatial attributes of reserve design. Four approaches that we think offer particular promise for elucidating these mechanisms are (a) emphasizing functional over structural connectivity, while incorporating (b) animal perception and a knowledge of sensory ecology that is (c) studied directly and deductively, while (d) purposely exploring variation among individuals, rather than seeking averages for populations. We illustrate some findings from our own studies that show how integrating these four approaches may reveal generalizations that are useful to conservation managers. Specifically, we show how animal movement through human-dominated landscapes can vary with context, perception, body condition and personality. We also show how the prevalence of sick animals might be related to both land use management and human–wildlife conflict. We conclude by suggesting that behavioral flexibility is an especially important conservation commodity that might be either augmented or reduced to support conservation goals.

Because of their training in behavioral theory, understanding of both sensory and motor systems, tendency to study behavior directly and deductively, and view of individual variation as the source of adaptation, behaviorists could play a much stronger role than they have in reserve design and management. A greater emphasis on behavior in this context is especially likely to support the ultimate conservation goal of coexistence that is championed by several visionary authors (e.g. Carson 1962, Wilson 1984, Walker 1992, Woodroffe *et al.* 2005). This emphasis may also extend other conceptual complements to the preservation of space as a cornerstone of conservation, such as the importance of maintaining ecological processes (Leopold 1949), the relevance of economic, political and social dimensions (West *et al.* 2006), and the necessity of engaging diverse stakeholders in conservation planning (Reed 2008).

## ACKNOWLEDGEMENTS

We thank Oded Berger-Tal and David Salz for inviting us to write this chapter and for many helpful comments on our ideas. Many individuals associated with our research projects supported our work and contributed to the ideas expressed here. The unpublished work we describe above was supported by Alberta Ingenuity Fund; Alberta Conservation Association; Alberta Ecotrust; Alberta Sport, Recreation, Parks, and Wildlife Foundation; Animal Damage Control; Canadian Foundation for Innovation; Canadian Wildlife Federation; Critical Ecosystem Partnership Fund; ESRI Canada; Natural Science and Engineering Research Council; Parks Canada Agency; TD Friends of the Environment Foundation; US Fish and Wildlife Service (Asian Elephant Conservation Fund); University of Alberta; and Wildlife Conservation Society.

## REFERENCES

Adriaensen, F., Chardon, J.P., De Blust, G., Swinnen, E., Villalba, S., Gulinck, H. and Matthysen, E. 2003. The application of "least-cost" modelling as a functional landscape model. *Landscape and Urban Planning,* 64:233–247.

Adriaenssens, B. and Johnsson, J.I. 2013. Natural selection, plasticity and the emergence of a behavioural syndrome in the wild. *Ecology Letters,* 16:47–55.

Afonso, P., Fontes, J., Holland, K.N. and Santos, R.S. 2008. Social status determines behaviour and habitat usage in a temperate parrotfish: implications for marine reserve design. *Marine Ecology Progress Series,* 359:215–227.

Ahearn, S.C., Smith, J.L.D., Joshi, A.R. and Ding, J. 2001. TIGMOD: an individual-based spatially explicit model for simulating tiger–human interaction in multiple use forests. *Ecological Modelling,* 140:81–97.

Arrignonon, F., Deconchat, M., Sarthou, J.-P., Balent, G. and Monteil, C. 2007. Modelling the overwintering strategy of a beneficial insect in a heterogeneous landscape using a multi-agent system. *Ecological Modelling,* 205:423–436.

Atwell, J.W., Cardoso, G.C., Whittaker, D.J., Campbell-Nelson, S., Robertson, K.W. and Ketterson, E.D. 2012. Boldness behavior and stress physiology in a novel urban environment suggest rapid correlated evolutionary adaptation. *Behavioral Ecology*, 23:960–969.

Austin, N.P. and Rogers, L.J. 2012. Limb preferences and lateralization of aggression, reactivity and vigilance in feral horses, *Equus caballus*. *Animal Behaviour*, 83:239–247.

Avens, L. and Lohmann, K.J. 2004. Navigation and seasonal migratory orientation in juvenile sea turtles. *Journal of Experimental Biology*, 207:1771–1778.

Baguette, M. and Van Dyck, H. 2007. Landscape connectivity and animal behavior: functional grain as a key determinant for dispersal. *Landscape Ecology*, 22:1117–1129.

Bakker, V.J. 2006. Microhabitat features influence the movements of red squirrels (*Tamiasciurus hudsonicus*) on unfamiliar ground. *Journal of Mammalogy*, 87:124–130.

Bakker, V.J. and Van Vuren, D.H. 2004. Gap-crossing decisions by the red squirrel, a forest-dependent small mammal. *Conservation Biology*, 18:689–697.

Baum, K.A., Haynes, K.J., Dillemuth, F.P. and Cronin, J.T. 2004. The matrix enhances the effectiveness of corridors and stepping stones. *Ecology*, 85:2671–2676.

Bayne, E.M., Van Wilgenburg, S.L., Boutin, S. and Hobson, K. 2005. Modeling and field-testing of ovenbird (*Seiurus aurocapillus*) responses to boreal forest dissection by energy sector development at multiple spatial scales. *Landscape Ecology*, 20:203–216.

Beier, P., Majka, D.R. and Spencer, W.D. 2008. Forks in the road: choices in procedures for designing wildland linkages. *Conservation Biology*, 22:836–851.

Beier, P., Spencer, W., Baldwin, R.F. and McRAE, B. 2011. Toward best practices for developing regional connectivity maps. *Conservation Biology*, 25:879–892.

Beier, P. and Gregory, A.J. 2012. Desperately seeking stable 50-year-old landscapes with patches and long, wide corridors. *PLoS Biology*, 10:e1001253.

Beier, P. and Noss, R.F. 1998. Do habitat corridors provide connectivity? *Conservation Biology*, 12:1241–1252.

Bélisle, M., Desrochers, A. and Fortin, M. J. 2001. Influence of forest cover on the movements of forest birds: a homing experiment. *Ecology*, 82:1893–1904.

Bélisle, M. and Desrochers, A. 2002. Gap-crossing decisions by forest birds: an empirical basis for parameterizing spatially-explicit, individual-based models. *Landscape Ecology*, 17:219–231.

Bélisle, M. and St. Clair, C.C. 2002. Cumulative effects of barriers on the movements of forest birds. *Conservation Ecology*, 5:9.

Bélisle, M. 2005. Measuring landscape connectivity: The challenge of behavioral landscape ecology. *Ecology*, 86:1988–1995.

Bender, D.J. and Fahrig, L. 2005. Matrix structure obscures the relationship between interpatch movement and patch size and isolation. *Ecology*, 86:1023–1033.

Bennett, A.F. 2003. *Linkages in the Landscape: The Role of Corridors and Connectivity in Wildlife Conservation*. International Union for Conservation of Nature, Switzerland. p.262.

Boyce, M.S., Mao, J.S., Merrill, E.H., Fortin, D., Turner, M.G., Fryxell, J.M. and Turchin, P. 2003. Scale and heterogeneity in habitat selection by elk in Yellowstone National Park. *Ecoscience*, 10:321–332.

Boyce, M.S. and McDonald, L.L. 1999. Relating populations to habitats using resource selection functions. *Trends in Ecology & Evolution*, 14:268–272.

Brown, C., Burgess, F. and Braithwaite, V.A. 2007. Heritable and experiential effects on boldness in a tropical *poeciliid*. *Behavioral Ecology and Sociobiology*, 62:237–243.

Calabrese, J.M. and Fagan, W.F. 2004. A comparison shoppers guide to connectivity metrics: trading off between data requirements and information content. *Frontiers in Ecology and Environment*, 2:529–536.

Campbell-Palmer, R. and Rosell, F. 2011. The importance of chemical communication studies to mammalian conservation biology: a review. *Biological Conservation*, 144:1919–1930.

Carpenter, S.R. 1996. Microcosm experiments have limited relevance for community and ecosystem ecology. *Ecology*, 77:677–680.

Carson, R. 1962. *Silent Spring*. Houghton Mifflin Harcourt, New York, USA.

Carter, A.J., Marshall, H.H., Heinsohn, R. and Cowlishaw, G. 2012. How not to measure boldness: novel object and antipredator responses are not the same in wild baboons. *Animal Behaviour*, 84:603–609.

Castellon, T.D. and Sieving, K.E. 2006. An experimental test of matrix permeability and corridor use by an endemic understory bird. *Conservation Biology*, 20:135–145.

Caughley, G. 1994. Directions in conservation biology. *Journal of Animal Ecology*, 63:215–244.

Convention on Biological Diversity 2010. Strategic plan for biodiversity 2011–2020. www.cbd.int/sp/ [accessed June 07, 2014].

Chetkiewicz, C.B. and Boyce, M.S. 2009. Use of resource selection functions to identify conservation corridors. *Journal of Applied Ecology*, 46:1036–1047.

Chetkiewicz, C.L.B., St. Clair, C.C. and Boyce, M.S. 2006. Corridors for conservation: integrating pattern and process. *Annual Review of Ecology, Evolution, and Systematics*, 37:317–342.

Clevenger, A.P. and Waltho, N. 2005. Performance indices to identify attributes of highway crossing structures facilitating movement of large mammals. *Biological Conservation*, 121:453–464.

Clobert, J., Danchin, E., Dhondt, A.A. and Nichols, J. (eds.) 2001. *Dispersal*. Oxford University Press, Oxford. pp. 452.

Colding, J. and Folke, C. 1997. The relations among threatened species, their protection, and taboos. *Conservation Ecology*, 1:6.

Creel, S. 2001. Social dominance and stress hormones. *Trends in Ecology & Evolution*, 16:491–497.

Cushman, S.A., McKelvey, K.S. and Schwartz, M.K. 2009. Use of empirically derived source-destination models to map regional conservation corridors. *Conservation Biology*, 23:368–376.

Daily, G.C., Ehrlich. P.R. and Sanchez-Azofeifa, A. 2001. Countryside biogeography: use of human-dominated habitats by the avifuna of southern Costa Rica. *Ecological Applications*, 11:1–13.

Damschen, E.I., Haddad, N.M., Orrock, J.L., Tewksbury, J.J. and Levey, D.J. 2006. Corridors increase plant species richness at large scales. *Science,* 313:1284–1286.

DeAngelis, D.L. and Mooij, W.M. 2005. Individual-based modeling of ecological and evolutionary processes. *Annual Review of Ecology, Evolution, and Systematics,* 36:147–168.

Debeffe, L., Morellet, N., Cargnelutti, B., Lourtet, B., Bon, R., Gaillard, J.-M. and Mark Hewison, A. 2012. Condition-dependent natal dispersal in a large herbivore: heavier animals show a greater propensity to disperse and travel further. *Journal of Animal Ecology,* 81:1327–1337.

Desrochers, A. and Hannon, S.J. 1997. Gap crossing decisions by forest songbirds during the post-fledging period. *Conservation Biology,* 11:1204–1210.

Desrochers, A., Hannon, S.J., Bélisle, M. and St. Clair, C.C. 1999. Movement of songbirds in fragmented forests: can we "scale up" from behaviour to explain occupancy patterns in the landscape? *International Ornithological Congress,* 22:2447–2464.

Diamond, J. 1975. The island dilemma: Lessons of modern biogeographic studies for the design of natural reserves. *Biological Conservation,* 7:129–146.

Dingemanse, N.J., Wright, J., Kazem, A.J., Thomas, D.K., Hickling, R. and Dawnay, N. 2007. Behavioural syndromes differ predictably between 12 populations of three-spined stickleback. *Journal of Animal Ecology,* 76:1128–1138.

Eads, D.A., Biggins, D.E., Livieri, T.M. and Millspaugh, J.J. 2014. Space use, resource selection and territoriality of black-footed ferrets: implications for reserve design. *Wildlife Biology,* 20:27–36.

Estrada, E. and Bodin, O. 2008. Using network centrality measures to manage landscape connectivity. *Ecological Applications,* 18:1810–1825.

Fahrig, L. 2003. Effects of habitat fragmentation on biodiversity. *Annual Review of Ecology, Evolution and Systematics,* 34:487–515.

Farrell, S.L., Morrison, M.L., Campomizzi, A.J. and Wilkins, R.N. 2012. Conspecific cues and breeding habitat selection in an endangered woodland warbler. *Journal of Animal Ecology,* 81:1056–1064.

Fernández-Juricic, E., Jimenez, M.D. and Lucas, E. 2001. Alert distance as an alternative measure of bird tolerance to human disturbance: implications for park design. *Environmental Conservation,* 28:263–269.

Ferraz, G., Nichols, J.D., Hines, J.E., Stouffer, P.C., Bierregaard, R.O. and Lovejoy, T.E. 2007. A large-scale deforestation experiment: effects of patch area and isolation on Amazon birds. *Science,* 315:238–241.

Flaherty, E.A., Smith, W.P., Pyare, S. and Ben-David, M. 2008. Experimental trials of the northern flying squirrel (*Glaucomys sabrinus*) traversing managed rainforest landscapes: perceptual range and fine-scale movements. *Canadian Journal of Zoology,* 86:1050–1058.

Fletcher, R.J. 2007. Species interactions and population density mediate the use of social cues for habitat selection. *Journal of Animal Ecology,* 76:598–606.

Forman, R.T.T. 1995. Some general principles of landscape and regional ecology. *Landscape Ecology,* 10:133–142.

Forman. R.T.T., Sperling, D., Bissonnette, J.A., Clevenger, A.P., Cutshall, C.D., Dale, V.H., Fahrig, L., France, R.L., Goldman, C.R., Heanue, K., Jones, J., Swanson, F., Turrentine, T. and Winter, T.C. 2006. *Road Ecology: Science and Solutions*. Island Press, Washington, DC, USA.

Forsman, J.T., Monkkonen, M., Korpimaki, E. and Thomson, R. L. 2012. Mammalian nest predator feces as a cue in avian habitat selection decisions. *Behavioral Ecology*, 24:262–266.

Fortin D., Beyer H.L., Boyce M.S., Smith D.W., Duchesne T. and Mao J.S. 2005. Wolves influence elk movements: behavior shapes a trophic cascade in Yellowstone National Park. *Ecology*, 86:1320–1330.

Di Franco, A., Gillanders, B.M., De Benedetto, G., Pennetta, A., De Leo, G.A. and Guidetti, P. 2012. Dispersal patterns of coastal fish: implications for designing networks of marine protected areas. *PloS one*, 7:e31681.

Gartner, M.C. and Powell, D. 2012. Personality assessment in snow leopards (*Uncia uncia*). *Zoo Biology*, 31:151–165.

Gascon, C., Williamson, G.B. and da Fonseca, G.A. 2000. Receding forest edges and vanishing reserves. *Science*, 288:1356–1358.

Gerkema, M.P., Davies, W.I.L., Foster, R.G., Menaker, M. and Hut, R.A. 2013. The nocturnal bottleneck and the evolution of activity patterns in mammals. *Proceedings of the Royal Society B*, 280:1–11.

Getz, W. M. and Saltz, D. 2008. A framework for generating and analyzing movement paths on ecological landscapes. *Proceedings of the National Academy of Sciences*, 105:19066–19071.

Gilbert-Norton, L., Wilson, R., Stevens, J.R. and Beard, K.H. 2010. A meta-analytic review of corridor effectiveness. *Conservation Biology*, 24:660–668.

Gillies, C.S., Hebblewhite, M., Nielsen, S.E., Krawchuk, M.A., Aldridge, C.L., Frair, J.L., Saher, D.J., Stevens, C.E. and Jerde, C.L. 2006. Application of random effects to the study of resource selection by animals. *Journal of Animal Ecology*, 75:887–898.

Gillies, C.S. and St. Clair, C.C. 2010. Functional responses in habitat selection by tropical birds moving through fragmented forest. *Journal of Applied Ecology*, 47:182–190.

Gonda, A., Välimäki, K., Herczeg, G. and Merilä, J. 2012. Brain development and predation: plastic responses depend on evolutionary history. *Biology Letters*, 8:249–252.

Groom, M.J., Meffe, G.K. and Carroll, C.R. 2006. *Principles of Conservation Biology* (pp. 174–251). Sinauer Associates, Sunderland, USA.

Guillette, L.M., Reddon, A.R., Hoeschele, M. and Sturdy, C.B. 2011. Sometimes slower is better: slow-exploring birds are more sensitive to changes in a vocal discrimination task. *Proceedings of the Royal Society B*, 278:767–773.

Gurevitch, J., Morrow, L.L., Wallace, A. and Walsh, J.S. 1992. A meta-analysis of competition in field experiments. *American Naturalist*, 140:539–572.

Haddad, N.M. 1999. Corridor and distance effects on interpatch movements: a landscape experiment with butterflies. *Ecological Applications*, 9:612–622.

Haddad, N.M., Bowne, D.R., Cunningham, A., Danielson, B.J., Levey, D.J., Sargent, S. and Spira, T. 2003. Corridor use by diverse taxa. *Ecology*, 84:609–615.

Haddad, N.M., Brudvig, L.A., Damschen, E.I., Evans, D.M., Johnson, B.L., Levey, D.J., Orrock, J.L., Resasco, J., Sullivan, L.L., Tewksbury, J.J., Wagner, S.A. and Weldon, A.J. 2014. Potential negative ecological effects of corridors. *Conservation Biology.* 28:1178–1187.

Haddad, N.M., Rosenberg, D.K. and Noon, B.R. 2000. On experimentation and the study of corridors: response to Beier and Noss. *Conservation Biology,* 14:1543–1545.

Hammond, J.I., Luttbeg, B. and Sih, A. 2007. Predator and prey space use: dragonflies and tadpoles in an interactive game. *Ecology,* 88:1525–1535.

Hanski, I. 1998. Metapopulation dynamics. *Nature,* 396:41–49.

Hanski, I. and Gilpin, M.E. 1997. *Metapopulation Biology: Ecology, Genetics and Evolution.* Academic Press, San Diego. pp. 580.

Harcourt, A.H., Coppeto, S. and Parks, S. 2002. Rarity, specialization and extinction in primates. *Journal of Biogeography,* 29:445–456.

Hebblewhite, M. and Haydon, D.T. 2010. Distinguishing technology from biology: a critical review of the use of GPS telemetry data in ecology. *Philosophical Transactions of the Royal Society B,* 365:2303–2312.

Herrero, S., Smith, T., DeBruyn, T.D., Gunther, K. and Matt, C.A. 2005. From the field: brown bear habituation to people – safety, risks, and benefits. *Wildlife Society Bulletin,* 33:362–373.

Hilty, J.A., Lidicker, W.Z.J. and Merenlender, A.M. 2006. *Corridor Ecology: The Science and Practice of Linking Landscapes for Biodiversity Conservation.* Island Press, Washington, DC, USA.

Hobbs, R.J. 1992. The role of corridors in conservation: solution or bandwagon? *Trends in Ecology & Evolution,* 7:389–392.

Hodgson, J.A., Thomas, C.D., Wintle, B.A. and Moilanen, A. 2009. Climate change, connectivity and conservation decision making: back to basics. *Journal of Applied Ecology,* 46:964–969.

Holyoak, M., Casagrandi, R., Nathan, R., Revilla, E. and Spiegel, O. 2008. Trends and missing parts in the study of movement ecology. *Proceedings of the National Academy of Sciences,* 105:19060–19065.

Hurlbert, S.H. 1984. Pseudoreplication and the design of ecological field experiments. *Ecological Monographs,* 54:187–211.

Ims, R.A. and Stenseth, N.C. 1989. Conservation biology: divided the fruitflies fall. *Nature,* 342:21–22.

Ise, J. 2011. *Our National Park Policy: A Critical History.* The Johns Hopkins University Press, New York, USA.

Itoh, K. 2002. Personality research with non-human primates: theoretical formulation and methods. *Primates* 43:249–261.

Kadoya, T. 2009. Assessing functional connectivity using empirical data. *Population Ecology,* 51:5–15.

Kerr, J.T. and Ostrovsky, M. 2003. From space to species: ecological applications for remote sensing. *Trends in Ecology & Evolution,* 18:299–305.

Kingsland, S. 2002. Designing nature reserves: adapting ecology to real-world problems. *Endeavour,* 26:9–14.

Knopff, A.A., Knopff, K.H., Boyce, M.S. and St. Clair, C.C. 2014. Flexible habitat selection by cougars in response to anthropogenic development. *Biological Conservation*, **178**:136–145.

Konno, A., Inoue-Murayama, M. and Hasegawa, T. 2011. Androgen receptor gene polymorphisms are associated with aggression in Japanese Akita Inu. *Biology Letters*, **7**:658–660.

Krebs, C.J., Keller, B.L. and Tamarin, R.H. 1969. *Microtus* population biology: demographic changes in fluctuating populations of *M.ochrogaster* and *M. pennsylvanicus* in southern Indiana. *Ecology*, **50**:587–607.

Krajl-Fiser, S., Weib, B.M. and Kotrschal, K. 2010. Behavioural and physiological correlates of personality in greylag geese (*Anser anser*). *Journal of Ethology*, **28**:363–370.

Kress, S.W. 1977. Establishing Atlantic puffins at a former breeding site. In *Endangered Birds: Management Techniques for Preserving Endangered Species* (ed. S. A. Temple), pp. 373–377. University of Wisconsin Press, Madison, USA.

Lamberson, R.H., Noon, B.R., Voss, C. and McKelvey, K.S. 1994. Reserve design for territorial species: the effects of patch size and spacing on the viability of the northern spotted owl. *Conservation Biology*, **8**:185–195.

Laursen, K., Kahlert, J. and Frikke, J. 2005. Factors affecting escape distances of staging waterbirds. *Wildlife Biology*, **11**:13–19.

Leopold, A. 1949. *A Sand County Almanac and Sketches Here and There*. Oxford University Press, Oxford, UK.

Liedvogel, M., Chapman, B.B., Muheim, R. and Akesson, S. 2013. The behavioral ecology of animal movement: reflections upon potential synergies. *Animal Migration*, **1**:39–46.

Lima S.L. and Zollner, P.A. 1996. Towards a behavioral ecology of ecological landscapes. *Trends in Ecology & Evolution*, **11**:131–35.

Lomolino, M.V. and Brown, J.H. 2009. The reticulating phylogeny of island biogeography theory. *The Quarterly Review of Biology*, **84**:357.

Luck, G.W. 2007. The relationships between net primary productivity, human population density and species conservation. *Journal of Biogeography*, **34**:201–212.

MacArthur, R.H. and Wilson, E.O. 1967. *The Theory of Island Biogeography*. Princeton University Press, Princeton, USA.

Machtans, C.S. 2006. Songbird response to seismic lines in the western boreal forest: a manipulative experiment. *Canadian Journal of Zoology*, **84**:1421–1430.

MacKenzie, D.I., Nichols, J.D., Royle, J.A., Pollock, K.H., Bailey, L.L. and Hines, J. E. 2006. *Occupancy Estimation and Modeling: Inferring Patterns and Dynamics of Species Occurrence*. Academic Press, San Diego, USA.

Manly B.F.J., McDonald L.L., Thomas D.L., McDonald T.L. and Erikson W.P. 2002. *Resource Selection by Animals: Statistical Design and Analysis for Field Studies*. Kluwer, New York, 221.

Margules, C.R. and Pressey, R. L. 2000. Systematic conservation planning. *Nature*, **405**:243–253.

Martin, G.R. 2012. Through birds' eyes: insights into avian sensory ecology. *Journal of Ornithology*, 153:S23–S48.

McDonald, W.R. and St. Clair, C.C. 2004. Elements that promote highway crossing structure use for small mammals in Banff National Park. *Journal of Applied Ecology*, 41:82–93.

McLellan, B.N. and Hovey, F.W. 2001. Natal dispersal of grizzly bears. *Canadian Journal of Zoology*, 79:838–844.

McRae, B.H., Dickson, B.G., Keitt, T.H. and Shah, V.B. 2008. Using circuit theory to model connectivity in ecology, evolution, and conservation. *Ecology*, 89:2712–2724.

McRae, B.H., Hall, S.A., Beier, P. and Theobald, D.M. 2012. Where to restore ecological connectivity? Detecting barriers and quantifying restoration benefits. *PloS one*, 7:e52604.

Minor, E.S. and Urban, D.L. 2008. A graph-theory framework for evaluating landscape connectivity and conservation planning. *Conservation Biology*, 22:297–307.

Moilanen, A. and Hanski, I. 2001. On the use of connectivity measures in spatial ecology. *Oikos*, 95:147–151.

Molles, M. and Cahill, J.C. 2011. *Ecology: Concepts and Applications*. McGraw-Hill Ryerson. 654 pp.

Moorcroft, P.R., Lewis, M.A. and Crabtree, R.L. 2006. Mechanistic home range models capture spatial patterns and dynamics of coyote territories in Yellowstone. *Proceedings of the Royal Society B: Biological Sciences*, 273:1651–1659.

Morrison, M.L. 2012. The habitat sampling and analysis paradigm has limited value in animal conservation: a prequel. *The Journal of Wildlife Management*, 76:438–450.

Mysterud, A. and Ims, R.A. 1998. Functional responses in habitat use: availability influences relative use in trade-off situations. *Ecology*, 79:1435–1441.

Nathan, R., Getz, W.M., Revilla, E., Holyoak, M., Kadmon, R., Saltz, D. and Smouse, P.E. 2008. A movement ecology paradigm for unifying organismal movement research. *Proceedings of the National Academy of Sciences*, 105:19052–19059.

Newmark, W.D. 1996. Insularization of Tanzanian parks and the local extinction of large mammals. *Conservation Biology*, 10:1549–1556.

Ovaskainen, O. 2004. Habitat-specific movement parameters estimated using mark-recapture data and a diffusion model. *Ecology*, 85:242–257.

Ovaskainen, O. and Cornell, S.J. 2003. Biased movement at a boundary and conditional occupancy times for diffusion processes. *Journal of Applied Probability*, 40:557–580.

Ottewell, K., Dunlop, J., Thomas, N., Morris, K., Coates, D. and Byrne, M. 2014. Evaluating success of translocations in maintaining genetic diversity in a threatened mammal. *Biological Conservation*, 171:209–219.

Owens, I.P. and Bennett, P.M. 2000. Ecological basis of extinction risk in birds: habitat loss versus human persecution and introduced predators. *Proceedings of the National Academy of Sciences*, 97:12144–12148.

Panzacchi, M., Van Moorter, B., Strand, O., Saerens, M., Kivimaki, I., St. Clair, C. C., Herfindal, I. and Boitani, L. 2016. Predicting the continuum between corridors and barriers to animal movements using Step Selection Functions and Randomized Shortest Paths. *Journal of Animal Ecology*, **85**:32–42.

Patt, A.J.M. and Sétamou, M. 2007. Olfactory and visual stimuli affecting host plant detection in *Homalodisca coagulate* (*Hemiptera: Cicadellidae*). *Environmental Entomology*, **36**:142–150.

Pe'er, G. and Kramer-Schadt, S. 2008. Incorporating the perceptual range of animals into connectivity models. *Ecological Modelling*, **213**:73–85.

Pressey, R.L. and Bottrill, M.C. 2008. Opportunism, threats and the evolution of systematic conservation planning. *Conservation Biology*, **22**:134–1345.

Prevedello, J. A., Forero-Medina, G. and Vieira, M. V. 2010. Movement behaviour within and beyond perceptual ranges in three small mammals: effects of matrix type and body mass. *The Journal of Animal Ecology*, **79**:1315–1323.

Proppe, D.S., Avey, M.T., Hoeschele, M., Moscicki, M.K., Farrell, T., St Clair, C.C. and Sturdy, C.B. 2012. Black-capped chickadees sing at higher pitches with elevated anthropogenic noise, but not with decreasing canopy cover. *Journal of Avian Biology*, **43**:1–8.

Pulliam, H.R. 1988. Sources, sinks, and population regulation. *American Naturalist*, **132**:652–661.

Reale, D., Martin, J., Coltman, D.W., Poissant, J. and Festa-Bianchet, M. 2009. Male personality, life-history strategies and reproductive success in a promiscuous mammal. *Journal of Evolutionary Biology*, **22**:1599–1607.

Reddon, A.R. and Hurd, P.L. 2008. Aggression, sex and individual differences in cerebral lateralization in a cichlid fish. *Biology Letters*, **4**:338–340.

Reed, M.S. 2008. Stakeholder participation for environmental management: a literature review. *Biological Conservation*, **141**:2417 2431.

Robins, A. and Phillips, C. 2009. Lateralized visual processing in domestic cattle herds responding to novel and familiar stimuli. *Laterality: Asymmetries of Brain, Body and Cognition*, **15**: 514–534.

Rogers, L.J. 2010. Relevance of brain and behavioural lateralization to animal welfare. *Applied Animal Behaviour Science*, **127**:1–11.

Rothermel, B.B. 2004. Migratory success of juveniles: a potential constraint on connectivity for pond-breeding amphibians. *Ecological Applications*, **14**:1535–1546.

Ruczynski, I. and Barton, K.A. 2012. Modelling sensory limitation: the role of tree selection, memory and information transfer in bats' roost searching strategies. *PloS one*, **7**:e44897.

Schmidt, K.A., Dall, S.R.X. and van Gils, J.A. 2010. The ecology of information: an overview on the ecological significance of making informed decisions. *Oikos*, **119**:304–316.

Schmiegelow, F.K.A., Machtans, C.S. and Hannon, S.J. 1997. Are boreal birds resilient to forest fragmentation: an experimental study of short-term community responses. *Ecology*, **78**:1914–1932.

Schooley, R.L. and Wiens, J.A. 2003. Finding habitat patches and directional connectivity. *Oikos*, **102**:559–570.

Scriber, J.M. 2010. Integrating ancient patterns and current dynamics of insect–plant interactions: taxonomic and geographic variation in herbivore specialization. *Insect Science*, 17:471–507.

Sellers, J.G., Mehl, M.R. and Josephs, R.A. 2007. Hormones and personality: Testosterone as a marker of individual differences. *Journal of Research in Personality*, 41:126–138.

Selonen, V. and Hanski, I.K. 2006. Habitat exploration and use in dispersing juvenile flying squirrels. *Journal of Animal Ecology*, 75:1440–1449.

Shumway, C.A. 1999. A neglected science: applying behavior to aquatic conservation. *Environmental Biology of Fishes*, 55:183–201.

Sieving, K.E., Willson, M.F. and De Santo, T.L. 2000. Defining corridor functions for endemic birds in fragmented south-temperate rainforest. *Conservation Biology*, 14:1120–1132.

Simberloff, D. and Abele, L.G. 1976. Island biogeography theory and conservation practice. *Science*, 191:285–286.

Simberloff, D. and Cox, J. 1987. Consequences and costs of conservation corridors. *Conservation Biology*, 1:63–71.

Simberloff, D., Farr, J.A., Cox, J. and Mehlman, D.W. 1992. Movement corridors: conservation bargains or poor investments? *Conservation Biology*, 6:493–504.

Siniff, D.B. and Jessen, C.R. 1969. A simulation model of animal movement patterns. *Advances in Ecological Research* 11:369–404.

Smith, B.R., and Blumstein, D.T. 2008. Fitness consequences of personality: a meta analysis. *Behavioral Ecology*, 19:448–455.

Sneddon, L.U. 2003. The bold and the shy: individual differences in rainbow trout. *Journal of Fish Biology*, 62:971–975.

Soule, M.E. and Terborgh, J. 1999. The policy and science of regional conservation. In: *Continental Conservation*. (eds. M.E. Soule & J. Terborgh), pp. 1–18. Island Press, Washington DC, USA.

Spear, S.F., Balkenhol, N., Fortin, M.-J., McRae, B.H. and Scribner, K. 2010. Use of resistance surfaces for landscape genetic studies: considerations for parameterization and analysis. *Molecular Ecology*, 19:3576–3591.

Stamps, J.A. 2012. The effect of conspecifics on habitat selection in territorial species. *Behavioral Ecology*, 28:29–36.

Stamps, J. and Swaisgood, R.R. 2007. Someplace like home: experience, habitat selection and conservation biology. *Applied Animal Behaviour Science*, 102:392–409.

Stankowich, T. 2008. Ungulate flight responses to human disturbance: a review and meta-analysis. *Biological Conservation*, 141:2159–2173.

St. Clair, C.C. 2003. Comparative permeability of roads, rivers, and meadows to songbirds in Banff National Park. *Conservation Biology*, 17:1151–1160.

St. Clair, C.C., Bélisle, M., Desrochers, A. and Hannon, S. 1998. Winter responses of forest birds to habitat corridors and gaps. *Conservation Ecology*, 2:13.

Steffen, W., Grinevald, J., Crutzen, P. and McNeill, J. 2011. The Anthropocene: conceptual and historical perspectives. *Philosophical Transactions of the Royal Society A*, 369:842–867.

Stevens, V.M., Polus, E., Wesselingh, R.A., Schtickzelle, N. & Baguette, M. 2005. Quantifying functional connectivity: experimental evidence for patch-specific resistance in the Natterjack toad (*Bufo calamita*). *Landscape Ecology,* 19:829–842.

Sutherland, W.J., Pullin, A.S., Dolman, P.M. and Knight, T.M. 2004. The need for evidence-based conservation. *Trends in Ecology & Evolution,* 19:305–308.

Tang, W. and Bennett, D.A. 2010. Agent-based modeling of animal movement: a review. *Geography Compass,* 4:682–700.

Taylor P.D., Fahrig L., Henein K. and Merriam G. 1993. Connectivity is a vital element in landscape structure. *Oikos,* 68:571–573.

Taylor P.D., Fahrig L. and With, K.A. 2006. Landscape connectivity: a return to the basics. In *Connectivity Conservation* (eds. K.R. Crooks and M. Sanjayan), pp. 29–43. Cambridge University Press, New York, USA.

Tracey, J.A., Zhu, J., Boydston, E., Fisher, R.N. and Crooks, K.R. 2013. Mapping behavioral landscapes for animal movement: a finite mixture modeling approach. *Ecological Applications,* 23:654–669.

Tremblay, M.A. and St Clair, C.C. 2009. Factors affecting the permeability of transportation and riparian corridors to the movements of songbirds in an urban landscape. *Journal of Applied Ecology,* 46:1314–1322.

Tremblay, M.A. and St. Clair, C.C. 2011. Permeability of a heterogeneous urban landscape to the movements of forest songbirds. *Journal of Applied Ecology,* 48:679–688.

Tommasi, L. 2009. Mechanisms and functions of brain and behavioural asymmetries. *Philosophical Transactions of the Royal Society B,* 364:855–859.

Turchin, P. 1998. *Quantitative Analysis of Movement: Measuring and Modeling Population Redistribution in Animals and Plants.* Sinauer Associates, MA, USA.

Turner, W., Spector, S., Gardiner, N., Fladeland, M., Sterling, E. and Steininger, M. 2003. Remote sensing for biodiversity science and conservation. *Trends in Ecology & Evolution,* 18:306–314.

Urban, D. and Keitt, T. 2001. Landscape connectivity: a graph-theoretic perspective. *Ecology,* 82:1205–1218.

Valente, D., Golani, I. and Mitra, P.P. 2007. Analysis of the trajectory of *Drosophila melanogaster* in a circular open field arena. *PloS one,* 2:e1083.

Vandermeer, J. and Perfecto, I. 2007. The agricultural matrix and a future paradigm for conservation. *Conservation Biology,* 21:274–277.

Walker, B.H. 1992. Biodiversity and ecological redundancy. *Conservation Biology,* 6:18–23.

Wenninger, E.J., Stelinski, L.L. and Hall, D.G. 2009. Roles of olfactory cues, visual cues, and mating status in orientation of *Diaphorina citri* Kuwayama (*Hemiptera: Psyllidae*) to four different host plants. *Chemical Ecology,* 38:225–234.

West, P., Igoe, J. and Brockington, D. 2006. Parks and peoples: the social impact of protected areas. *Annual Reviews in Anthropology,* 35:251–277.

Whittington, J., St. Clair, C.C. and Mercer, G. 2004. Path tortuosity and the permeability of roads and trails to wolf movement. *Ecology and Society,* 9:4.

Wiens, J.A., Chr, N., Van Horne, B. and Ims, R.A. 1993. Ecological mechanisms and landscape ecology. *Oikos,* 66:369–380.

Wiens, D.J., Reynolds, R.T. and Noon, B.R. 2006. Juvenile movement and natal dispersal of Northern Goshawks in Arizona. *The Condor*, **108**:253–269.

Wijnstekers, W. 2003. *The Evolution of CITES*. CIC – International council for game and wildlife conservation, Budakeszi, Hungary.

Wilcox, B.A. and Murphy, D.D. 1985. Conservation strategy: the effects of fragmentation on extinction. *The American Naturalist*, **125**:879–887.

Williams, J.C., ReVelle, C.S. and Levin, S.A. 2005. Spatial attributes and reserve design models: a review. *Environmental Modeling and Assessment*, **10**:163–181.

Wilson, E.O. 1984. *Biophilia*. Harvard University Press. Cambridge, USA.

Wilson, E.O.and Willis, E.O. 1975. *Applied Biogeography. Ecology and Evolution of Communities*. Belknap Press, Cambridge, MA.

Wiltschko, W. and Wiltschko, R. 1996. Magnetic orientation in birds. *The Journal of Experimental Biology*, **38**:29–38.

Wisenden, B.D. 2000. Olfactory assessment of predation risk in the aquatic environment. *Philosophical Transactions of the Royal Society of London Series B*, **355**:1205–1208.

Wolf, M. and Weissing, F.J. 2012. Animal personalities: consequences for ecology and evolution. *Trends in Ecology & Evolution*, **27**:452–461.

Woodroffe, R. and Ginsberg, J.R. 1998. Edge effects and the extinction of populations inside protected areas. *Science*, **280**:2126–2128.

Woodroffe, R, Thirgood, S. and Rabinowitz, A. (eds.) 2005. *People and Wildlife, Conflict or Co-existence?* Cambridge University Press, Cambridge, UK.

Zeller, K.A., McGarigal, K. and Whiteley, A.R. 2012. Estimating landscape resistance to movement: a review. *Landscape Ecology*, **27**:777–797.

Zollner, P.A. 2000. Comparing the landscape level perceptual abilities of forest sciurids in fragmented agricultural landscapes. *Landscape Ecology*, **15**:523–533.

Zollner, P. and Lima, S. 1999. Illumination and the perception of remote habitat patches by white-footed mice. *Animal Behaviour*, **58**:489–500.

# Behavior-based management: conservation translocations

BEN D. BELL

## 8.1 INTRODUCTION

The translocation of organisms is defined as human-mediated movement of living organisms from one area, with release in another (IUCN 2012). Translocated animals must go through a process that their evolutionary history has not prepared them for. They are transported from their native range, often boxed and/or sedated and/or blindfolded, placed in a completely unfamiliar landscape and, in the case of captive-bred animals, altogether a completely novel environment. Surviving the translocation process depends to a large extent on the individual's behavior and decision-making during the time following the release. This behavior is a derivative of the species' evolutionary history, the individual's past experience, and conditions during transfer and at the release site. Most importantly, if the animal is capable of learning (and most translocated animals have at least some learning capabilities), this behavior will change as the animal gains experience in its new environment. This change can be termed "post release behavioral modification" (PRBM; Berger-Tal & Saltz 2014). Barring the case of ecological traps (see Chapter 4), as knowledge accumulates over time, behaviors will be modified accordingly to become more adaptive. Thus, PRBM is expected to increase the future fitness of the translocated animal. The novel environment dictates a need for rapid learning, while minimizing risk (the extent of which is unknown to a newly released animal). The need to learn a novel environment is most probably stressful (Dickens et al. 2010), making the animals susceptible to other types of threat, such as diseases (Harrington et al. 2013) and predators (Griffin et al. 2000). A key goal of the manager is, therefore, to shorten as much as possible the time necessary for the animals to become well acquainted with their new environment and learn to obtain resources while minimizing risk.

*Conservation Behavior: Applying Behavioral Ecology to Wildlife Conservation and Management*, eds. O. Berger-Tal and D. Saltz. Published by Cambridge University Press. © Cambridge University Press 2016.

"Conservation translocations" are translocations carried out for conservation purposes (IUCN 2012). Such translocations can be categorized according to the area into which the organisms are released. The term "population restoration" embraces any conservation translocation to site within the taxon's indigenous range and comprises two types of releases: (1) Reinforcement – intentional movement and release of an organism into an existing population of conspecifics to enhance the viability of the extant population. The existence of conspecifics may, in some cases, speed up the learning process of translocated individuals; and (2) Reintroduction – intentional movement and release of an organism to a site within its indigenous range from which it has disappeared to establish a new population and promote the persistence of a species. Conservation introduction is the intentional movement and release of an organism outside its indigenous range to prevent its extinction or to fulfill the function of an extinct species. Translocated animals typically come from two very different sources: captivity and the wild. The different translocation types and sources of animals result in different forms of inexperience in translocated animals, and, therefore, different learning challenges and processes. In addition, the age and social status of the individuals, and the sex ratio of the released group, may impact the PRBM process.

Conservation translocations have been an important aspect of conservation management for many years, but have met with varying success (e.g. IUCN 1998, Bell & Merton 2002, Griffiths & Pavajeau 2008, Germano & Bishop 2009, Jones & Merton 2012, Ottewell et al. 2014). Because they often fail (Griffith et al. 1989, Short et al. 1992, Fischer & Lindenmayer 2000), and because they often lack adequate post-release monitoring and reporting (Short et al. 1992, Fischer & Lindenmayer 2000, Short 2009), translocations are subject to a fair amount of criticism (Short 2009). Given that behavior is a key element in the fitness of animals, it is safe to assume that the success of translocations is strongly influenced by the behavior of the animals released (but see Linklater et al. 2012). However, as Seddon et al. (2007) noted, behavior (e.g. social behavior or foraging) was listed as the topic category for only 5% of 454 translocation papers published between 1990 and 2005, indicating that this aspect of focal species biology has had relatively less emphasis in the past. Animals in a novel landscape must make key decisions concerning their foraging, learning and movement strategies with little information to go on, for example: to be bold or cautious, to be exposed or to hide, to invest more in searching and learning or in exploiting and exhausting resources already located (Berger-Tal & Avgar 2012, Berger-Tal *et al.* 2014). Although presented as binary decisions, these options represent extremes on a continuous scale.

The animal's ability to make a correct (i.e. adaptive) decision depends on:

(a) Learning capabilities which are inherited traits stemming from the species' evolutionary history, but can vary among individuals.
(b) Current knowledge or behavior patterns (either learned or inherited) when released.
(c) The release protocol and conditions at the new site.

These three elements interact, and it is this interaction that determines the rapidity of PRBM. Thus, these elements should form the skeleton of the behavioral considerations that must be integrated into translocation and release protocols. By considering these issues, a good translocation protocol can be designed to fit the evolutionary constraints on the animals' behavior and enhance the rapidity and directions of the PRBM. Specifically, the learning abilities, the state of knowledge the animal has when released, the conditions at the release site and the release protocol can be manipulated so as to shorten the time frame of the PRBM by either speeding up learning, reducing the amount of modification neces- sary or by reducing the threats that are associated with the PRBM period (e.g. increased risk of predation). In terms of the conservation behavior framework (Chapter 1, Figure 1.1) these elements fall within the behavior- based management theme. The other two themes in the conservation behavior framework – behavioral response of animals to anthropogenic disturbances and behavioral indicators – obviously have a bearing on translocations (e.g. anthropogenic disturbances such as roads might influ- ence the PRBM, and PRBMs can be used as indicators) but are not directly linked to the procedure itself. However, patterns of PRBM can be used in an adaptive management approach to improve future translocations (Berger-Tal & Saltz 2014).

The animal's state of knowledge when released can be manipulated by pre-release training. This approach is commonly applied in training captive- bred individuals designated for translocations, for example, pre-release aversion training in the California condor (*Gymnogyps californianus*) to avoid power-lines and humans, or post-release training to induce migration in the whooping crane (*Grus americana*) (Lewis 1995, 1997; Bell & Merton 2002). This topic is covered in Chapter 10. By contrast, in the behavior-sensitive management realm, conservation decision-making and protocols are based on the species' behavior and learning abilities, as part of the overall suite of options available, and focus on providing and maintaining the environment necessary to enhance animal fitness, rather than manipulating the animals' behavior.

Translocations, therefore, can present two unique challenges: (1) Pre-release training, which is an active behavioral modification while the animals are still in captivity, and (2) Design and implementation of the translocation protocols that will facilitate rapid learning once the animal is in its new environment. In this chapter I focus on the latter, dealing with behavior-based management of translocation for conservation. Specifically, I focus on the key factors affecting PRBM, including landscape and habitat issues, threats, translocation protocols and how these have been or should be considered and managed when preparing translocations.

## 8.2 BEHAVIORAL FLEXIBILITY AND LEARNING IN TRANSLOCATIONS

Learning behavior can be central to conservation and, therefore, to translocation management (e.g. Reed & Merton 1991). The survival of individual animals and the consequent success of the translocation (as determined by the viability and growth rate of the population) are dependent on how well and how rapidly the animals can learn to function in their new environment. The manager's goal is to minimize the learning period that the animal has to go through after its release into an unfamiliar landscape. Species differ in their learning capabilities and learning ontogeny (Chapter 3). At one extreme are species that have little behavioral flexibility. In such species, behaviors may be inherently fixed or may become fixed at a specific time during the early ontogeny of the individual, termed the "critical period." The "critical period" is a period occurring at the early stages of development during which an organism is sensitive to key exogenous stimuli that are important for its future development; if the organism does not receive the appropriate stimulus during this period it may be slow, or unable, to develop these functions later in life. At the other extreme are animals that exhibit high behavioral flexibility, including social aspects, such as parental teaching and social learning. These animals are capable of modifying their behavior as the environment changes at all stages of their life, although the capacity to learn may decline with age. Behavioral flexibility, learning abilities and sociality are generally associated with brain size, and, as it turns out, brain size can predict translocation success; evidently because it is associated with behavioral flexibility that provides advantages when animals are placed in novel environments (Sol & Lefebvre 2000, Sol et al. 2002). Thus, the consideration of the interaction between behavior, age and ontogeny must be a part of any translocation program.

Animals that tend to have fixed behaviors and reduced learning capabilities cannot be managed to improve their behavioral responses, and if raised in unsuitable conditions their behavior cannot be modified or rectified later. Instead, the release procedure and the environment into which they are introduced must be selected and tailored to fit their behavioral characteristics. Important components of behaviors that become fixed during the critical period (such as kin recognition, sexual imprinting and parental effects such as a flight response to novel occurrences) must be acquired before release into the wild. If landscape-related behaviors are learned during the critical period, translocations may succeed only with the release of sub-adults that will then undergo the critical period learning experience in the wild (this probably dictates a low probability of success).

In species that tend to be more flexible in their behavior, learning becomes a fundamental process of adapting to a new environment and can be enhanced through various procedures. A translocated animal has three sources of knowledge that can be integrated into its learning process and decision-making:

(a) Past experience.
(b) Learning from conspecifics already present at the site of release.
(c) Learning from its own experience at the site of release.

Learning is especially important in highly social species (Thornton & Clutton-Brock 2011) and PRBM associated with social structure may be a strong indicator of the future fitness of the translocated population (Strum 2005). Sociality may produce behavior traditions that will modify selection pressures and influence genetic evolution. Such traditions are likely to play an important part in translocation of social animals, which may best be moved as a group in which survival-related traditions are well established, or only after pre-release training where such survival traditions are addressed (e.g. Griffith et al. 2000).

Many species requiring translocation have become rare due to being specialized – including reduced behavioral flexibility. Species become behaviorally specialized either because they evolved in a low-disturbance environment or due to intense competition. Such species often have limited abilities to respond behaviorally to pressures from invasive predators and/or competitors, e.g. the endemic terrestrial fauna in New Zealand following human settlement (Bell 1991, Gibbs 2006, King et al. 2009). With an evolutionary history essentially without mammalian predators, many New Zealand birds are (or were) mammal predator-naïve, being tame and/or flightless. With the lack or reduction of an innate response to novel threatening situations and little ability to discern threatening situations, these species were able to survive only in environments where invasive mammals are either absent or in low numbers, such as islands or predator-excluded sanctuaries. Island endemics are particularly prone to endangerment and extinction, exhibiting low flexibility and specialized behavior (Hairston et al. 1970, Pianka 1970, King 1980, Temple 1985, Rodda et al. 2002). Bell and Merton (2002) drew attention to difficulties some threatened birds face due to inherent evolutionary traits, many of which fall under the category of inflexible behavior, noting that situations in which critically endangered bird species occur may have made them inherently vulnerable. A disproportionately high number of threatened birds, about half, occur on islands where they evolved with little disturbance. This is particularly true for isolated islands far from land. Such species are less able to learn and, therefore, are ill-adapted to behaviorally deal with recently emerging, human-driven threats (Temple 1985, 1986; BirdLife International 2008). A behavioral characteristic that is typical of these island species is behavioral naïvety or tameness, mentioned above. Behavioral naïvety involves loss of defensive behaviors and adaptations required to survive incursions of exotic predators, such as carnivores and rodents. Most of these species evidently do not possess the genetic ability to learn and modify their behavior. This has led to decline or extinction of many species (e.g. birds like kakapo [Strigops habroptilus] and Stephens Island wren [Xenicus (Traversia) lyalli]) (Bell 1991, Gill & Moon 1999, Bell & Merton 2002, Tennyson & Martinson 2006). Conservation translocation for these species dictates stringent planning, including, for example, anti-predator training, predator control or constructing barriers to exclude predators (Bell & Merton 2002, Campbell-Hunt 2002, Moore et al. 2008).

## 8.3 WHAT ARE THE COMPONENTS OF PRBM

PRBM can be categorized according to the behavioral domains described in the conservation behavior framework: Foraging and vigilance, movement and space use, reproductive behavior and social organization. As mentioned earlier, the change over time in these behaviors will, in most cases, increase the fitness of the individual, and it is in the interest of the practitioner to hasten/shorten this process. Upon release, animals are expected to show increased levels of vigilance necessary to buffer the risks associated with an unfamiliar landscape. Foraging may be less efficient and intermittent due to apprehension (Kotler *et al.* 2002), or the absence of a cognitive map of the area, which dictates the need for exploration for resource exploitation (Berger-Tal *et al.* 2014). The extent of this exploration is limited by the efficiency of the exploitation – which initially is low. This, in turn, will be expressed in the animal's movement patterns at all scales: from the finest, single-step movement, to course scale movement patterns such as dispersal and home range establishment. As knowledge accumulates, a shift will occur in the exploration/exploitation relationship. In general terms, and depending on the species, movement patterns of animals are expected to shift from cautious local movements necessary to identify nearby resources that will enable exploration, to increased exploration, and finally settling down into regular movement patterns within a well-defined home range, with local foraging and inter- and intra-patch movement including repeated returns to preferred locations (Berger-Tal & Saltz 2014). Social structure and landscape attributes may interact with the aforementioned patterns, depending on whether there are resident conspecifics at the site of release (Dolev et al. 2002) and the stress induced by the process on the cohesiveness of the group released.

## 8.4   BEHAVIORAL ISSUES AT SUCCESSIVE STAGES OF THE TRANSLOCATION PROCESS

The restoration of endangered populations involves a series of stages: know your species, diagnose causes of population decline and test remedial action, intensive management (maximizing productivity and survival of each individual if population is at critically low level to increase numbers and genetic diversity as soon as possible, including translocation), population management (species not at such critically low levels that nevertheless require restoration management, including translocation) and, finally, monitoring during and after such restoration efforts (Jones 2004). Further treatment of these stages is given specifically for translocations by the IUCN (2012): deciding when translocation is an acceptable option, planning a translocation, feasibility and design, risk assessment, release and implementation, monitoring and continuing management, and dissemination of information. In terms of behavior-based management I focus on three phases where behavioral considerations are most vital: (1) The feasibility study, (2) The planning and preparation phase, and

(3) The transfer and release. Although post-release monitoring often includes behavioral studies, these studies are oriented toward assessing the success of the translocation and, therefore, tend more to fall under the topic of behavioral indicators (see Part IV of this book).

### 8.4.1 Feasibility

Assessing the feasibility of a translocation requires knowledge of the focal species' biology, and in the current context, its behavior patterns in relation to a possible translocation, as well as its conservation status, range, abundance and population status. Behavioral aspects might include, for example, behavioral flexibility and learning abilities, social relationships, foraging needs and abilities, vulnerability to predators or competitors, and implications of these behavioral characteristics before, during and after envisaged translocation. While I recognize the importance of such an extensive assessment, sometimes the urgency of intervention (e.g. Butler & Merton 1992) precludes an extensive preliminary study (Bell & Merton 2002). The conservation manager of endangered species faces a dilemma caused by the need for knowledge on one hand, and the urgency of attending to the species, on the other. Nevertheless, behavioral considerations, even if based on limited knowledge available on the ontogeny, behavior and ecology of a species (or its relatives), can contribute significantly to probability of success.

From a behavioral perspective, the feasibility study must consider two key issues: (1) Where can animals be obtained, i.e. what is the source of animals? (2) Where are they going, namely the attributes of the release area. These considerations also follow through the preparation and release phases. Risk analysis is also an important component of the feasibility study and should include behavioral components. Integrating behavior into risk analyses is a complex task and is addressed in Chapter 9.

#### 8.4.1.1 The source of the animals

The main issue here is whether the animals come from a captive-bred stock or from the wild. If the animals targeted for translocation will be of captive stock, many behavioral traits necessary for survival in the wild may be missing, although the extent of this depends on the species concerned. In general, predators raised in captivity are likely to have deficiencies in hunting techniques and will lack any hunting experience. Predators affected most by this lack of experience will be those that tend to be flexible in their behavior (e.g. mammalian predators) and are typified by a strong parental affect in their development (i.e. when offspring learn to hunt from the parents). Captive predators released in the wild must become successful hunters quickly or they will starve. Prey species raised in captivity tend to lose anti-predator behavioral attributes, and, similar to predators, must acquire these quickly or they will be subject to predation. Thus, a feasibility study must take into account the need and costs of pre-release training. Studies have shown that, aside from training, the attributes of the captive environment affect the speed at which learning takes place in the wild (Zidon *et al.* 2009).

Using multiple sources of animals may have behavioral consequences. For example, in New Zealand, translocated North Island kokako (*Callaeas wilsoni*) transferred to Kapiti Island were from multiple sources. Adults from each source had distinctive dialects and the translocated female kokako preferentially chose males whose repertoire was typical of the acoustic environment they experienced before translocation (Rowe & Bell 2007). Song analysis and pair formation of kokako born on Kapiti Island indicated that such assortative mating was a temporary phenomenon in the years following translocation, which did not continue following juvenile recruitment, but points to the possibility of social problems developing when the translocated population comes from different sources.

Allee effects (Stephens & Sutherland 1999) should also be taken into account when the source of the animals is considered. The size of the group released must be sufficient to produce a positive growth rate. Such Allee effects are expected in highly social animals where cooperation between group members contributes to their fitness. The feasibility study should therefore consider the possibility of obtaining the necessary number of animals for release. If a breeding nucleus is planned, then the breeding facility should be large enough to house a population that can produce the necessary nucleus for release without being degraded (Saltz 1998). If the animals will be captured out of a wild population then the feasibility study should address the feasibility of trapping the necessary number of animals out of the wild within an acceptable time frame, the ability of the source population to sustain a removal of such a number, the need and costs of a large enough habituation enclosure, and of maintaining the release group within this enclosure until a cohesive group is formed. The feasibility of obtaining the age and sex structure desired for the released group must be assessed as well.

Social behavior must also be considered. This is especially true for highly social species with a strong hierarchical group structure. In such species, the severance of group bonds can potentially impact the fitness of the entire group or specific members of it. Although some studies suggest that there is little effect of social bonds on post-release survival (Bly-Honness *et al.* 2004), others found that strong social bonds enhance survival (Silk *et al.* 2009, 2010) and reproductive success (Cameron *et al.* 2009) of individuals in highly social species. Consequently translocations should attempt to release "organic" groups (Kleiman 1989, 1994; Rohan *et al.* 2002). If a translocation from a wild population is considered, the feasibility study must address the feasibility of actually being able to trap an entire social group or calculate the costs of forming one by an extended stay within an adequate habituation enclosure at the site of release, or modify the number of animals needed for release to overcome the reduced survival and reproductive success (based on risk analysis).

### 8.4.1.2   Attributes of the release area

Does the translocation involve reintroduction, reinforcement or a conservation introduction? Each one of these entails different behavioral considerations.

In the case of reintroduction, the feasibility study must evaluate the adequateness of the target area in terms of the behavior of the animal. Specifically, changes in community structure (e.g. alien predators the species is not familiar with) and anthropogenic modifications that may act as disturbing factors or ecological traps must be considered. Due to changes in those former habitats, or the arrival of invasive competitors or predators, the species considered for translocation may no longer be able to survive there. Many threatened species survive in relictual habitats that represent a fragment of their former distribution and these are not necessarily optimal. In the case of the flightless takahe (*Porphyrio hochstetteri*), for example, can one say that its behavioral repertoires and ecological requirements are best suited to its current relictual subalpine habitat in the Fiordland region of New Zealand? This issue was widely debated (Mills *et al.* 1982, 1984, 1988; Beauchamp & Worthy 1988; Atkinson & Millener 1991; Gray & Craig 1991; Clout & Craig 1995). The species' success when translocated to lowland habits, such as islands, suggests that takahe may indeed be better suited to lowland habitats rather than the subalpine habitat in which they survived (Jamieson & Ryan 2001), despite earlier skepticism (Mills *et al.* 1982).

With reinforcement (supplementation), the behavioral interactions between the native populations and the newly released animals must be considered as well. In some cases, the newly released individuals will benefit from the native individuals by learning from them (Dolev *et al.* 2002). In other cases, such as territorial species, the newly released animals may face intra-specific aggression with local individuals having a home-court advantage (Metcalfe *et al.* 2003). In highly social species, the release of a new group within the range of a resident group may destabilize both the groups, reducing survival and fertility.

There are several examples where problems might have been avoided if a behavioral focus was included in the feasibility study. Predatory behavior of Laysan finches (*Telespiza cantans*) was not anticipated when they were introduced to Midway Island (Bailey 1956, Halliday 1978), for while they initially survived, this was at the expense of some resident birds, whose eggs they ate, and the predatory impact of resident New Zealand falcons (*Falco novaeseelandiae*) appears to have been a factor in the limited success of translocating critically endangered orange-fronted parakeets (*Cyanoramphus malherbi*) to Maud Island (BirdLife International 2013).

### 8.4.2 Planning and preparation

The two components of translocation identified here – planning and preparation – are complementary and both contribute to the translocation program, once considered feasible. Planning sets out the overall formulation, evaluation and selection of considerations relating to a proposed translocation, and the behavior of the species targeted for translocation should be an important focus. Specifically, the behavior of the translocated animals may be impacted by these plans and consequently affect the PRBM process and the outcome of the translocation. Thus, particular emphasis must be placed on behavioral issues during this phase.

The preparation phase should focus on releasing the animals with behavioral attributes and experience that will shorten the process of PRBM. It should include the behavioral preparation of animals for translocation (this applies mostly to captive-bred animals and includes both considerations regarding the captive environment and active training. Here I focus on the former as the latter is covered in Chapter 10). It also includes the decision on the group composition that is to be released, selecting the habitat and target site/s for translocation, considering what site management might be necessary (e.g. to manage risks or to enhance the new environment), what provisions need to be made to transport animals there (e.g. to minimize risk and distress) and how best to release animals (e.g. soft or hard release?).

#### 8.4.2.1 Preparing animals for translocation

Whether wild or captive stocks, or a combination of both, are used in a conservation translocation depends on the species under consideration and availability. There are costs and benefits to each approach, but with respect to behavior, wild stock is generally preferable. Although wild animals may be exposed to anthropogenic activity and disturbances, wild stock will not have experienced the intense behavioral modification that typifies domestically bred and raised animals (Price 1999). Typically, translocations from the wild require far fewer considerations and complications than when captive stock is involved. Short (2009) noted that use of captive breeding is often seen as a high-risk strategy of last resort. Key issues that are perceived of higher risk are disease, and behavioral or genetic modification (Ehrlich & Ehrlich 1981, Chivers 1991, Dodd & Seigel 1991, Lindburg 1992, Viggers et al. 1993, Snyder et al. 1996, Sigg 2006). Seddon et al. (2007) identified poor health, individuals lacking fearfulness and no opportunity to learn key behaviors, such as predator recognition, as the main problems with captive-reared animals. So while there have been successes, many captive-breeding-for-translocation attempts have failed due to behavioral deficiencies of released animals, often because captive animals lack natural parental care or other environmental influences during critical learning periods (Shumway 1999). Despite problems that might arise through captive rearing and maintenance, release of captive stock may be the only resort, or may offer advantages including captive propagation or rearing of large numbers of animals, such as in artificial incubation programs of reptiles (e.g. in the tuatara, *Sphenodon punctatus*; Keall et al. 2010).

The species' behavioral characteristics have a strong bearing on the extent of behavioral modification in captivity (Price 1999). Species that characteristically have strong parental behavioral effects, such as teaching the offspring to hunt (e.g. large predatory social mammals as opposed to herpetofauna), will exhibit pronounced post-release key behavioral deficiencies. In the wild, animals may need to learn a range of behaviors that are difficult to replicate in captive settings, such as foraging or prey capture skills, selecting secure roosting or breeding sites, learning predator avoidance, and various intraspecific relationships and traditions (Conway 1980, Tudge 1992). Captive breeding programs can inadvertently alter behaviors of animals, including interfering with normal patterns of mate selection, creating inappropriate social conditions, absence of anti-predator behavior and conditioning to humans. The loss or alteration of behaviors such as these can have marked effects on released animals.

In most captive rearing settings anthropogenic activity provides a buffer between the animal and its environment. Consequently, captive rearing forms a domestic behavioral phenotype. This experience includes the presence or absence of key stimuli, changes in intraspecific aggressive interactions and anthropogenic presence. Typically, captive rearing reduces the sensitivity of animals to environmental changes and novel stimuli.

Consequently, captive rearing often results in animals ill-prepared for release to the wild.

The ability of captive-raised wild animals to survive and reproduce in nature may depend on how much their natural behavior has changed and become inflexible. The behavioral flexibility of an individual animal depends, amongst other things, on the environment in which it was raised (Fagen 1982). Many mammals raised in a rich environment that encourages play show greater flexibility as adults. A richer environment is also less stressful (Carlstead & Shepherdson 2000). On the other hand, if the stimuli are excessive and consist mostly of elements related to anthropogenic presence, habituation and a general tendency to ignore novel situations may result. Thus, reintroduced Persian fallow deer (*Dama mesopotamica*) reared in a public zoo had shorter flush and flight distances, inappropriate flight modes (walking as opposed to running), and tended to bed in open areas, as compared with individuals originating from a captive breeding facility with only limited access to the public (Zidon *et al.* 2009). The individuals from the zoo also exhibited slower learning curves (PRBMs) for these behaviors than those originating from the captive-breeding facility, and, interestingly, also exhibited higher levels of stress (as measured by fecal cortisol) in the wild. Such behavioral deficiencies and the resulting stress may impact post-release reproductive success and survival (Saltz & Rubenstein 1995, Bar-David *et al.* 2005, Zidon 2009).

Imprinting is a rapid learning process that occurs early in the life of many animals and establishes behavioral patterns, including recognition of and attraction to conspecifics (commonly a parent) by attempting to stay near it (Abercrombie *et al.* 1979, Lorenz 1981). Imprinting can be a problem during captive rearing if a young animal mal-imprints on anything other than its own species, such as a human. In cases where bird eggs and/or chicks may be collected from the wild and then incubated/reared or fostered in captivity or in the wild in association with management and translocation (Bell & Merton 2002), care is needed to avoid their mal-imprinting on humans they come into contact with (Lorenz 1981). In New Zealand, the problem of mal-imprinting was evident in young black robins (*Petroica [Miro]traversi*) that mal-imprinted on and hybridized with their Chatham Island tomtit (*Petroica macrocephala chathamensis*) foster parents (Butler & Merton 1992). Similarly, when a captive-reared male kakapo ("Sirocco") mal-imprinted on humans, it developed unusual "muttering" vocalizations, having probably learnt these calls when captive-reared as a young bird, hearing human voices from a nearby hut (Bell *et al.* 2013).

To avoid the deleterious effects of mal-imprinting, young black robins that were mal-imprinted on Chatham Island tomtit foster parents were translocated to tomtit-free islands (Butler & Merton 1992). In the recovery of some critically endangered birds, such as the Mauritius parakeet (*Psittacula eques*) (Greenwood 1996, Bell & Merton 2002), eggs and/or chicks were collected from the wild and then incubated/reared in captivity, with care being taken to use models of adults to avoid imprinting on humans. Such techniques proved valuable for birds such as takahe, kakapo and black stilt (*Himantopus novaezelandiae*) in New Zealand as well as, for instance, Mauritius kestrel (*Falco punctatus*) and whooping crane. To avoid mal-imprinting on humans, captive-reared takahe chicks have been routinely hand-fed using takahe-head puppets with internal speakers playing adult feeding calls (Eason & Williams 2001). Steps have also now been taken to avoid captive-reared young kakapo mal-imprinting on humans (Balance 2010).

As part of their conservation and rehabilitation, some captive-reared individuals soon to be released into the wild are provisioned with food at the intended release site (termed "hacking" in falconry) while they gradually become independent, such as captive breeding and release programs for the peregrine (*Falco peregrinus*) in North America (Enderson *et al.* 1998). Another technique used in captive-rearing-for-later-release management is "head-starting". Here, eggs or young collected from the wild are given a "head-start" by rearing them in a captive environment free of risks that are in the wild (e.g. predation). Head-starting has been successfully undertaken in a range of species, including the rowi (*Apteryx rowi*; Abbott 2014), marine turtles (Pritchard 1979; Eckert *et al.* 1999), tuatara (Nelson *et al.* 2002, Keall *et al.* 2010) and Caribbean rock iguanas (genus *Cyclura*; Alberts 2007). However, head-starting in some species, such as Kemp's ridley sea turtle (*Lepidochelys kempi*), is not necessarily effective (e.g. Woody 1990).

The extent to which the gene pool of the population has been altered during the domestication process may also impact the flexibility in behavioral development. Inbreeding and genetic bottlenecks (especially if the source of animals is captive breeding – Bowker-Wright *et al.* 2012) can have behavioral implications affecting viability (e.g. reduced diversity in vocal behavior; Corfield *et al.* 2008, Ramstad *et al.* 2010, Digby *et al.* 2013). Such genetic considerations and how to deal with them in captivity have been addressed elsewhere (e.g. Ballou 1984).

To summarize, captive conditions during the preparation phase (to differ from focused training) are important. During this period managers should focus on minimizing the loss of important behaviors necessary to maintain fitness in the wild. Specifically, animals should preferably be provided with an environment that is rich in stimulations resembling as much as possible the ones they would encounter in the wild. Care should be taken to ensure that what is novelty in the wild will remain as such in captivity so animals will not become habituated. This translates into minimizing human contact, minimizing exposure to human-related noises (traffic etc.) and minimizing exposure to domestic animals that resemble wild predators but are unable to approach the captive animals (e.g. dogs or cats walking along the fence of the enclosure).

8.4.2.2 Deciding on the size and composition of the group to be released
The structure of the released group of animals (including size, age and sex composition) is a crucial component in the viability of a translocation. Group size, age and sex composition are the basic elements that affect demographic stochasticity and are the basis of any population growth model used for risk assessment. Such models are also used to assess the impacts of stochastic processes on a given group structure in translocations and, therefore, the size and group structure necessary to achieve a desired probability of success (Saltz 1996). The size, age and sex structure of a released group also have behavioral ramifications, and their impacts are a function of how they interact with social structure.

In terms of size, as a rule, larger releases are expected to increase the probability of translocation success (Germano & Bishop 2009) by reducing the impacts of stochasticity. However, even large releases result in small populations to start with, and the small population paradigm may also have a deterministic element, namely the demographic Allee effect. Armstrong and Wittmer (2011) pointed out that Allee effects can potentially lead to reintroduction failure despite habitat quality being sufficient to allow long-term persistence if the population survived the establishment phase. The components of an Allee effect are often behavioral (Stephens & Sutherland 1999) and are potentially more pronounced in highly social species. Knowledge of the diverse behavioral mechanisms that drive demographic Allee effects must be assessed before animals are translocated and should be used as a basis for setting targets for the size of the nucleus to be released. Regretfully, and unavoidably, the number of animals translocated is often dictated by availability. A good example is the case of the translocation program for the rarest kiwi in New Zealand – the rowi. Abbott et al. (2013) report on effects of early rearing experience and optimal release group size. Rowi are flightless ratites that form monogamous, highly territorial pairs with extended periods of parent–offspring association. Restocking the sole remaining rowi population involved rearing chicks on predator-free islands isolated from adult conspecifics. To reflect adult social organization, releases into the wild population traditionally took place in pairs or small groups. Abbott *et al.* (2013) hypothesized that as a result of behavioral mechanisms induced by pre-release experience, individuals in larger groups would have a higher survival rate than those in small release groups. They tested this experimentally by manipulating release group size over three years. Modeling revealed that of all variables tested, group size was the only factor with significant influence on post-release survival, with survival of individuals in small groups being significantly lower than that of individuals released in large groups. They suggested that social attraction and increased conspecific tolerance resulting from an individual's rearing environment were reasons for this. These findings induced a change in release protocols and triggered further research into behavioral plasticity and long-term effects of rearing conditions in conservation management (Abbott *et al.* 2013, Abbott 2014).

By contrast, when solitary territorial animals are translocated, a simultaneous release of a large group at the same site may result in elevated aggression and reduced survival (Linklater & Swaisgood 2008, Linklater *et al.* 2011). Successive releases of small groups are recommended at a rate that will allow territorial establishment of one group before the other is released. This in turn will result in reduced aggression and will emulate the range expansion of a growing wild population (Nolet & Rosell 1994).

Age and sex structure are also behaviorally important, especially in social species. Specifically, the interaction between behavioral traits, age, sex and social structure can impact effective population size (Anthony & Blumstein 2000). Possible traits include reproductive suppression, sexually selected infanticide, mechanisms of mate choice, mating systems, social plasticity, dispersal, migration, conspecific attraction and reproductive behaviors. For example, in species where males have access to fixed-membership female groups (either by harem or territorial behavior), a small release will result in a single male controlling the female group for an extended period of time, significantly reducing effective population size (Saltz *et al.* 2000). A release of excess males is unlikely to alleviate this problem and will only drive increased levels of aggression and destabilization. The solution to this problem is probably the repeated releases of existing stable groups in different areas.

Another aspect of age and sex structure is dominance relationships. In social groups with dominance hierarchy, when dominance relationships are not well established aggressive interactions may result. Specifically, aggression is expected between closely ranked individuals (Heitor & Vicente 2010) and if the position of the alpha male or female is not well established or can be challenged by equally powerful individuals, the group may destabilize, reducing fertility and even resulting in a split followed by Allee effects. In the reintroduction of the Arabian oryx (*Oryx leucoryx*) in Israel, social groups are isolated before release to form a stable social structure that typically consists of one older male and female with the rest of the herd being sub- and young adults (D. Saltz, pers. comm.).

Age and sex play an important role in the dispersal behavior of animals (Johnson & Gaines 1990), and dispersal is often a factor in the failure of translocations (Le Gouar *et al.* 2011). Dispersal is often sex biased, but the direction of the bias changes with species. Under natural conditions it is often the sub-adults that show a tendency to disperse, either for inbreeding avoidance, due to local resource completion, or for other reasons, or other reasons. Whatever the reason, in many species, sub-adults have an innate tendency to disperse while adults exhibit strong fidelity to their established home ranges and will abandon this range only under extreme conditions. Alternatively, animals released in an unknown landscape may attempt to home (i.e. to disperse back to their previous home range; Dickens *et al.* 2009). Given the above, this tendency may be higher in adult animals that already have an established home range. This is one of the reasons that soft releases are favored (see next section). The key point is that dispersal behavior in reintroduced animals can be the result of either an innate behavioral tendency to disperse or a (behavioral) drive to return to the previous range, and these, in turn, are affected by the age and gender of the animal, the release protocol and conditions at the release site.

### 8.4.2.3 Selecting the habitat and target site/s for translocation

In general, the consideration of the choice of an animal's habitat – the place with a particular kind of environment inhabited by that organism (Abercrombie *et al.* 1979) – has led to the development of much theory on the behavioral mechanisms involved in animal habitat selection. Cody (1985) noted that habitat selection had been central in avian behavioral ecology for several decades, and that substantial work had also been done on habitat selection in other groups, e.g. Wecker's deermouse studies (1964) and work on *Anolis* lizards by Liester *et al.* (1975). As conservation managers need to select suitable habitats and target site/s for translocation, it follows that habitat selection theory could, in principle, be applied to conservation and management of wildlife (Morris 2003). Translocated animals often do not behave according to "rules," however (Buechner 1989), and may be attracted to dispersal sinks (low-quality habitats).

A lack of resources may also drive long-distance dispersal of animals in different directions (Negro et al. 1997), increasing the probability of failure. This suggests that the site targeted for release should be carefully selected, providing the best available habitat (in terms of resources, sink patches, and other risks) within the former range of the species. A preferred release area in terms of the behavior of the animal is an area that can promote quick learning regarding the whereabouts of resources and shelter. One might consider preferable landscapes within the former range as those that have a rich representation of different types of habitat that provide good refuge and easy access to a variety of resources, and fewer nuisance disturbances (e.g. humans). Artificial structures already in the intended site may be potential hazards to species not familiar with them – e.g. power lines, wind turbines and other physical structures have posed threats to critically endangered bird species, especially to relatively naïve individuals used in captive-release experiments, e.g. California condor and whooping crane (Bell & Merton 2002). Habitat enhancement may also carry risks, as unwanted species may also be attracted – e.g. rock-piles constructed to enhance herpetofauna habitats may also attract unwanted predators or competitors (Bell et al. 2010, Karst 2013).

The habitat should support and channel, as much as possible, the PRBM, while reducing levels of stress and risk. In translocations of prey species some form of predator control at the site may be essential, not only to increase survival but also to reduce apprehension so that animals can learn faster (Harthoorn 1962). Alternatively, predators should be released at times of peak prey abundance, such as in the Canada lynx (Lynx canadensis) reintroduction in Colorado (Devineau et al. 2011). Other timing issues, such as avoiding the breeding season, should also be considered. Training of animals and provision of good refuge near the release site are possible, but require a good understanding of the predator–prey relationship (Esque et al. 2010) if they are to be beneficial. Regarding refuge, breeding or foraging behaviors, many birds and mammals, as well as reptiles (e.g. tuatara) and insects (e.g. the tree weta Hemideina crassidens), may benefit from construction of artificial retreat or nest sites in their new habitats, as well as provision of supplementary foods, or the planting of food trees, if such resources are limited in the intended translocation site.

Released animals prefer to settle in familiar types of habitat, and areas where learning the essentials is easy and is less risky. Such habitat selection may cause animals to reject novel areas lacking cues similar to those in their habitat of origin. A preference by released animals for familiar cues may encourage them to seek out inappropriate, low-quality habitats following release at a new location, invoking the concept of the ecological trap. Ecological traps reduce the fitness of individuals to the point where the average population growth rate may be negative. Although this phenomenon has received comprehensive theoretical treatment in the recent literature, the corollary of the ecological trap, i.e. when animals choose to avoid good-quality habitats (termed "perceptual trap"; Patten & Kelly 2010) is discussed less often. This failure to recognize high-quality sites may generate an Allee effect (of sorts), which would subside only when densities in the poorer habitat are high enough to force individuals into the better habitats. Until this threshold is reached, the probability of failure of the reintroduction due to stochastic factors will be high. If the growth rate in the poorer habitat is negative (a strong Allee effect) the reintroduction is destined to fail. Improved understanding of factors determining the colonization of high-quality sites could aid conservationists in mitigating the damaging effects of maladaptive habitat selection (Gilroy & Sutherland 2007). In field experiments with the lesser prairie-chicken (*Tympanuchus pallidicinctus*) in New Mexico, Patten and Kelly (2010) found empirical support for the concept, and cite several other potential examples suggesting that these perceptual traps are perhaps more prevalent than has been appreciated. They also point out that while an ecological trap may be negated by improving habitat quality, biologists will be hard pressed to negate a perceptual trap, which will require determining which cues an animal uses to select a habitat and then devising a means of managing those cues so that an animal is lured to the better habitat.

Other insights from habitat selection theory may help conservation managers encourage released animals to settle in appropriate habitats. Stamps and Swaisgood (2007) considered situations in which free-living dispersers prefer new habitats that contain stimuli comparable to those in their natal habitat, a phenomenon called natal habitat preference induction (NHPI). Theory predicts NHPI when dispersers experienced favorable conditions in their natal habitat, and have difficulty estimating the quality of unfamiliar habitats. They stated that NHPI is especially likely to occur when performance in a given habitat is enhanced if an animal developed in that same habitat type. Animals exhibiting NHPI are expected to rely on conspicuous cues that can be quickly and easily detected during search, and to prefer new habitats possessing cues that match those encountered in their natal habitat. A major obstacle to successful relocations is that newly released animals may reject habitat near the release site and rapidly travel long distances away before settling (Stamps & Swaisgood 2007). An NHPI perspective argues that long-distance movements away from release sites occur because releasees prefer to settle in familiar types of habitat, and reject novel areas lacking cues similar to those in their habitat of origin. Stamps and Swaisgood (2007) note that if tests for familiar cues in captivity indicate that the members of a species respond positively to particular stimuli, providing these stimuli in captivity and at the destination site might increase the chances that captive-born animals would "feel at home" in the otherwise-unfamiliar location.

### 8.4.3 Release

This phase involves three stages: The collection and transfer of animals from the source site (the wild, or in captivity) to the release site, habituation at the release site (not always implemented) and the release into the wild. A wide diversity of animals have been translocated for reinforcement and/or reintroduction as part of their conservation management, including insects, fishes, amphibians, reptiles, birds and mammals, so that a varied range of transportation techniques have been devised. The key goal is reducing stress (including injury), or more specifically distress (see Linklater 2010, Linklater *et al.* 2010). There is substantial material available on how to transport wildlife (e.g. Dmytryk 2012) so I will not expand anymore on this issue.

The IUCN guidelines for translocation (2012) recommend that animals become familiarized with their new environment using a soft release (but see Rohan *et al.* 2002 and Hardman & Moro 2006). This is most often done in some sort of an (habituation) enclosure. The role of such an enclosure is strictly behavioral. The enclosure has two main purposes: (1) to familiarize the animals with the habitat and the surroundings in a safe predator-free setting and (2) to prevent homing attempts by the animals by fencing them in at the site of release until they develop an affinity to the location.

As mentioned earlier, the ability of a translocated individual to make the correct decisions after its release depends, in part, on its current knowledge. In this respect the enclosure enhances the current knowledge by providing a learning experience regarding the types of resources and the structure of the landscape in the vicinity of the enclosure. Thus, habituation enclosures should include within them, as best they can, a good representation of available resources and habitat structure. The location and type of fence, cage or aviary should enable the animals to view the surroundings so when they are released the landscape is not entirely novel. Finally, the enclosure should be large enough to sustain the released group in good conditions and so intraspecific aggression is minimized (Cassinello & Pieters 2000). By doing so, the release is expected to be far less stressful for the animal and the process of PRBM will be shorter, because the animal has familiarized itself with part of the new landscape.

As mentioned above, dispersal is often a factor in the failure of reintroductions (Le Gouar et al. 2011), and in many cases this so-called dispersal may actually be a homing response by the animals. To prevent this homing, the animals must be re-trained to recognize the release area as "home" (Bright & Morris 1994), a process termed "anchoring." The effectiveness of the enclosure in preventing the homing response is dependent on the amount of time the animals spend in the enclosure and the conditions the enclosure provides (Parker et al. 2008). Thus, the tenure in the enclosure is basically a process by which the familiarity and affinity to the current home range (enclosure) overrides the previous home range. While this is important, it does generate two problems: First, if the animals were originally from captive stock and had undergone pre-release training, part or all of this training may be lost unless it is refreshed in the enclosure. Second, the longer the stay in the enclosure, the stronger the affinity to it may be and the eventual exiting from the enclosure may become stressful and animals may restrict their activity to the vicinity of the enclosure, becoming more predictable in their movements and, thus, easier targets for predators.

Given the problems noted above, a key question regarding habituation enclosures is what is the optimal length of stay of the animals in them? Devineau et al. (2011) addressed this question for the Canada lynx reintroduction in Colorado. They found that survival increased with time in the release enclosure, and reached an asymptote at 5–6 weeks, after which there was no additional fitness gain. The optimal length in the release enclosure may also depend on the age of the animals. Adult female Persian fallow deer released after a three-month stay in a habituation enclosure exhibited very conservative movement, while sub-adults dispersed several km away (Dolev et al. 2002). Arabian oryx reintroduced in Israel appeared to show increased homing behavior if they remained less than 3 months in the habituation enclosure (D. Saltz, pers. comm.). An extreme example is that of a failed attempt to reintroduce ostriches (Struthio camelus) in Israel (Rinat 2007). The movements of these individuals after spending 3 months in a habituation enclosure were erratic during the first week following the release, after which they could not be tracked again. Presumably, ostrich, which evidently have limited learning abilities and may have most of their behavioral patterns established during critical period/s, were unable to comprehend a new landscape and attempted homing. The conclusion was the ostrich must be raised in an enclosure at the site of release from hatching, and only those hatched on site can be released.

Another method of anchoring is the provision of scattered feeding stations in the vicinity of the release site. This was done for several bird species in New Zealand, including: kaka (*Nestor meridionalis*), stitchbird (*Notiomystis cincta*), bellbird (*Anthornis melanura*) and red-crowned parakeet (*Cyanoramphus novaezelandiae*). Other foraging examples relate to the use of supplementary feeding regimes to enhance food supply in translocated populations, e.g. in two threatened New Zealand birds: kakapo and stitchbird (Clout *et al.* 2002, Armstrong *et al.* 1999, 2007). Additional forms of anchoring include, for example, use of live captive animals as decoys, or even realistic artificial decoys – a practice long used by duck-shooters! Auditory playback of calls is another means of behavioral anchoring, used with some success in encouraging return of translocated seabird chicks to sites around New Zealand (e.g. Miskelly & Taylor 2004, Bell *et al.* 2005), while realistic artificial decoys and calls have been used to attract the Australasian gannet (*Morus serrator*) there.

In Africa, black rhinoceros (*Diceros bicornis*) are variously sedentary or wide-ranging in their movements after release (Hamilton & King 1969) and differences between rhinos suggests that temperament may play a role in responses to the stress of novel environments (Linklater & Swaisgood 2008). Linklater *et al.* (2006) conducted an experimental trial on the effect of scent broadcasting on black rhinoceros post-release behavior. Rhino were captured and held in individual enclosures (bomas) for 31–61 days, fitted with horn-implant radio transmitters and released near-simultaneously over three days at different individual locations spaced throughout the Mun-ya-Wana Game Reserve, South Africa. This release strategy appeared to have benefits compared with translocation techniques that instead release rhinoceros from bomas at the same site but separately over many days. Before release the dung and urine soaked substrate from the bomas of seven rhinos (treatment group) was collected and spread within 2 km of their future release site. The control group, comprising eight rhinos, did not have scent broadcast about their release site. They hypothesized that the presence of a rhino's scent would reduce its post-release movement and predicted that rhinos would settle around or near their own scent. Contrary to their prediction, individuals in the treatment group moved significantly farther than those from the control group. Although they attempted to hold all factors constant between the two groups, retrospective hormone analyses showed that the control rhinos had higher levels of reproductive hormones, but despite this potential confound, the results contributed toward understanding black rhino behavioral ecology and improving translocation strategies. Rhino did not avoid the dung from other rhino, and appear to have even been attracted by it, indicating that conspecific scent might play a role in black rhinoceros movements after release and be a tool for managing home range establishment. They note that improving our understanding of this effect will depend on studies of black rhinoceros olfactory behavior, scent chemistry and further experimental trials of the scent broadcasting technique.

### 8.4.4  Post-release monitoring

IUCN (2012) note that monitoring behavior of translocated individuals can be a valuable, early indicator of translocation progress, but its value depends on comparative data from either comparable natural populations or the same individuals before removal from their source population. There are many reports of post-release monitoring covering a wide range of animals, including those that emphasize behavioral aspects. For example, Shier and Owings (2006, 2007) studied effects of social learning on predator training and post-release survival in juvenile black-tailed prairie dogs (*Cynomys ludovicianus*), while Bar-David *et al.* (2008) found that models based on space-use patterns of reintroduced species can be used as projection models in the context of landscape planning, using a case study of Persian fallow deer reintroduced in northern Israel. Competitive exclusion of one reintroduction by another can also occur. For example, Empson and Fastier (2013) report on transfers of New Zealand robin and New Zealand tomtit (*Petroica macrocephala*) into the Zealandia sanctuary. Although successful tomtit breeding was observed both within and outside the sanctuary, predation pressure was higher outside, while a progressive move of tomtit territories out of the sanctuary was thought to have been a response to increasing aggression from the expanding robin population inside. As well as interspecific competition, intraspecific competition operates. For example, two successive releases of the threatened Maud Island frog (*Leiopelma pakeka*) were monitored (Bell *et al.* 2004, Bell 2010), suggesting that chemical cues (Lee & Waldman 2002, Waldman & Bishop 2004) may be important as territorial signals, identifying individuals and possibly deterring new arrivals from settling (Trewenack *et al.* 2007). Using a continuum multi-species model framework describing dispersal and settling of transferred animals, Trenwenack *et al.* (2007) suggested that settling occurred at a constant rate, with deterrence from chemical signals probably playing a significant role.

## 8.5  MANAGEMENT IMPLICATIONS

The responsibility of the wildlife conservation manager is to maximize the success of a translocation, including shortening as much as possible the time necessary for the animals to become well acquainted with their new environment and learn to obtain resources while minimizing risk. Drawing on the Berger-Tal *et al.* (2011) conceptual framework and on other parts of this review chapter, implications for wildlife manager actions relating to conservation translocations are summarized as follows:

## General

(1) Recognize that every wildlife translocation is an experiment, and, whenever possible, translocations should be designed and evaluated to test hypotheses that will further improve our understanding and success through adaptive management.

(2) Assemble available information on the ecology and behavior of the focal species, and its distributional and numerical status, as important preliminaries to consideration of a translocation.

(3) Be aware of risks that may be impacting remnant populations, and minimize these impacts at the release site and source site where possible.

(4) Consider the interactions among behavior, age and ontogeny when designing translocation programs.

## Feasibility

(1) Determine the type of conservation translocation envisaged, reinforcement or reintroduction, and whether it is feasible – refer to the *IUCN Guidelines* (IUCN 2012) for full advice on the process and refine it based on the species' known or expected behavioral ecology and learning capabilities.

## Planning and preparation

(1) Determine *a priori* specific goals, overall ecological purpose, and inherent technical and biological limitations of a given translocation, and note that evaluation processes incorporate behavioral, experimental and modeling approaches.

(2) Develop a translocation plan that incorporates behavioral, demographic and genetic aspects of translocating the species, including selection of animals to transfer (individual behavioral traits), their possible behavioral preparation for translocation, selection of the release site, and behavioral monitoring and care of animals during and following release.

(3) Where possible, ensure that the source population can sustain harvesting for translocation, undertaking modeling as necessary including behavioral, demographic and genetic considerations (e.g. Allee effects).

(4) Behavioral modification, such as anti-predator training, may be applicable to both wild and captive-reared animals, but particularly the latter, and may involve classical conditioning procedures, e.g. where animals learn that model predators are predictors of aversive events.

(5) Be aware that problems such as emergence of inappropriate responses might arise during aversion training. Be prepared to modify protocols to eliminate or minimize these inappropriate responses.

(6) Be aware that head-starting programs offer opportunities for those species that can be mass-reared in safe conditions prior to release, but that any unwanted behavioral impacts of captive rearing must be mitigated.

### Release phase

(1) Ensure harmonious and demographically appropriate social grouping (e.g. by age, sex and dominance) during selection for translocation and during transportation to the release site.

(2) Ensure the selected animals are fit and healthy and have received necessary veterinary checks, to reduce stress (that may lead to maladaptive behaviors) and to minimize introduction of diseases and parasites into the release site.

(3) Consider premedication of animals (e.g. using sedatives) prior to transportation for translocation to reduce behavioral and physiological distress.

(4) Where possible, methods of individual identification of released animals should be utilized (e.g. ear tags, leg bands, radio collars), so that behavior, location, history and fate of individuals can be documented post-release.

(5) Try to minimize the learning period that the animal has to go through when released into a novel environment, as it is most vulnerable then.

(6) Minimize chances of detrimental impacts on focal species at the release site, in terms of immediate needs (e.g. shelter, nest-sites, den-sites, food supply) by providing necessary resources (e.g. nest-boxes, feeding stations, roosting sites), and over the longer term be mindful of possible cascading effects of translocation on other species in the community.

(7) Review and consider possible anchoring methods at the release site, including use of soft versus hard release protocols, habituation enclosures, holding during sensitive periods early in life, supplementary feeding stations and acoustic lures.

**Post-release monitoring**

(1) To learn from the translocation, monitoring post-release behavior of translocated individuals is a valuable, early indicator of the translocation progress, providing an opportunity to optimize chances of success and to ascertain the extent to which the animal's behavior may change. If the released animal is capable of learning, its behavior will change as the animal gains experience (post-release behavioral modification or "PRBM"). PRBM is an important indicator of success.

(2) Comparative behavioral data from comparable natural populations and from temporal trends in the released individuals can be of value in ascertaining the extent of the behavioral modifications.

## ACKNOWLEDGEMENTS

I wish to express my sincere thanks to David Saltz and Oded Berger-Tal for their patience and assistance in the preparation of this chapter and I am grateful to the late Don Merton, and to George Gibbs, Jen Moore, Nicky Nelson, Wayne Linklater, Andrew Digby and Rachael Abbott for further information and background material.

## REFERENCES

Abbott, R. 2014. Behavioural mechanisms affecting the success of translocations: an investigation using New Zealand's rarest ratite, the rowi. Unpublished Ph.D. Thesis, Victoria University of Wellington, NZ.

Abbott, R., Bell, B. and Nelson, N. 2013. Improving conservation management of New Zealand's rarest kiwi (Apteryx rowi): effects of early rearing experience and optimal release group size. Abstracts of Behaviour 2013: joint meeting of the 33rd International Ethological Conference (IEC) and the Association for the Study of Animal Behaviour (ASAB), 4–8 August, 2013, The Sage, Gateshead, Newcastle, UK.

Abercrombie, M., Hickman, C.J. and Johnson, M.L. 1979. The Penguin Dictionary of Biology. Penguin Books, UK.

Alberts, A.C. 2007. Behavioral considerations of headstarting as a conservation strategy for endangered Caribbean rock iguanas. Applied Animal Behavior Science, 102:380–391.

Anthony, L.L. and Blumstein, D.T. 2000. Integrating behavior into wildlife conservation: the multiple ways that behavior can reduce Ne. Biological Conservation, 95:303–315.

Armstrong, D.P., Castro, I., Alley, J.C., Feenstra, B. and Perrott, J.K. 1999. Mortality and behavior of hihi, an endangered New Zealand honeyeater, in the establishment phase following translocation. Biological Conservation, 89:329–339.

Armstrong D.P., Castro I. and Griffiths R. 2007. Using adaptive management to determine requirements of re-introduced populations: the case of the New Zealand hihi. *Journal of Applied Ecology*, **44**:953–962.

Armstrong, D.P and Wittmer, H.U. 2011. Incorporating Allee effects into rein-troduction strategies. *Ecological Research*, **26**:687–695.

Atkinson, I.A.E. and Millener, P.R. 1991. An ornithological glimpse into New Zealand's pre-human past. *Acta XX Congressus Internationalis Ornithologici*, **2**:129–192.

Bailey, A.M. 1956. *Birds of Midway and Laysan Islands. Museum Pictorial*, 12. Denver: Denver Museum of Natural History.

Balance, A. 2010. *Kakapo: Rescued from the Brink of Extinction*. Nelson: Craig Potton Publishing.

Ballou, J. D. 1984. Strategies for maintaining genetic diversity in captive populations through reproductive technology. *Zoo Biology*, **3**:311–323.

Bar-David, S., Saltz, D., Dayan, T., Perelberg, A. and Dolev, A. 2005. Demographic models and reality in reintroductions: the Persian fallow deer in Israel. *Conservation Biology*, **19**:131–138.

Bar-David, S., Saltz, D., Dayan, T. and Shkedy, Y. 2008. Using spatially expanding populations as a tool for evaluating landscape planning: the reintroduced Persian fallow deer as a case study. *Journal for Nature Conservation*, **16**:164–174.

Beauchamp, A.J. and Worthy, T.H. 1988. Decline in distribution of the Porphyrio (=Notornis) mantelli: a re-examination. *Journal of the Royal Society of New Zealand*, **18**:103–112.

Bell, B.D. 1991. Recent avifaunal changes and the history of ornithology in New Zealand. *Acta XX Congressus Internationalis Ornithologici*, **2**:193–230.

Bell, B.D. 2010. The threatened Leiopelmatid frogs of New Zealand: natural history integrates with conservation. *Herpetological Conservation and Biology*, **5**:515–528.

Bell, B.D., Bishop, P.J.and Germano, J.M. 2010. Lessons learned from a series of translocations of the archaic Hamilton's frog and Maud Island frog in central New Zealand. pp. 81–87 in Soorae, P. S. (ed.) *Global Re-introduction Perspectives: Additional Case-studies from Around the Globe*. IUCN/SSC Re-introduction Specialist Group, Abu Dhabi, UAE.

Bell, B.D., Carpenter, J.K., Dewhurst, P.L., Karst, T.M. and Browning, S. 2013. Unusual vocalisations from a male kakapo (*Strigops habroptilus*) imprinted on humans. *Notornis*, **60**:265–268.

Bell, B.D. and Merton, D.V. 2002. Critically endangered bird populations and their management. pp. 103–138 in K. Norris and D. J. Pain, (eds.) *Conserving Bird Biodiversity: General Principles and Their Application*. Cambridge: Cambridge University Press.

Bell, B.D., Pledger, S. and Dewhurst, P. 2004. The fate of a population of the endemic frog Leiopelma pakeka (*Anura: Leiopelmatidae*) translocated to restored habitat on Maud Island, New Zealand. *New Zealand Journal of Zoology*, **31**:123–131.

Bell, M., Bell, B.D. and Bell, E.A. 2005. Translocation of fluttering shearwater (*Puffinus gavia*) chicks to create a new colony. *Notornis*, **52**:11–15.

Berger-Tal, O., Polak, T., Oron, A. *et al.* 2011. Integrating animal behavior and conservation biology: a conceptual framework. *Behavioral Ecology*, 22:236–239.

Berger-Tal, O. and Avgar, T. 2012. The glass is half full: overestimating the quality of a novel environment is advantageous. *PLoS ONE*, 7:e34578.

Berger-Tal, O., Nathan, J., Meron, E. and Saltz, D. 2014. The exploration-exploitation dilemma: a multidisciplinary framework. *PLoS ONE*, 9:e95693.

Berger-Tal, O. and Saltz, D. 2014. Using the movement patterns of reintroduced animals to improve reintroduction success. *Current Zoology*, 60:515–526.

BirdLife International. 2008. *Many threatened birds are restricted to small islands.* Presented as part of the BirdLife State of the world's birds website. Available from: www.birdlife.org/datazone/sowb/casestudy/173. Accessed December 7, 2013.

BirdLife International. 2013. Species factsheet: Cyanoramphus malherbi. www.birdlife.org/datazone/speciesfactsheet.php?id=1477 Accessed September 11, 2013.

Bly-Honness K., Truett, J.C. and Long, D.H. 2004. Influence of social bonds on post-release survival of translocated black-tailed prairie dogs (*Cynomys ludivicianus*). *Ecological Restoration*, 22:204–209.

Bowker-Wright, G., Bell, B.D., Williams, M.J. and Ritchie, P. 2012. Captive breeding and release diminishes genetic diversity in brown teal, *Anas chlorotis*, an endangered New Zealand duck. *Wildfowl*, 62:174–187.

Bright, P. W. and Morris, P.A. 1994. Animal translocation for conservation: performance of dormice in relation to release methods, origin and season. *Journal of Applied Ecology*, 31:699–708.

Buechner, M. 1989. Are small-scale landscape features important factors for field studies of small mammal dispersal sinks? *Landscape Ecology*, 2:191–199.

Butler, D. and Merton, D.V. 1992. *The Black Robin – Saving the World's Most Endangered Bird.* Oxford: Oxford University Press.

Cameron, E. Z., Setsaasa, T.H. and Linklater, W.L. 2009. Social bonds between unrelated females increase reproductive success in feral horses. *Proceedings of the National Academy of Science USA*, 106:13850–13853.

Campbell-Hunt, D. 2002. *Developing a Sanctuary.* Wellington: Victoria Link Ltd.

Carlstead, K. and Shepherdson, D. 2000. Alleviating stress in zoo animals with environmental enrichment. pp. 337–349 in Moberg, G.P. and Mench, J.Y.A. (eds.). *The Biology of Animal Stress: Basic Principles and Implications for Animal Welfare.* Wallingford: CABI Publishing.

Cassinello, J. and Pieters, I. 2000. Multi-male captive groups of endangered dama gazelle: social rank, aggression, and enclosure effects. *Zoo Biology*, 19:121–129.

Chivers, D.J. 1991. Guidelines for re-introductions: procedures and problems. *Symposium Zoological Society London*, 62:89–99.

Clout, M.N. and Craig, J.L. 1995. The conservation of critically endangered flightless birds in New Zealand. *Ibis*, 137:S181–S190.

Clout, M.N., Elliott, G.P. and Robertson, B.C. 2002. Effects of supplementary feeding on the offspring sex ratio of kakapo: a dilemma for the conservation of a polygynous parrot. *Biological Conservation*, 107:13–18.

Cody, M.L. (ed.) 1985. *Habitat Selection in Birds*. Orlando: Academic Press Inc.

Conway, W.G. 1980. An overview of captive propagation. pp. 199–208. In: M.E. Soulé and Wilcox B.A. (eds.) *Conservation Biology: An Evolutionary-Ecological Perspective*. Sunderland: Sinauer Associates.

Corfield, J., Gillman, L. and Parsons, S. 2008. Vocalisations of the North Island brown kiwi (*Apteryx mantelli*). *Auk*, 125:326–335.

Devineau, O., Shenk, T.M., Doherty Jr., P.F., White, G.C. and Kahn, R.H. 2011. Assessing release protocols for Canada lynx reintroduction in Colorado. *Journal of Wildlife Management*, 75:623–630.

Dickens, M.J., Delehanty, D.J., Reed, J.M. and Romero, L.M. 2009. What happens to translocated game birds that "disappear"? *Animal Conservation*, 12:418–425.

Dickens M.J., Delehanty, D.J. and Romero, L.M. 2010. Stress: an inevitable component of animal translocation. *Biological Conservation*, 143:1329–1341.

Digby, A., Bell, B.D. and Teal, P.D. 2013. Vocal cooperation between the sexes in *little spotted kiwi Apteryx owenii*. *Ibis*, 155:229–245.

Dmytryk, R. 2012. Transporting wildlife. pp. 177–178 in *Wildlife Search and Rescue: A Guide for First Responders*. Chichester: John Wiley & Sons, Ltd.

Dodd, C.K. and Seigel, R.A. 1991. Relocation, repatriation, and translocation of amphibians and reptiles – are they conservation strategies that work. *Herpetologica*, 47:336–363.

Dolev, A., Saltz, D., Bar-David, S. and Yom-Tov, Y. 2002. The impact of repeated releases on the space-use patterns of reintroduced Persian fallow deer (*Dama dama mesopotamica*) in Israel. *Journal of Wildlife Management*, 66:737–746.

Eason, D.K. and Williams, M.J. 2001. Captive rearing: a management tool for the recovery of the endangered takahe. pp. 80–95 in Lee, W.G. and Jamieson, I.G. (eds.) *The Takahe: Fifty Years of Conservation Management and Research*. Dunedin: University of Otago Press.

Eckert, K.L., Bjorndal, K.A., Abreu-Grobois, F.A. and Donnelly, M. (eds.) 1999. *Research and Management Techniques for the Conservation of Sea Turtles*. IUCN/ SSC Marine Turtle Specialist Group Publication No. 4.

Ehrlich, P.R. and Ehrlich, A. 1981. *Extinction: The Causes and Consequences of the Disappearance of Species*. New York: Random House.

Empson, R. and Fastier, D. 2013. Translocations of North Island tomtits (*Petroica macrocephala toitoi*) and North Island robins (*P. longipes*) to Zealandia-Karori Sanctuary, an urban sanctuary. What have we learned? *Notornis*, 60:63–69.

Enderson, J.H., White, C.M. and Banasch, U. 1998. Captive breeding and releases of peregrines *Falco peregrinus* in North America. pp. 437–444 in Chancellor, R. D., Meyburg, B-U and Ferrero, J.J. (eds.). *Holarctic Birds of Prey*. ADENEX-WWGBP: Mérida & Berlin.

Esque, T.C., Nussear, K.E., K. Drake, K.K., *et al.* 2010. Effects of subsidized predators, resource variability, and human population density on desert tortoise populations in the Mojave Desert, USA. *Endangered Species Research*, 12:167–177.

Fagen, R. 1982. Evolutionary issues in development of behavioral flexibility. *Ethology*, 5:365–383.

Fischer, J. and Lindenmayer, D.B. 2000. An assessment of the published results of animal relocations. *Biological Conservation*, 96:1–11.

Germano, J.M. and Bishop, P.J. 2009. Suitability of amphibians and reptiles for translocation. *Conservation Biology*, 23:7–15.

Gibbs, G. 2006. *Ghosts of Gondwana: The History of Life in New Zealand*. Nelson: Craig Potton Publishing.

Gill, B. and Moon, G. 1999. *New Zealand's Unique Birds*. Auckland: Reed.

Gilroy, J.J. and Sutherland, W.J. 2007. Beyond ecological traps: perceptual errors and undervalued resources. *Trends in Ecology & Evolution*, 22:351–356.

Gray, R.D. and Craig, J.L. 1991. Theory really matters: hidden assumptions in the concept of "habitat requirements." *Acta XX Congressus Internationalis Ornithologici*, 4:2553–2560.

Greenwood, A.G. 1996. The echo responds – a partnership between conservation biology, aviculture and veterinary science. *Proceedings of the International Aviculturists Society*, January 1996, pp. 6–7. Orlando, Florida.

Griffin, A.S., Blumstein, D.T. and Evans, C.S. 2000. Training captive-bred or translocated animals to avoid predators. *Conservation Biology*, 14:1317–1326.

Griffith, B., Scott, J. M., Carpenter, J.W. and Reed, C. 1989. Translocation as a species conservation tool – status and strategy. *Science*, 245:477–480.

Griffiths, R.A. and Pavajeau, L. 2008. Captive breeding, reintroduction, and the conservation of amphibians. *Conservation Biology*, 22:852–861.

Hairston, N.G., Tinkle, D.W. and Wilbur, H.M. 1970. Natural selection and the parameters of population growth. *Journal of Wildlife Management*, 34:681–690.

Halliday, T. 1978. *Vanishing Birds – Their Natural History and Conservation*. New Zealand: Holt, Rinehart and Winston.

Hamilton, P. and King, J. 1969. The fate of black rhinoceros released in Nairobi National Park. *East African Wildlife Journal*, 7:73–83.

Hardman, B. and Moro, D. 2006. Optimising reintroduction success by delayed dispersal: is the release protocol important for hare-wallabies? *Biological Conservation*, 128:403–411.

Harrington, L.A., Moehrenschlager, A., Gelling, M. *et al.* 2013. Conflicting and complementary ethics of animal welfare considerations in reintroductions. *Conservation Biology*, 27:486–500.

Harthoorn, A.M. 1962. Translocation as a means of preserving wild animals. *Oryx*, 6:215–227.

Heitor, F. and Vicente, L. 2010. Dominance relationships and patterns of aggression in a bachelor group of Sorraia horses (*Equus caballus*). *Journal of Ethology*, 28:35–44.

IUCN. 1998. *IUCN Guidelines for Re-introductions*. IUCN, Gland, Switzerland & Cambridge, UK. 10 pp.

IUCN. 2012. *IUCN Guidelines for Reintroductions and Other Conservation Translocations*. IUCN, Gland, Switzerland

Jamieson, I.G. and Ryan, C.J. 2001. Closure of the debate over the merits of translocating takahe to predator-free islands. pp. 96–113 in Lee, W.G. and Jamieson, I.G. (eds.) *The Takahe: Fifty Years of Conservation Management and Research*. Dunedin: University of Otago Press.

Jones, C.G. 2004. Conservation management of endangered birds. In *Bird Ecology and Conservation: A Handbook of Techniques*. Sutherland, W.J., Newton, I. and Green R. (eds.) Oxford: Oxford University Press (reprinted twice in 2005).

Jones, C.G. and Merton, D.G. 2012. A tale of two islands: the rescue and recovery of endemic birds in New Zealand and Mauritius. Chapter 2 (pp. 33–72) in Ewen, J.G., Armstrong, D.P., Parker, K.A. and Seddon, P.J. (eds.) *Reintroduction Biology: Integrating Science and Management.* Wiley-Blackwell.

Johnson, M. L. and Gaines, M.S. 1990. Evolution of dispersal: theoretical models and empirical tests using birds and mammals. *Annual Review of Ecology and Systematics,* 21:449–480.

Karst, T.M. 2013. Mortality mitigation of a translocated rare New Zealand frog *Leiopelma pakeka.* Unpublished MSc. Thesis, Victoria University of Wellington, NZ.

Keall, S.N., Nelson, N.N. and Daugherty, C.H. 2010. Securing the future of threatened tuatara populations with artificial incubation. *Herpetological Conservation and Biology,* 5:555–562.

King C. M., Roberts, C.D., Bell, B.D. *et al.* 2009. Phylum Chordata: lancelets, fishes, amphibians, reptiles, birds, mammals. In Gordon, D.P. (ed.), *New Zealand Inventory of Biodiversity, Volume 1, Kingdom Animalia: Radiata, Lophotrochozoa, Deuterostomia,* pp. 431–551. Christchurch: Canterbury University Press.

King, W.B. 1980. Ecological basis of extinctions in birds. *Acta XVII Congressus Internationalis Ornithologici,* 905–911.

Kleiman, D.G. 1989. Reintroduction of captive mammals for conservation. *BioScience,* 39:152–161.

Kleiman, D.G. 1994. Criteria for reintroductions. Chapter 14 (pp. 287–303) in Olney, P.J.S, Mace, G.M. and Feistner, A.T.C (eds.), *Creative Conservation: Interactive Management of Wild and Captive Animals.* Chapman & Hall.

Kotler, B.P., Brown, J.S., Dall, S.R.X. *et al.* 2002. Foraging games between gerbils and their predators: temporal dynamics of resource depletion and apprehension in gerbils. *Evolutionary Ecology Research,* 4:495–518.

Lee, J. and Waldman, B. 2002. Communication by faecal chemo signals in an archaic frog, *Leiopelma hamiltoni. Copeia,* 2002:679–686.

Le Gouar, P., Mihoub, J.B. and Sarrazin, F. 2011. Dispersal and habitat selection: behavioural and spatial constraints for animal translocations. pp.138–162 in Ewen, J. G., Armstrong, D.P., Parker, K.A. and Seddon, P.J. (eds.). *Reintroduction Biology: Integrating Science and Management.* New York: John Wiley and Sons.

Liester, A.R., Gormn, G.C. and Arroyo, D.C. 1975. Habitat selection behavior of three species of *Anolis* lizards. *Ecology,* 56:220–225.

Lewis, J.C. 1995. Whooping crane (*Grus americanus*). pp. 153 in Poole, A. and Gill, F. (eds.), *The Birds of North America.* Philadelphia and Washington, DC: The Academy of Natural Sciences & the American Ornithologists Union.

Lewis, J.C. 1997. Alerting the birds. *Endangered Species Bulletin,* 22:22–23.

Lindburg, D.G. 1992. Are wildlife reintroductions worth the cost? *Zoo Biology,* 11:1–2.

Linklater W. 2010. Distress – an underutilised concept in conservation and missing from Busch and Hayward (2009). *Biological Conservation,* 143:1037–1038.

Linklater, W.L., Adcock, K., du Preez, P. *et al.* 2011. Guidelines for large herbivore translocation simplified: black rhinoceros case study. *Journal of Applied Ecology*, 48:493–502.

Linklater, W.L., Flamand, J., Rochat, Q. *et al.* 2006. Preliminary analyses of the free-release and scent-broadcasting strategies for black rhinoceros reintroduction. *Conservation Corporation Africa Ecological Journal* 7:26–34.

Linklater, W.L., Gedir, J.V., Law, P.R. *et al.* 2012. Translocations as experiments in the ecological resilience of an asocial mega-herbivore. *PLoS ONE* 7:e30664. doi:10.1371/journal.pone.0030664.

Linklater, W.L., MacDonald, E., Flamand, J. and Czekala, N. 2010. Declining and low fecal corticoids are associated with distress, not acclimation to stress, during the translocation of African rhinoceros. *Animal Conservation*, 13:104–111.

Linklater, W.L. and Swaisgood, R.R. 2008. Reserve size, conspecific density, and translocation success for black rhinoceros. *Journal of Wildlife Management*, 72:1059–1068.

Lorenz, K. 1981. *The Foundations of Ethology*. New York: Springer-Verlag.

Metcalfe, N. B., Valdimarsson, S. K. and Morgan, I. J. 2003. The relative roles of domestication, rearing environment, prior residence and body size in deciding territorial contests between hatchery and wild juvenile salmon. *Journal of Applied Ecology*, 40:535–544.

Mills, J.A., Lavers, R.B., Lee, W.G. and Garrick, A.S. 1982. *Management Recommendations for the Conservation of Takahe*. Internal Report, New Zealand Wildlife Service, Department of Internal Affairs, Wellington.

Mills, J.A., Lavers, R.B. and Lee, W.G. 1984. The takahe: a relict of the Pleistocene grassland avifauna of New Zealand. *New Zealand Journal of Ecology*, 7:57–70.

Mills, J.A., Lavers, R.B. and Lee, W.G. 1988. The post-Pleistocene decline of the takahe (*Notornis mantelli*): a reply. *Journal of the Royal Society of New Zealand*, 18:122–118.

Miskelly, C.M. and Taylor, G.A. 2004. Establishment of a colony of common diving petrels (*Pelecanoides urinatrix*) by chick transfers and acoustic attraction. *Emu*, 104:205–211.

Moore, J.A., Bell, B.D. and Linklater, W.L. 2008. The debate on behavior in conservation: New Zealand integrates theory with practice. *BioScience*, 58:454–459.

Morris, D.W. 2003. How can we apply theories of habitat selection to wildlife conservation and management? *Wildlife Research*, 30:303–319.

Negro, J. J., Hiraldo, F. and Donázar, J.A. 1997. Causes of natal dispersal in the lesser kestrel: inbreeding avoidance or resource competition? *Journal of Animal Ecology*, 66:640–648.

Nelson, N.J., Keall, S.N., Brown, D. and Daugherty, C.H. 2002. Establishing a new wild population of tuatara (*Sphenodon guntheri*). *Conservation Biology*, 16:887–894.

Nolet, B. A. and Rosell, F. 1994. Territoriality and time budgets in beavers during sequential settlement. *Canadian Journal of Zoology*, 72:1227–1237.

Ottewell, K., Dunlop, J., Thomas, N. *et al.* 2014. Evaluating success of transloca-tions in maintaining genetic diversity in a threatened mammal. *Biological Conservation*, 171:209–219.

Parker, I.D., Watts, D.E., Lopez, R.R. *et al.* 2008. Evaluation of the efficacy of Florida Key deer translocations. *Journal of Wildlife Management*, 72:1069–1075.

Patten, M.A. and Kelly, J.F. 2010. Habitat selection and the perceptual trap. *Ecological Applications*, 20:2148–2156.

Pianka, E.R. 1970. On r and K selection. *American Naturalist*, 104:592–597.

Price, E.O. 1999. Behavioral development in animals undergoing domestication. *Applied Animal Behaviour Science*, 65:245–271.

Pritchard, P.C.H. 1979. Head-starting and other conservation techniques for marine turtles *Cheloniidae* and *Dermochelyidae*. *International Zoo Yearbook*, 19:38–42.

Ramstad, K.M., Pfunder, M., Robertson, H.A. *et al.* 2010. Fourteen microsatellite loci cross-amplify in all five kiwi species (*Apteryx* spp.) and reveal extremely low genetic variation in little spotted Kiwi (*A. owenii*). *Conservation Genetics Resources*, 2:333–336.

Reed, C. and Merton. D. 1991. Behavioral manipulation of endangered New Zealand birds as an aid towards species recovery. *Acta XX Congressus Internationalis Ornithologici*, 4:2514–2522.

Rinat, Z. 2007. The bitter fate of ostriches in the wild. *Haaretz Daily Newspaper*. Dec.25, 2007.

Rodda, G.H., Fritts, T.H., Campbell III, E.W. et al. 2002. Practical concerns in the eradication of island snakes. Pp. 260–265 in Veitch, C. R. & Clout, M. N. (eds.), *Turning the Tide: the Eradication of Invasive Species*. IUCN SSC Invasive Species Specialist Group. IUCN, Gland, Switzerland and Cambridge, UK.

Rohan, C., Boulton, R. and Clarke, M. 2002. Translocation of the socially complex black-eared miner *Manorina Melanotis*: a trial using hard and soft release techniques. *Pacific Conservation Biology*, 8:223–234.

Rowe, S. J. and Bell, B.D. 2007. The influence of geographic variation in song dialect on post-translocation pair formation in North Is kokako (*Callaeas cinerea wilsoni*). *Notornis*, 54:28–37.

Saltz, D. 1996. Minimizing extinction probability due to demographic stochasti-city in a reintroduced herd of Persian fallow deer. *Biological Conservation*, 75:27–33.

Saltz, D. 1998. A long-term systematic approach to reintroductions: the Persian fallow deer and Arabian oryx in Israel. *Animal Conservation*, 1:245–252.

Saltz, D., Rowen, M. and Rubenstein, D.I. 2000. The effect of space-use patterns of reintroduced Asiatic wild ass on effective population size. *Conservation Biology*, 14:1852–1861.

Saltz, D., Rubenstein, D.I. 1995. Population dynamics of a reintroduced Asiatic Wild Ass (*Equus hemionus*) herd. *Ecological Applications*, 5:327–335.

Seddon, P.J., Armstong, D.P. and Maloney, R.F. 2007. Developing the science of reintroduction biology. *Conservation Biology*, 21:303–312.

Shier, D.M. and Owings, D.H. 2006. Effects of predator training on behavior and post-release survival of captive prairie dogs (*Cynomus ludovicianus*). *Biological Conservation*, 132:126–135.

Shier, D.M. and Owings, D.H. 2007. Effects of social learning on predator training and postrelease survival in juvenile black-tailed prairie dogs (*Cynomus ludovicianus*). *Animal Behaviour*, 73:567–577.

Short, J. 2009. The characteristics and success of vertebrate translocations within Australia. A final report to Department of Agriculture, Fisheries and Forestry. Wildlife Research and Management Pty. Ltd., Kalamunda, West Australia.

Short, J., Bradshaw, S.D., Giles, J.R., Prince, R.I.T. and Wilson, G.R. 1992. Reintroduction of macropods (Marsupialia: Macropodoidea) in Australia – a review. *Biological Conservation*, 62:189–204.

Shumway, C.A. 1999. A neglected science: applying behavior to aquatic conservation. *Environmental Biology of Fishes*, 55:183–201.

Sigg, D.P. 2006. Reduced genetic diversity and significant genetic differentiation after translocation: comparison of the remnant and translocated populations of bridled nailtail wallabies (*Onychogalea fraenata*). *Conservation Genetics*, 7:577–589.

Silk, J.B., Beehner, J.C., Bergman, T.J. *et al.* 2009. The benefits of social capital: close social bonds among female baboons enhance offspring survival. *Proceedings of the Royal Society B*, 276:3099–3104.

Silk, J. B., Beehner, J.C., Bergman, T.J. *et al.* 2010. Strong and consistent social-bonds enhance the longevity of female baboons. *Current Biology*, 20:1359–1361.

Snyder, N.F.R., Derrickson, S.R., Beissinger, S.R. *et al.* 1996. Limitations of captive breeding in endangered species recovery. *Conservation Biology*, 10:338–348.

Sol, D. and Lefebvre, L. 2000. Behavioural flexibility predicts invasion success in birds introduced to New Zealand. *Oikos*, 90:599–605.

Sol, D., Timmermans, S. and Lefebvre, L. 2002. Behavioral flexibility and invasion success in birds. *Animal Behaviour*, 63:495–502.

Stamps, J. A. and Swaisgood, R. R. 2007. Someplace like home: experience, habitat selection and conservation biology. *Applied Animal Behaviour Science*, 102:392–409.

Stephens, P.A. and Sutherland, W.J. 1999. Consequences of the Allee effect for behaviour, ecology and conservation. *Trends in Ecology & Evolution*, 14:401–405.

Strum, S.C. 2005. Measuring success in primate translocation: a baboon case study. *American Journal of Primatology*, 65:117–140.

Temple, S.A. 1985. Why endemic island birds are so vulnerable to extinction. *Bird Conservation*, 2:3–6.

Temple, S.A. 1986. The problem of avian extinctions. In *Current Ornithology*, pp. 453–485. New York: Plenum Publishing Corp.

Tennyson, A. and Martinson, P. 2006. *Extinct Birds of New Zealand*. Wellington: Te Papa Press.

Thornton, A. and Clutton-Brock, T. 2011. Social learning and the development of individual and group behaviour in mammal societies. *Philosophical Transactions of the Royal Society B: Biological Sciences*, 366:978–987.

Trewenack, A.J., Landman, K.A. and Bell, B.D. 2007. Dispersal and settling of translocated populations: a general study and a New Zealand amphibian case study. *Journal of Mathematical Biology*, 55:575–604.

Tudge, C. 1992. *Last Animals at the Zoo: How Mass Extinction Can Be Stopped.* Washington, DC: Island Press.

Viggers, K.L., Lindenmayer, D.B. and Spratt, D.M. 1993. The importance of disease in reintroduction programmes. *Wildlife Research*, 20:678–698.

Waldman, B. and Bishop, P.J. 2004. Chemical communication in an archaic anuran amphibian. *Behavioral Ecology*, 14:88–93.

Wecker, S.C. 1964. Habitat selection. *Scientific American*, 211:109–116.

Woody, J.B. 1990. Guest editorial: Is "headstarting" a reasonable conservation measure? "On the surface, yes; in reality, no." *Marine Turtle Newsletter*, 50:8–11.

Zidon, R., Daltz, D., Shore, L.S. and Motro, U. 2009. Behavioral changes, stress, and survival following reintroduction of Persian fallow deer from two breeding facilities. *Conservation Biology*, 23:1026–1035.

# From individual behavior to population viability: implications for conservation and management

CARMEN BESSA GOMES AND FRANÇOIS SARRAZIN

## 9.1 INTRODUCTION

In this chapter we explore how behavior can be taken into account in quantitative population management. Indeed, incorporating behavior in management should take place at all levels of the management process, including in quantitative population management. Quantitative tools are often used to compare management scenarios and can play an integrative role in adaptive management strategies. Whether we envision behavior as a process amenable to management or a constraint impairing population processes, its integration into quantitative tools is likely to make these tools more efficient and more resilient to uncertainty.

The interplay between behavior and population management and, in particular, conservation, is well-rooted in the late 1990s conservation science literature, a period where many behavioral ecologists came forward to discuss the potential contributions of behavioral ecology for conservation (Caro 1998, Sutherland 1998, Gosling & Sutherland 2000). They argued that behavioral ecology could contribute to conservation science not only as a tool for the management of particular species, as ex-situ conservation programs had often done, but by offering conceptual tools that could contribute to better understanding the extinction process. Behavioral ecology could, in particular, help us understand species' capability to adapt to anthropogenic disturbance (Sutherland & Norris 2002, Sutherland 2006) and predict their vulnerability (Anthony & Blumstein 2000, Sutherland & Norris 2002).

*Conservation Behavior: Applying Behavioral Ecology to Wildlife Conservation and Management*, eds. O. Berger-Tal and D. Saltz. Published by Cambridge University Press. © Cambridge University Press 2016.

Behavior is a fundamental element in the dynamics of animal populations, as many processes affecting these dynamics are rooted in individual behavior (Sutherland 1996). Animal behavior can be defined as being the range of actions made by organisms in response to various stimuli or inputs, whether internal or external, conscious or subconscious, overt or covert, and voluntary or involuntary. Behavior therefore serves as a fundamental link between the single organism and population processes. Population management typically focuses on factors that determine presence, survival and reproductive rates, which are largely mediated by behavior. Hence, behavior may be integrated in management by identifying the key behavioral responses underlying the population processes that managers seek to address.

Many behavioral responses are indeed already part of the manager's toolbox. If we focus on the three population processes mentioned above (presence, survival and reproduction), we easily find a myriad of examples almost as diverse as the study cases and the management goals (conservation, exploitation, eradication). Managers can influence animals' settling decisions through specific management protocols, including using decoys or odor cues to simulate the presence of conspecifics, or establishing temporary on-site enclosures to ensure habitat familiarity prior to releases of animals into the wild (Chapter 8). Managers also routinely manage animal dispersal in fragmented landscapes based on behavioral principles (Chapter 7). Indeed, species occupancy patterns are likely to depend on behavioral process underlying animal movement. The functional connectivity of a given landscape depends not only on its structure, but also on how animals respond to this structure: i.e. how the behavior of a dispersing organism is affected by landscape structure and elements (Van Dyck & Baguette 2005). Such behavior has been the object of specific management through the creation of movement corridors that offer adequate vegetation and cover. Managers may also use behavior to counter animal movement by devices that dissuade animals from given sites. Another management domain with strong behavioral implications is ex situ conservation, particularly the persistence of captive populations of endangered species through captive breeding (Chapter 10). Undeniably, the reproductive success of captive individuals depends on a wide range of behavioral traits, spanning from mate choice to parental care, and on their interaction with the environment.

Unaccounted behavioral traits have at times proven themselves as major management setbacks. For example, many selectively harvested populations where exploitation was restricted to adult males found themselves exposed to a surplus infant mortality due to infanticide. Although it was long known that brown bear (*Ursus arctos*) cub survival was lower in hunted populations (e.g. McCullough 1981, Stringham 1983), only recently has this phenomenon been related to male reproductive strategies (Wielgus *et al.* 2001, Swenson 2003). The same phenomenon has also been observed in lion (*Panthera leo*) populations subject to trophy hunting, and dominance status proxies were suggested to identify animals that should not be targeted, as their disappearance would result in infanticide by a new incoming male (Whitman *et al.* 2004). Increased infanticide rate in response to habitat disturbance was also documented in blue monkey (*Cercopithecus mitis*) troops, but direct causes of this phenomenon remain largely unknown (Butynski 1990).

The relation between behavior and management should be a key consideration in conservation science. Many behavioral traits play a key role in the vulnerability of species to anthropogenic activities. These mostly fall within behavioral disciplines such as animal movement, habitat selection for reproduction or foraging, breeding strategies and social organization. Identifying behaviors that are correlates of extinction risk allows us to go beyond a case-by-case approach. However, assessing the impact of behavioral traits on population viability will depend on our knowledge of trait reaction norms. The reaction norm describes the pattern of phenotypic expression across a range of environments. Hence, behavioral trait reaction norms will influence individual and, consequently, population performance under different environmental conditions. Traits become vulnerability factors if maladapted to the environment in which the population occurs. This is likely to be true when the environment is rapidly changing because of increasing human activities. Nevertheless, empirical evidence of a connection between behavioral traits and risk of extinction remains scant. Several studies of different taxa have highlighted a relationship between reproductive strategies to extinction risk proxies. Brashares (2003) examined mammal species loss from African protected areas and found that monogamous and polygynous species living in small groups were prone to local extinction. Likewise, passerine species exhibiting biparental care are less likely to establish long-term populations when introduced in insular systems (Bessa-Gomes *et al.* 2003) than species where parental care can be ensured by a single parent. As for sexual selection, sexual dichromatism has been shown to correlate with lower introduction success in birds (Bessa-Gomes *et al.* 2003), and with local extinction risks in North American birds (Doherty *et al.* 2003). However, no evidence was found to support this hypothesis when assessing global extinction risk in either birds or mammals (Morrow & Pitcher 2003, Morrow & Fricke 2004).

Here we examine the link between animal behavior and population dynamics in order to highlight how behavior affects responses to particular anthropogenic disturbance and management practices. We will examine behavior correlates of population response to disturbance and management practices so that we can identify recurrent patterns regarding the role of behavior in population dynamics. In particular, we want to examine which categories of behavior are most commonly identified as being correlated to population response to conservation threats (e.g. logging, habitat loss and fragmentation, climate change), conservation practices (e.g. translocation success), harvesting and eradication. Having identified the categories of behavior more commonly identified as being correlated to population processes, we will discuss how to move from case-by-case integration of behavior to generic integration of behavior in population dynamics, conservation and management. Additionally, we will focus on situations where behavior results in frequency- or density-dependence, thus inducing non-linear population dynamics. Such non-linearity can be used to better understand extinction dynamics (e.g. Allee effect) as well as population growth and regulation. We will illustrate how behavior can be integrated into common quantitative tools, namely when assessing population viability and minimum viable population size, and when estimating sustainable harvesting. We will finally examine the limits and potential of such tools, highlighting their potential contributions not only to conservation science, but also to other domains such as the management of exploited populations and the control of alien and pest species.

## 9.2 THE BEHAVIOR–POPULATION DYNAMICS NEXUS

### 9.2.1 Basic concepts of population dynamics

Population conservation and management primarily relies on the ability to drive population dynamics toward predefined targets. The targets will vary from critical population abundance, or density in a given area, to desired population performance in terms of growth or persistence. Nevertheless, they all depend on population dynamics. These dynamics result from the combination of three basic demographic processes: survival, reproduction and dispersal, whatever the focal population. Whether a population declines to extinction, increases and invades its environment, is subject to any kind of regulation, or responds either positively or negatively to conservation or management processes, depends at each instant on any factors influencing these three demographic processes. Quantifying demography is thus central to diagnose and project population dynamics.

Survival is the product of a combination of behavioral outputs. It involves the ability to implement relevant foraging strategies to gather water, nutrients and energy to maintain metabolism and allow development, and the ability to escape abiotic and biotic threats including predation or parasitism through vigilance, learning and movement. It may equally include behavioral processes by which organisms alter their own (or other species') environment, a phenomenon termed niche construction (Laland *et al.* 2000). Reproduction entails a large set of behaviors from breeding habitat selection, mate selection and mate guarding resulting from sexual selection and parental investment. In contrast to survival and reproduction, dispersal is not always considered as a pure demographic process. Indeed it is mostly seen as the results of decision-making behaviors regarding the risks and benefits of remaining philopatric versus crossing a somewhat suboptimal habitat matrix and selecting a new breeding habitat. Each of these decisions may involve cues and public information concerning habitat quality, the presence of conspecifics, or their ability to breed successfully in a given patch (Doligez *et al.* 2002, Dall *et al.* 2005). However, since the outcome of such decision processes determines emigration and immigration, dispersal can be a major determinant of both local population and overall metapopulation growth rate.

Although the population growth rate is the direct consequence of the combination of survival, reproduction and dispersal, quantifying either the overall population growth rate or its three main forces *per se* may not be useful for management if we don't understand the drivers of these forces. Each of these parameters, estimated at the population level, is the result of individuals' fate, which, in turn, results from the interactions between individuals' phenotype and the environment that varies over time and space. The expression of individuals' genotypes in the successive environments they face shapes individuals' phenotype through epigenetic, development, plasticity and behavioral processes. The ability of these phenotypes to survive, disperse and reproduce defines their capacity to spread genes within their population, i.e. their fitness. From a demographic point of view, a direct connection exists between fitness and the estimated population growth rate associated with a given set of life history traits.

The complexity of environments is of first importance because it includes abiotic and biotic factors and their own dynamics. Among biotic factors, direct interspecific interactions, including predation, competition, parasitism, mutualism or symbiosis, are well known, but indirect interactions through ecological engineering are increasingly considered to significantly affect habitat suitability for numerous organisms (Wright & Jones 2006). All these forces have acted through evolution and resulted in the selection of life history traits that link a given life cycle, i.e. a biodemographic strategy, to the multidimensionality of its niche. These strategies are largely structured along a gradient of generation times from short-lived to long-lived species (Gaillard *et al.* 2005) that is at least partly connected to a gradient of environments from fluctuating to more stable ones. The study of life cycles has generated a highly comprehensive theory (see Stearns [1992] for a comprehensive review) that has been supported by the extensive analyses of the properties of structured populations (Caswell 2001). The structured population approach simply considers classes of individuals sharing similar demographic rates within a given population. A high diversity of classes can be identified from ages to stages, such as size, sex, color, shape or any phenotypic trait. Hence, the structured population approach can easily be used to account for behavior by classing individuals according to behavioral similarities, such as social groups, dominance ranks, helping strategies, foraging abilities or dispersal rates. The proportion of individuals in each class is the result of the life cycle and the values of each life history trait or demographic parameter. In the same way, the contributions of the individuals of each class to the future population (i.e. their reproductive value) are the outputs of the life cycle and its parameters.

One of the main properties of structured population models is that the response of fitness or population growth rate to a change in a given life history trait, or vital rate, is directly linked to the reproductive values and proportion of the classes concerned by the changing vital rate. This sensitivity of fitness or population growth rate to vital rates is a key point to understand the potential or actual impact of any behavioral trait on fitness in an evolutionary context, or on population dynamics in a demographic one. Indeed, numerous studies on wild populations have shown a gradient of sensitivities of population growth rates from high sensitivities to reproduction or juvenile survival in short-lived species, to high sensitivities to adult survival in long-lived ones. From an evolutionary point of view, selective pressures on life history traits to which fitness is highly sensitive have a direct impact compared to selective pressure on other traits. One consequence in population dynamics is that such traits with high fitness sensitivity are likely to be buffered and less variable in response to normal environmental fluctuations. Variation of such traits would induce a significant variation in fitness that will be counter selected. In a demographic context, the response of vital rates to density dependence should be negatively correlated to the sensitivity of population growth rate to these vital rates. This was quantified in various ungulates where proportion of breeders declines and age at first breeding increases with density (Gaillard *et al.* 1998). Such basic consequences of fitness sensitivity to vital rates variation in natural populations may have huge consequences in human-modified environments. The speed of anthropogenic environmental change at global or local scales may disrupt the adaptive links between life cycle, vital rates and environmental selective pressures (Chevin *et al.* 2010). Through the main forces of habitat changes, overexploitation, invasive species and climate changes, biotic and abiotic pressures are likely to act on vital rates independently from the sensitivity of fitness or population growth rate to them. A small impact on a key vital rate can then quickly induce an extinction vortex if negative or an invasion dynamic if positive (Caswell 2001, Beissinger & McCullough 2002, Morris & Doak 2002). This conceptual framework is thus also a very powerful way to evaluate *a priori* the potential response of a population to conservation and management measures through the identification of target vital rates and quantification of necessary efforts to restore or regulate their value. It is important to underline that a high sensitivity of fitness or population growth rate to a given vital rate does not necessarily mean that such vital rate will be more amenable to management than a vital rate of lower sensitivity. However, these sensitivity analyses are valuable because they allow us to compare and identify the most efficient strategy of conservation or management.

The impacts of behavior can be included in this framework to evaluate their consequences on population dynamics, conservation and management. Most of these approaches have been developed for stable populations or constant vital rates. This must be seen as a proxy of temporary population dynamics regimes. Nevertheless the extension of sensitivity analyses to stochastic dynamics has also been considered. Indeed, demographic stochasticity, the explicit realization of vital rates at individual levels, and environmental stochasticity, the variation of mean vital rates in response to environmental fluctuations, play major roles in population dynamics particularly for those subject to conservation and management (Beissinger & McCullough 2002, Morris & Doak 2002). Stochastic processes must be accounted for to estimate extinction probabilities or mean extinction times. Their randomness may, however, be partly artificial. These stochastic processes can, at times, act as conceptual black boxes encompassing a large number of micro and macro events that are largely deterministic but often too difficult to make explicit and quantify, such as individual heterogeneity. Among them, behaviors are largely involved in the actual results of these apparently stochastic processes.

### 9.2.2 The impact of behavior on demographic processes

As we stated in the previous section, behavior is a fundamental element in understanding population dynamics. The abiotic or biotic components of the environment will often elicit behavioral responses aimed at enhancing animal performance as conditions change. The pattern of behavioral response across a range of environments can be considered as a behavioral norm of reaction that describes how organisms respond to varying environments. This norm of reaction is a key element to understand how organisms will perform under variable environmental conditions. This is well illustrated by the particular example of the Allee effect. The Allee effect is a form of density dependence, in which there exists a positive relation between population size or density and population growth. Such relation was first described by Warder C. Allee in 1931 (in Courchamp *et al.* 1999), and was initially defined as a demographic phenomenon that affected the overall population growth rate. However, Allee (Allee *et al.* 1949) evoked behavioral mechanisms such as mate finding to justify the existence of such form of density-dependence. The Allee effect remained elusive and was even decried in the population dynamics literature until the mid-1990s because it is difficult to document (Stephens & Sutherland 1999, Stephens *et al.* 1999). But acknowledging its relation to behavioral traits that encompass positive interactions among congenerics has shed a new light on Allee effects. Other than the demographic Allee effect described by Allee, it is now current to consider the existence of component Allee effects, which are defined as positive correlations between low population density and any fitness component. The existing knowledge from behavioral ecology is now used to examine the relation between density and vital rates and this has strongly improved our understanding of the Allee effect phenomenon (Mooring *et al.* 2011, Bateman *et al.* 2012, Luque *et al.* 2013).

To successfully account for behavior in quantitative management we need to identify the behavioral traits that are likely to impact the population processes we are addressing, and the major environmental drivers of these traits (Figure 9.1). As the focus here is on population dynamics, we will concentrate on specific classes of behavior that have a well-established link to variation in vital rates. In this context, behavioral traits can be grouped into three broad classes, depending on the vital rates they influence: (1) traits associated with reproduction (reproductive system and strategies); (2) traits

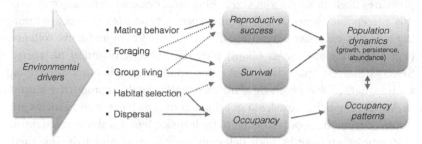

**Figure 9.1:** Behavioral traits affect population processes through their effect on life-history traits, particularly on reproduction and survival, as well as through their effect on occupancy patterns, and through the interaction of both.

associated with survival (foraging, social organization, collaboration and competition); (3) traits associated with dispersal (breeding habitat selection and movement). Although these categories are largely artificial, they help us to build a systematic approach to integrate behavior into a population-level framework, while integrating many of the previously identified behavior impacts on population dynamics.

### 9.2.2.1 Behavior and reproductive success

Social-reproductive systems, including mating systems, social organization and sexual selection, have received particular attention in studies linking dynamics and behavior. These traits may affect reproductive success through various mechanisms, ranging from the reproductive status of individuals to the probability of both finding and accepting a potential mate, as well as the modulation of parental investment in response to partner or socio-reproductive context. Many of these mechanisms have been associated with population density and composition (sex-ratio and age structure), and are considered as component Allee effects (Stephens et al. 1999). The probability of finding an acceptable breeding partner at low density is postulated to depend on the social organization (Jager et al. 2006), as well as on mating systems (e.g. monogamous species are more vulnerable to stochastic variation of the sex-ratio: Legendre et al. 1999, Bessa-Gomes et al. 2010) and mating behavior in general (mate choice: Møller & Legendre 2001; sexual selection: Kokko & Brooks 2003). Theoretical work indicates that the mating system is likely to affect access to reproduction but its impact depends on population sex-ratio (Bessa-Gomes et al. 2004). Accordingly, deciding whether to account for the mating system in quantitative management must be based on our knowledge of sex-ratio variation in the population, particularly of the selective pressures likely to act on both sexes. Empirical evidence generally supports this perception. Hence, in polygynous species submitted to selective hunting of males, excessive harvesting of males has led to reproductive collapse (e.g. saïga antelope (*Saiga tatara*) harems: Milner-Gulland *et al.* 2003). Selective harvest may also affect populations through other behaviors, such as mate choice. Poaching of males has affected the reproductive output in African elephant (*Loxodonta sp.*) populations, as large mature males are removed from the population due to their tusks, females show maladaptive mate choice strategy by both delaying reproduction and reducing their maternal investment (Dobson & Poole 1998). Selective harvesting may also strongly impact the reproductive success of species where males may kill infants to gain access to reproduction (sexual selected infanticide, see introduction).

Reproductive success can also depend on habitat choice. The selection of a breeding habitat can be based on diverse mechanisms and entail the use of different information sources, ranging from environmental cues to personal information or public information. If habitat choice relies on environmental cues, then it may constitute an obvious link between the external environment and the reproductive output. This is true whether the choice is adaptive or, even more so, when the choice is maladaptive. Indeed, if the environment has changed, particularly in response to human disturbance, some cues may no longer correspond to suitable breeding habitat and animals may choose an inadequate breeding site, i.e. an ecological trap (see Chapter 4). This phenomenon can be illustrated by the example of grasshopper sparrow (*Ammodramus savannarum*), a granivorous bird that builds its nest at the proximity of cereal fields, anticipating a bonanza of feeding resource for its offspring (Shochat *et al.* 2005). Nevertheless, upon harvest, such bonanza will be cut short before nestlings reach their independence, resulting in high nestling mortality. Other species base their choice on public information, i.e. on how well their congeners are faring, and will settle in areas where previous-year breeders were successful. If habitat quality is predictable, such strategy reduces the risk of making a poor choice, but can be problematic in low density populations because stochasticity in small groups reduces the relevance of the available information (Mihoub *et al.* 2009).

### 9.2.2.2 Behavior and survival

Behavioral strategies affecting resource acquisition, foraging success or predator avoidance play a key role in survival. The demographic consequences of behavior is well-illustrated by group living species, where survival of group members often depends on mechanisms requiring a threshold number of conspecifics (Courchamp *et al.* 1999). African wild dog (*Lycaon pictus*) group members need to ensure pup protection from predators, while maintaining a hunting group size large enough to ensure the capture of large prey and to reduce kleptoparasitism by larger predators. Consequently, pups born in groups containing less than six adults and sub-adults are unable to survive (Courchamp *et al.* 2000).

Species that do not hunt collectively may still need conspecifics for resource acquisition, such as when species rely on social information on resource location. Such dependence on social information is well illustrated by social foragers exploiting unpredictable resources such as obligate raptor scavengers feeding on ungulate carcasses. In many vulture conservation programs, the dependence of individuals on social information has been surpassed through the creation of feeding sites (Figure 9.2). Because feeding sites render resources predictable, we might expect individual vultures to rely on their own experience regarding resource location. This seems to be the case, and when vulture home range characteristics and feeding habitat selection were analyzed (Monsarrat *et al.* 2013), it showed that feeding stations were always preferred compared to the rest of the habitat, where resources are unpredictable. Vulture survival rates are very high and

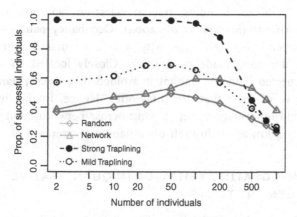

**Figure 9.2:** The proportion of successful vultures foraging on predictable resources only depends of the presence of congeners if individuals rely on public information (strategies "random" – gray line with diamonds – and "Network" – gray line with triangles). If individuals dispose of personal information on resource location (strategies traplining – dashed lines), which is likely to be the case when resources are predicable, vultures will be able to locate resources even when at low density (adapted from Deygout *et al.* [2010]).

have remained almost constant over the last 30 years (Chantepie *et al.* 2015). Nevertheless they still keep foraging on unpredictable resources even after more than 30 years of feeding in feeding stations. This information is crucial for the management of vultures in regions where carrion may be intentionally poisoned or unintentionally toxic due to veterinary chemicals used on livestock. Intensive programs to provide safe food in feeding places cannot totally prevent threats to the survival of these long-lived birds.

Survival can also be influenced by habitat selection strategies. Behavioral ecologists generally agree that animals derive benefits from familiarity with spaces that they inhabit or visit, yet site familiarity is rudimentary or lacking in most models of habitat selection (Berger-tal & Saltz 2014).

### 9.2.2.3 Behavioral determinants of dispersal patterns

Species distribution is largely determined by a range of biotic and abiotic factors, such as temperature, rain, altitude, resources and ecological interactions. Fine scale occupancy, however, depends on behavioral processes as well, such as habitat choice and dispersal. Considering the behavioral processes that underlie the decision of leaving the current site, choosing a given route, and settling in a given new site, may strongly contribute to our understanding of the overall population dynamics and occupancy patterns. This is well illustrated by the facultative colonial lesser kestrels (*Falco naumanni*), where occupancy patterns depend on breeding habitat selection behavior, leading to local aggregations (Serrano *et al.* 2004). Occupancy patterns are also likely to depend on site connectivity, a variable that depends on the movement decisions made by animals. Closely located sites may be poorly connected if matrix habitat is avoided by animals, namely, due to the degree of cover that they may offer. Hence, Iberian lynx (*Lynx pardinus*) will on average travel 30 km to cross between habitat patches located only 4 km apart through open habitat (Ferreras *et al.* 2001).

## 9.3 INTEGRATING BEHAVIOR IN QUANTITATIVE MANAGEMENT TOOLS

### 9.3.1 Construction of behavior-based models of population dynamics

The construction of behavior-based models is based upon the behavior/life history nexus, focusing on the relation between phenotype, performance and life table. Behavior can be accounted for in quantitative management tools using several different approaches. The actual approach implemented in any particular case depends on the information available regarding population dynamics and behavioral processes, their interaction and the goal of the model being built. We have chosen to focus here on four modeling approaches: (1) functional; (2) mechanistic; (3) behavioral; and (4) individual-based models.

### 9.3.1.1 Functional models

The integration of behavior into quantitative management tools through functional models is based on generic population-dynamics models (e.g. the logistic growth curve). Information on behavior is taken into account by making vital rates vary in response to environmental drivers, including local density of focal species or other interacting species. Hence, many density-dependent models are indeed functional models in that the mechanisms underlying the relation between density and vital rates remain implicit. This is the case of many models examining the Allee effect (see Boukal & Berec [2002] for a review of the functions most often used). Several functions have been proposed to integrate the impact of positive intra-specific interactions into population dynamics regardless of the behavioral mechanism considered. Such approach has several advantages and can be used if some information on the mechanisms linking behavior to population dynamics is available. Also, it allows us to examine the consequences of generic behavior traits using models that are amenable to numerical analysis, and are likely to generate general results that can improve our overall understanding of population dynamics.

Functional models were used to examine the impact of meerkat (*Suricata suricatta*) social systems on their population dynamics (Bateman *et al.* 2011, 2012). As with the African wild dogs described previously, meerkats are obligate cooperators, and their need for helpers could generate an Allee effect at the group level. Using functional population models, modified to incorporate environmental conditions and potential Allee effects, the authors observed that although per capita meerkat mortality is subject to a component Allee effect, it contributes relatively little to observed variation in group dynamics. Indeed, conventional density-dependent factors affecting other demographic rates, particularly emigration, govern group dynamics. Such findings highlight the need to consider demographic processes before drawing conclusions about how behavior affects population processes in socially complex systems.

### 9.3.1.2 Mechanistic models

As in functional models, mechanistic models take behavior into account through vital rates variation in response to environmental drivers. They differ from the previous in that the behavioral mechanism underlying the relation between the driver and vital rate is no longer implicit; the shape of the response is the direct consequence of this relation. This can be illustrated by comparing two alternative mating functions that are currently used, the harmonic mean and the minimum function (Figure 9.3). Independently of the function chosen, a mating function influences the population growth rate because it influences the degree of access to partners, which in turn determines the probability of breeding. Two-sex models can account for the mating systems by explicitly modeling female access to reproduction (i.e. the probability that females will breed) as a function of male availability (Caswell 2001). As females can only produce young if they mate, the probability that they will breed is given by the ratio between the numbers of mated females over the total number of potentially reproductive females. When males and females do not mate one-to-one, the relative abundance of each sex in the mating population is biased in favor of the sex that is able to establish more pair bonds. Therefore, when males can acquire several partners, the maximum number of pair bonds that males can establish is greater than one. This maximum value sets an upper limit on the number of breeding pairs, and, concurrently, of breeding events, per male (termed potential reproductive rate [Clutton-Brock 1989]). When we consider the minimum function as the mating function, the number of pairs is set by the less abundant sex after correction for the potential reproductive rate. Hence, this mating function clearly identifies the pairing mechanism and the major constraint acting on mating, the scarcity of receptive partners. The alternative functional approach uses the harmonic mean of the number of males and females as a mating function. This functional approach results in the same probability of mating when males and females are functionally equally abundant (Caswell 2001). Nevertheless, any departure from equilibrium results in higher reproductive rates than the mechanistic model and can overestimate population growth (Bessa-Gomes *et al.* 2010).

Mechanistic models of behavior can also be used to examine occurrence patterns. Boyer *et al.* (2006) were able to examine primate spatial distribution patterns as a function of foraging rules and resource distribution. Using a simple foraging model where individual primates follow mental maps and choose their displacements according to a maximum efficiency criterion, the authors showed that tree-size frequency distribution can induce non-Gaussian movement patterns with multiple spatial scales (Levy walks). These results are consistent with field observations of tree-size

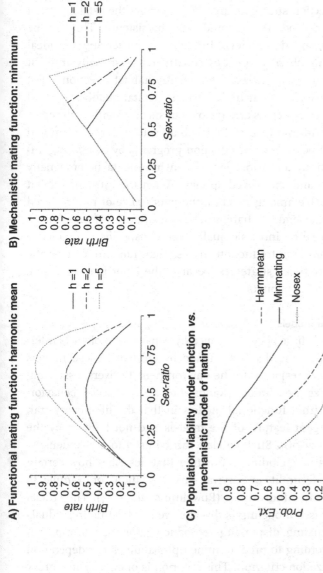

**Figure 9.3:** When modeling the mating process using either a functional (A) or mechanistic (B) approach, we observe qualitatively relatively similar patterns of behavioral performance (here accessed as mating rate at population level). Nevertheless, population viability depends on the modeling approach used (C), being significantly lower when estimated using a mechanistic model. Indeed, although more amenable to numerical analysis, the functional model fails to capture the limitation in partners when the operational sex-ratio is not at equilibrium (adapted from Bessa-Gomes *et al.* [2010]).

variation and spider monkey (*Ateles geoffroyi*) foraging patterns and contribute to our understanding of the patterns of seed dispersal by foraging primates.

Breeding habitat selection strategies are another class of behavior that is expected to influence the occurrence pattern of animals. In particular, breeding habitat selection strategies are likely to affect the establishment success of reintroduced populations and their persistence. This can be assessed using mechanistic models. Instead of considering an ideal free distribution, Mihoub *et al.* (2009) constructed a spatially implicit demographic model that considered five breeding habitat selection rules: (1) random, (2) choose sites with high intrinsic habitat quality, (3) avoid conspecifics, (4) join conspecifics and (5) join conspecifics with high reproductive success. The impact of breeding habitat selection was examined within the context of possible reintroduction programs by comparing different release methods under various levels of environmental heterogeneity levels, for both long- and short-lived species. When the intrinsic habitat quality is highly variable among patches, the persistence of reintroduced populations strongly depends on habitat selection strategies (Mihoub *et al.* 2009). Strategies based on intrinsic quality and conspecific reproductive success lead to a higher reintroduction success than random, conspecific presence or avoidance-based strategies, because the latter may aggregate individuals in suboptimal habitats.

### 9.3.1.3 Behavioral models

Behavior-based models have been developed by behavioral ecologists interested in understanding the decisions made by animals. These models examine the behavioral response to the environmental drivers, assuming that animals maximize their fitness (Rands *et al.* 2003). Hence, behavior-based models differ from functional models in that the life-history trait variation is an emergent feature of the models and not built in by the modeler (Sutherland 2006). Such models can be used to derive density-dependent processes by including a function that describes how certain components, such as fecundity or survival, vary in relation, for example, to the resources acquired by individuals (Rowcliffe *et al.* 2004). The crucial feature of behavior-based modeling is that the events affecting individuals in a given state (i.e. mating, dispersing, reproducing, joining a group, etc.) are not modeled according to predetermined probabilities but depend on some kind of optimization criterion. This criterion is often a simple proxy for fitness (e.g. individuals behave in a way that maximizes their foraging intake rate, or minimizes competition with close relatives), but may also be based on residual reproductive value. The residual reproductive value represents an organism's future reproduction and incorporates expected survival and future reproductive success. Residual reproductive value approaches have the intuitive appeal of simulating behaviors of evident evolutionary stability but may not differ greatly from more obvious and tractable proxies. Given that they are based on the solid ground of natural selection, behavior-based models are likely to produce patterns of individual behavior (and therefore patterns of population behavior) of far greater realism than patterns predicted by simpler functional population models that do not explicit behavioral processes.

Behavioral models can be used for predictive purposes as long as the principles acting at the individual level remain constant (Sutherland & Norris 2002). Hence, other than its contribution to our current understanding of population biology, the integration of ecological, behavioral and evolutionary biology knowledge at the population level may reveal itself an important tool to tackle global change issues (Sutherland 2006). By explicitly integrating the selective value of individual strategies, behavioral population models are hypothesized to help predict population response to novel, human-originated conditions. This should help us to project expansion, regulation or decline of populations.

Behavioral models often rely on dynamic programming or dynamic games to assess the evolutionary stable strategy (ESS) of the behaviors under examination (Houston & McNamara 1999). Hence, by defining foraging rules at the individual level, Rands and colleagues were able to examine not only individual foraging success but also emergent population processes, such as synchronized foraging (Rands et al. 2003, 2008). Likewise, behavioral models based on dynamic programming were used to examine oystercatchers (*Haematopus ostralegus*) foraging behavior, particularly the modes of competition likely to affect foraging success (Stillman et al. 2000b, 2002). These models proved to be parameterized much more quickly than the alternative of measuring interference directly. A thorough sensitivity analysis, as well as a model validation procedure using historical records, showed that the models were robust and accurately predicted mortality patterns, proving their potential as valuable tools for population modeling and management (Stillman et al. 2000a).

Simulation experiments using a behavioral model explored the role of brown trout individual adaptive behavior in response to food limitation (Railsback & Harvey 2011). This was used to demonstrate that behavior can affect managers' understanding of conservation problems. The model includes many natural complexities in habitat, fish physiology and behavior. The simulated population was always food limited, irrespective of food concentration increase over a wide range of food supply. In fact, as food availability increased, the population also increased at a higher growth rate and consumed a higher percentage of the food supply, apparently because higher food concentrations increase the relative amount of habitat that is profitable for drift feeders. The behavior responsible for this response was activity selection: when food was abundant, fish chose to feed less frequently and more nocturnally, thereby reducing predation and increasing survival. Hence, food availability continued to limit population size notwithstanding its high level. Contrary to what was expected, it was impossible to detect an upper threshold of food supply beyond which food was no longer the limiting resource. In this example, explicit consideration of adaptive behavior produced a novel but reliable understanding of food effects on salmonid populations.

### 9.3.1.4 Individual-based models

Another way to integrate behavior in population dynamics models is to build in the "first principles" operating at the individual level within the framework of Individual-Based Models (IBMs). IBMs of behavior allow us to examine the effects of individual actions both on its performance and on the emergent behavior at the level of the population. This allows generating scenarios and predictions according to our basic knowledge on the behavioral process under analysis. The impact of individual relatedness and population genetic structure on individual behavior and consequent performance can be explored through demo-genetic versions of the IBMs that specifically take into account these factors.

The model constructed by Deygout et al. (2009, 2010) to examine scavenger foraging behavior can illustrate this approach. By defining foraging rules at the individual level, authors observed that foraging strategies based on personal information (as opposed to social information) could lead to high aggregation around resources (Deygout et al. 2010) and strongly affect scavengers' functional role due to reduced scavenging efficiency (Deygout et al. 2009). Such results have important consequences for griffon vulture management, indicating that the current management action of establishing vulture "restaurants" may affect not only vulture behavior, but also population dynamics and ecosystem processes.

Behavioral models can be used to parameterize the reaction norms within behavioral IBMs. Hence, the foraging rules derived from oyster-catchers behavioral models were integrated into IBMs and used to examine both oystercatchers' individual performance and population dynamics under varying environmental conditions (Stillman *et al.* 2000a, 2000b, 2002), as well as their response to human disturbance (West *et al.* 2002). Stillman and colleagues highlighted the importance of integrating fine detail behavioral information by showing that current mortality patterns are better explained by models that account for both dominance relations and interference (Stillman *et al.* 2000b). These models resulted in a body of knowledge that was mobilized to build easily assessed predictors of oystercatcher mortality and have ultimately been used for the sustainable management of shore areas (Stillman *et al.* 2003; West *et al.* 2003; Ditdurell *et al.* 2005).

Another good example is an analysis of the population viability of the alpine marmot using IBMs (Grimm *et al.* 2003). This species displays marked social behavior, living in social groups of up to twenty individuals. To account for these characteristics, the model was spatially explicit at the scale of clusters of neighboring territories, and spatially implicit at larger scales. The decisive aspect of marmot life history – winter mortality – was described by a logistic regression where mortality increases with age and winter severity, and decreases with the number of subdominant individuals present in a group (i.e. incorporating a functional approach into the IBM). Model predictions of group size distribution are in good agreement with the results of the field study. The model shows that the effect of sociality on winter mortality is very effective in buffering environmental harshness and fluctuations.

### 9.3.2 Parameterization

Although incorporating behavior population models is potentially promising, the integration of behavior into population studies dedicated to conservation and management remains a difficult task. Indeed, it involves several stages: (1) identifying *a priori* behavioral traits likely to impact population dynamics, (2) obtaining the data in the field necessary for parameterization of behavioral mechanisms and their demographic consequences, (3) assessing the significance and magnitude of behavioral effects on demographic parameters.

These three stages are challenging because of the discrepancy in the time scales at which behavioral and demographic processes actually occur. Indeed the occurrence and duration of behavior can range across seconds, minutes, hours, days or weeks, depending on whether we consider behaviors such as predation or predator avoidance, mate choice and mate guarding, or seasonal migration and natal or breeding dispersal. By contrast, vital rates are typically measured at an annual scale, or alternatively scaled according to generation time or seasonality (Armbruster & Lande 1993, Caswell 2001). Additionally, behavior can be observed and measured at the individual level whereas vital rates have to be estimated at the population level or at the intra-population level (e.g. for a given age or stage class).

A *priori* identification of key behaviors likely to impact population dynamics relies on various forms of knowledge. Of course, good knowledge of the natural history of the target species or of phylogenetically and ecologically related species is vital. However, it is clear that when emergency issues arise with threatened and invasive species, these species have rarely been studied intensively enough for the appropriate decision-making process and this decision must be undertaken by assembling a mosaic of information. In this constrained context, using life cycle modeling and sensitivity and elasticity analyses is fruitful to identify potential key vital rates and understand which behavioral process is likely to affect these values. When the population growth rate is highly sensitive to a given vital rate, the estimate of this parameter has to be accurate to allow relevant projection of population trajectories. This is classically the case for adult survival in long-lived species (Ferriere *et al.* 1996). Understanding and quantifying the demographic consequences of behavior on such parameter is, thus, of great importance.

From a methodological point of view, estimating the demographic parameters of animal populations has received a lot of attention from biometricians (Lebreton & Clobert 1991, Williams *et al.* 2001, Choquet *et al.* 2013). The development of field methods and statistics is still improving to account for low detectability and movement of individuals (e.g. Duriez *et al.* 2009). If breeding success is generally estimated from ad hoc methods defined for each breeding system, other parameters such as age at maturity, proportion of breeders, survival and natal or breeding dispersal are commonly estimated within the framework of mark-recapture methods (e.g. Rivalan *et al.* 2005, Schaub & von Hirschheydt 2009, Balkiz *et al.* 2010, Hernández-Matías *et al.* 2010, Péron *et al.* 2010). This body of field techniques, data analyses and modeling, is increasingly used to address a range of questions and particularly the link between behavior and demography. While demographic parameters were typically estimated without explicit connections to the underlying processes that drive their values, recent approaches (such as multi-event models: Pradel 2005) allow us to discriminate between the actual state of the individuals (e.g. alive, dead, in place A or B, with status x or y) and the various events (e.g. stay or move, breed or not), that led them to their state. The potential of these methods are large and promising. It is likely that they will be limited by field constraints more than statistic or modeling ones. Additionally, the parameterization of behavioral effects on demography should not only consider mean values but also the contribution of the potential variation in behavioral events (between individuals and over time) to the temporal variation of demographic parameters. Taking behavior into account may contribute to better understand individual heterogeneity that will otherwise be ascribed to stochasticity. Once again, field constraints remain strong since long time series are rarely available for specific individuals that are members of populations at risk. Even when long time series of demographic data are available, they are rarely accompanied by behavioral data.

Clearly, the ultimate quantification of the impact of behavior may range from relatively easy for an annual all-or-nothing behavior (e.g. to breed or not, stay or disperse) to highly challenging for repetitive short-term events such as daily foraging or competitive interactions. Indeed, the final demographic balance of these events mostly results from metabolic efficiency that is hard to quantify over the long term, in the field and for numerous individuals. Even the behavioral parameters of a functional response that limits the ability of predators to consume more prey in a simple prey–predator system are non-accessible in most cases in the real world of conservation and management. It is nevertheless of great importance to predict the numerical response of this system. The discrepancy between the quantitative units used to measure behavioral efficiency and the demographic consequences of these behaviors remains high and leaves space for conceptual and technical innovation.

## 9.4 PERSPECTIVES

The long-term and wide-scale conservation of wildlife in human-dominated landscapes largely relies on our capacity to identify and cope with the negative impacts of human activities on these populations as well as the negative impact of some of these populations on human activities. The recoveries of wild animal populations still often face hostility due to their actual or supposed impacts on, for example, crops, livestock or even game species. Both effects from and on animal populations can involve behavioral processes. Despite the fact that the history of behavioral-based population conservation and management is already long and rich, numerous challenges remain to be addressed. These challenges are both conceptual and technical.

At the individual and population levels, understanding how behavior responds to drastic reduction in population density is crucial to avoid the proximal phase of extinction vortices (Henle *et al.* 2004). In particular, in very small populations, assessing the relevance of the concept of personality will be necessary to discriminate anecdotal and stochastic patterns from structured patterns (Dingemanse *et al.* 2010, Wolf & Weissing 2012). Likewise, understanding how behavior responds to high population density will remain crucial to smoothly regulate overabundant populations (Tavecchia *et al.* 2007).

In a longer-term perspective, continuing researches on the genetic and epigenetic basis of behavior, reaction norms and plasticity, as well as learning and information transfer through sociality and culture, are still relevant in the context of increasing global changes. The speed of these changes, including climatic changes, is such that even short-lived populations will have huge difficulties to respond by selection (but see Sasaki *et al.* 2009). The capacity of individuals to maintain the adequacy of their phenotypes to their changing environments will largely rely on their behavioral flexibility: their capacity to forage on new resources, to avoid new predators, competitors or parasites, or to breed in new habitats (see Chapter 5). Behavior will play a major role in the species' ability to interact with their environment and thus is an essential aspect of species ability to modulate their niche in changing environments. Alternatively, dispersal behavior will be a crucial response for numerous organisms. But moving toward a changing habitat will probably enhance the risk for ecological traps.

The conservation strategy of assisted colonization (i.e. the intentional movement and release of an organism outside its indigenous range to avoid extinction of populations of the focal species [IUCN/SSC 2013]), is still the object of dispute (Seddon *et al.* 2014). The critiques range from poor feasibility and efficiency to high risk of uncontrolled invasive dynamics. But even the most optimistic and positive views about these proactive conservation actions agree that understanding behavioral plasticity and learning of translocated individuals and their offspring is central if we are to expect any positive output on the long-term viability of these new populations. Once again, linking behavior to population viability is crucial in this context.

There is a global tendency in biodiversity conservation and management to move from strictly individual and population approaches to more integrative approaches at the level of communities and ecosystems. This does not mean that research at the population level is no longer needed. Population remains the most operational and powerful conceptual level for many species conservation and management plans.

The population level provides a central framework for keeping a link between evolutionary and ecological concepts and processes, which is of first importance in behavioral ecology. But accounting for interspecific interactions that largely rely on behavioral interactions remains to be developed in population-centered studies and particularly in conservation and management. Indeed, behavioral traits may be major drivers of population dynamics within trophic networks, and host-parasites or plant-pollinators systems, constituting key factors for conservation and management. This is true for present populations and communities that suffer perturbations, and even more so under the perspective of creating new assemblages through assisted colonization. In community and ecosystem ecology functional traits have been identified and are now used to set up management strategies (Lavorel & Garnier 2002). These functional traits have been mostly studied in plants, but their equivalents in animals should be explored and should necessarily consider behavioral traits or strategies. This would allow us to reconnect the main ecological processes of predation, competition, parasitism, mutualism, symbiosis or even ecosystem engineering to individual behavior and population dynamics. It is therefore extremely important to reinforce the strength, the relevance and the efficiency of behavioral-based population conservation and management.

## REFERENCES

Allee, W.C., Park, O., Emerson, A.E., Park, T. and Schmidt, K.P. 1949. *Principles of Animal Ecology*. Philadelphia: WB Saunders Co.

Anthony, L.L. and Blumstein, D.T. 2000. Integrating behaviour into wildlife conservation: the multiple ways that behaviour can reduce Ne. *Biological Conservation*, 95:303–315.

Armbruster, P. and Lande, R. 1993. A population viability analysis for African elephant (*Loxodonta africana*): how big should reserves be? *Conservation Biology*, 7:602–610.

Balkiz, Ö., Bechet, A., Rouan, L., Choquet, R., Germain, C., Amat, J.A., *et al.* 2010. Experience-dependent natal philopatry of breeding greater flamingos. *Journal of Animal Ecology*, 79:1045–1056.

Bateman, A.W., Coulson, T. and Clutton-Brock, T.H. 2011. What do simple models reveal about the population dynamics of a cooperatively breeding species? *Oikos*, 120:787–794.

Bateman, A.W., Ozgul, A., Coulson, T. and Clutton-Brock, T.H. 2012. Density dependence in group dynamics of a highly social mongoose, *Suricata suricatta*. *Journal of Animal Ecology*, 81:628–639.

Beissinger, S. and McCullough, D.R. 2002. *Population Viability Analysis*. Chicago: Chicago University Press.

Berger-tal, O. and Saltz, D. 2014. Using the movement patterns of reintroduced animals to improve reintroduction success. *Current Zoology*, 60:515–526.

Bessa-Gomes, C., Danek-Gontard, M., Cassey, P., Møller, A.P., Legendre, S. and Clobert, J. 2003. Mating behaviour influences extinction risk: insights from demographic modelling and comparative analysis of avian extinction risk. *Annales Zoologici Fennici*, 40:231–245.

Bessa-Gomes, C., Legendre, S. and Clobert, J. 2004. Allee effects, mating systems and the extinction risk in populations with two sexes. *Ecology Letters*, 7:802–812.

Bessa-Gomes, C., Legendre, S. and Clobert, J. 2010. Discrete two-sex models of population dynamics: on modelling the mating function. *Acta Oecologica*, 36:439–445.

Boukal, D. and Berec, L. 2002. Single-species models of the Allee effect: extinction boundaries, sex ratios and mate encounters. *Journal of Theoretical Biology*, 2018:375–394.

Boyer, D., Ramos-Fernández, G., Miramontes, O., Mateos, J.L., Cocho, G., Larralde, H., *et al.* 2006. Scale-free foraging by primates emerges from their interaction with a complex environment. *Proceedings of the Royal Society B: Biological Sciences*, 273:1743–50.

Brashares, J.S. 2003. Ecological, behavioral, and life-history correlates of mammal extinctions in West Africa. *Conservation Biology*, 17:733–743.

Butynski, T. 1990. Comparative ecology of blue monkeys (*Cercopithecus mitis*) in high- and low-density subpopulations. *Ecological Monographs.*, 60:1–26.

Caro, T. 1998. *Behavioral Ecology and Conservation Biology*. Oxford: Oxford University Press.

Caswell, H. 2001. *Matrix Population Models. Construction, Analyses and Interpretation*. 2nd Edition. Sunderland: Sinauer.

Chantepie, S., Teplitsky, C., Pavard, S., Sarrazin, F., Descaves, B., Lecuyer, P., *et al.* 2015. Age-related variation and temporal patterns in the survival of a long-lived scavenger. *Oikos*, in Press.

Chevin, L.M., Lande, R. and Mace, G.M. 2010. Adaptation, plasticity, and extinction in a changing environment: towards a predictive theory. *PLoS Biology*, 8.

Choquet, R., Sanz-Aguilar, A., Doligez, B., Nogué, E., Pradel, R., Gustafsson, L., *et al.* 2013. Estimating demographic parameters from capture-recapture data with dependence among individuals within clusters. *Methods in Ecology and Evolution*, 4:474–482.

Clutton-Brock, T.H. 1989. Mammalian mating systems. *Proceedings of the Royal Society B: Biological Sciences*, 236: 339–372.

Courchamp, F., Clutton-brock, T. and Grenfell, B. 1999. Inverse density dependence and the Allee effect. *Trends in Ecology & Evolution*, 14:405–410.

Courchamp, F., Clutton-Brock, T. and Grenfell, B. 2000. Multipack dynamics and the Allee effect in the African wild dog, *Lycaon pictus*. *Animal Conservation*, 3:277–285.

Dall, S.R.X., Giraldeau, L.-A., Olsson, O., McNamara, J.M. and Stephens, D.W. 2005. Information and its use by animals in evolutionary ecology. *Trends in Ecology & Evolution*, 20:187–93.

Deygout, C., Gault, A., Duriez, O., Sarrazin, F. and Bessa-Gomes, C. 2010. Impact of food predictability on social facilitation by foraging scavengers. *Behavioral Ecology*, 21:1131–1139.

Deygout, C., Gault, A., Sarrazin, F. and Bessa-Gomes, C. 2009. Modeling the impact of feeding stations on vulture scavenging service efficiency. *Ecological Modelling*, 220:1826–1835.

Dingemanse, N.J., Kazem, A.J.N., Réale, D. and Wright, J. 2010. Behavioural reaction norms: animal personality meets individual plasticity. *Trends in Ecology & Evolution* 25, 81–89

Ditdurell, S., Stillman, R., Triplet, P., Aulert, C., Ditbiot, D., Bouchet, A, *et al.* 2005. Modelling the efficacy of proposed mitigation areas for shorebirds: a case study on the Seine estuary, France. *Biological Conservation* 123:67–77.

Dobson, A. and Poole, J. 1998. Conspecifics aggregation and conservation biology. In Caro, T. (ed.) *Behavioral Ecology and Conservation Biology*. Oxford: Oxford University Press.

Doherty, P.F., Sorci, G., Royle, J.A., Hines, J.E., Nichols, J.D. and Boulinier, T. 2003. Sexual selection affects local extinction and turnover in bird communities. *Proceedings of the Royal Society B: Biological Sciences*, 100:5858–62.

Doligez, B., Danchin, E. and Clobert, J. 2002. Public information and breeding habitat selection in a wild bird population. *Science*, 297:1168–1170.

Duriez, O., Saether, S. a., Ens, B.J., Choquet, R., Pradel, R., Lambeck, R.H.D., *et al.* 2009. Estimating survival and movements using both live and dead recoveries: a case study of oystercatchers confronted with habitat change. *Journal of Applied Ecology*, 46:144–153.

van Dyck, H. and Baguette, M. 2005. Dispersal behaviour in fragmented landscapes: routine or special movements? *Basic and Applied Ecology*, 6:535–545.

Ferreras, P., Gaona, P., Palomares, F. and Delibes, M. 2001. Restore habitat or reduce mortality? Implications from a population viability analysis of the Iberian lynx. *Animal Conservation* 4:265–274.

Ferriere, R., Sarrazin, F., Legendre, S. and Baron, J.-P. 1996. Matrix population models applied to viability analysis and conservation: theory and practice using the ULM software. *Acta Oecologica*, 17:629–656.

Gaillard, J.-M., Yoccoz, N.G., Lebreton, J.-D., Bonenfant, C., Devillard, S., Loison, A., *et al.* 2005. Generation time: a reliable metric to measure life-history variation among mammalian populations. *American Naturalist*, 166:119–23.

Gaillard, J.M., Festa-Bianchet, M. and Yoccoz, N.G. 1998. Population dynamics of large herbivores: variable recruitment with constant adult survival. *Trends in Ecology & Evolution*, 13:58–63.

Gosling, L. and Sutherland, W.J. 2000. *Behaviour and Conservation*. Cambridge: Cambridge University Press.

Grimm, V., Dorndorf, N. and Frey-Roos, F. 2003. Modelling the role of social behavior in the persistence of the alpine marmot *Marmota marmota. Oikos*, 102:124–136.

Henle, K., Sarre, S. and Wiegand, K. 2004. The role of density regulation in extinction processes and population viability analysis. *Biodiversity and Conservation*, 13:9–52.

Hernández-Matías, A., Real, J., Pradel, R., Ravayrol, A., Vincent-Martin, N., Bosca, F., *et al.* 2010. Determinants of territorial recruitment in Bonelli's eagle (*Aquila fasciata*) populations. *Auk*, 127:173–184.

Houston, A.I. and McNamara, J.M. 1999. *Models of Adaptive Behaviour: An Approach Based on State.* Cambridge: Cambridge University Press.

IUCN/SSC. (2013). *Guidelines for Reintroductions and Other Conservation Translocations.* UICN Species Survival Commission, Gland, Switzerland.

Jager, H.I., Carr, E.A. and Efroymson, R.A. 2006. Simulated effects of habitat loss and fragmentation on a solitary mustelid predator. *Ecological Modelling*, 191:416–430.

Kokko, H. and Brooks, R. 2003. Sexy to die for? Sexual selection and the risk of extinction. *Annales Zoologici Fennici*, 40:207–219.

Laland, K.N., Odling-Smee, J. and Feldman, M.W. 2000. Niche construction, biological evolution, and cultural change. *Behavioral and Brain Sciences*, 23:131–146.

Lavorel, S. and Garnier, E. 2002. Predicting changes in community composition and ecosystem functioning from plant traits: revisiting the Holy Grail. *Functional Ecology*, 16:545–556.

Lebreton, J.D. and Clobert, J. 1991. Bird population dynamics, management and conservation: the role of mathematical modelling. In C.M. Perrins and Leberton, J.D. (eds.) *Bird Population Studies: Relevance to Conservation and Management* Oxford: Oxford University Press.

Legendre, S., Clobert, J., Møller, A. and Sorci, G. 1999. Demographic stochasticity and social mating system in the process of extinction of small populations: the case of passerines introduced to New Zealand. *American Naturalist*, 153:449–463.

Luque, G.M., Giraud, T. and Courchamp, F. 2013. Allee effects in ants. *Journal of Animal Ecology*, 82:956–965.

McCullough, D.R. 1981. Population dynamics of the Yellowstone grizzly bear. In Fowler, C.W. and Smith T.D. (eds.) *Dynamics of Large Mammal Populations.* New York: John Wiley and Sons, New York.

Mihoub, J.-B., Le Gouar, P. and Sarrazin, F. 2009. Breeding habitat selection behaviors in heterogeneous environments: implications for modeling reintroduction. *Oikos*, 118:663–674.

Milner-Gulland, E.J., Bukreeva, O.M., Coulson, T., Lushchekina, A.A., Kholodova, M. V. and Grachev, I.A. 2003. Reproductive collapse in saiga antelope harems. *Nature*, 422:2003.

Møller, A. and Legendre, S. 2001. Allee effect, sexual selection and demographic stochasticity. *Oikos*, 92:27–34.

Monsarrat, S., Benhamou, S., Sarrazin, F., Bessa-Gomes, C., Bouten, W. and Duriez, O. 2013. How predictability of feeding patches affects home range and foraging habitat selection in avian social scavengers? *PLoS ONE*, 8:e53077.

Mooring, M.S., Fitzpatrick, T.A., Nishihira, T.T. and Reisig, D.D., 2011. Vigilance, predation risk, and the Allee effect in desert bighorn sheep. *Journal of Wildlife Management*, 68:519–532.

Morris, W.F. and Doak, D.F. 2002. *Quantitative Conservation Biology. Theory and Practice of Population Viability Analyses.* Sunderland: Sinauer.

Morrow, E.H. and Fricke, C. 2004. Sexual selection and the risk of extinction in mammals. *Proceedings of the Royal Society B: Biological Sciences*, 271:2395–2401.

Morrow, E.H. and Pitcher, T.E. 2003. Sexual selection and the risk of extinction in birds. *Proceedings of the Royal Society B: Biological Sciences*, 270:1793–1799.

Péron, G., Lebreton, J.-D. and Crochet, P.-A. 2010. Breeding dispersal in black-headed gull: the value of familiarity in a contrasted environment. *Journal of Animal Ecology*, 79:317–326.

Pradel, R. 2005. Multievent: an extension of multistate capture-recapture models to uncertain states. *Biometrics*, 61:442–447.

Railsback, S.F. and Harvey, B.C. 2011. Importance of fish behaviour in modelling conservation problems: food limitation as an example. *Journal of Fish Biology*, 79:1648–1662.

Rands, S.A, Cowlishaw, G., Pettifor, R.A, Rowcliffe, J.M. and Johnstone, R.A. 2003. Spontaneous emergence of leaders and followers in foraging pairs. *Nature*, 423:432–434.

Rands, S.A, Cowlishaw, G., Pettifor, R.A, Rowcliffe, J.M. and Johnstone, R.A. 2008. The emergence of leaders and followers in foraging pairs when the qualities of individuals differ. *BMC Evolutionary Biology*, 8:51.

Rivalan, P., Prevot-Julliard, A., Choquet, R., Pradel, R., Jacquemin, B. and Girondot, M. 2005. Trade-off between current reproductive effort and delay to next reproduction in the leatherback sea turtle. *Oecologia*, 145:564–574.

Rowcliffe, J.M., Pettifor, R. a. and Carbone, C. 2004. Foraging inequalities in large groups: quantifying depletion experienced by individuals in goose flocks. *Journal of Animal Ecology*, 73:97–108.

Sasaki, K., Fox, S.F. and Duvall, D. 2009. Rapid evolution in the wild: changes in body size, life-history traits, and behavior in hunted populations of the Japanese mamushi snake. *Conservation Biology*, 23:93–102.

Schaub, M. and von Hirschheydt, J. 2009. Effect of current reproduction on apparent survival, breeding dispersal, and future reproduction in barn swallows assessed by multistate capture-recapture models. *Journal of Animal Ecology*, 78:625–635.

Seddon, P.J., Griffiths, C.J., Soorae, P.S. and Armstrong, D.P. 2014. Reversing defaunation: restoring species in a changing world. *Science*, 345:406–412.

Serrano, D., Forero, M., Donázar, J. and Tella, J. 2004. Dispersal and social attraction affect colony selection and dynamics of lesser kestrels. *Ecology*, 85:3438–3447.

Shochat, E., Patten, M. and Morris, D. 2005. Ecological traps in isodars: effects of tallgrass prairie management on bird nest success. *Oikos*, 111:159–169.

Stearns, S.C. 1992. *The Evolution of Life Histories*. Oxford: Oxford University Press.

Stephens, P., Sutherland, W. and Freckleton, R. 1999. What is the Allee effect? *Oikos*, 87:185–190.

Stephens, P.A. and Sutherland, W.J. 1999. Consequences of the Allee effect for behaviour, ecology and conservation. *Trends in Ecology & Evolution*, 14:401–405.

Stillman, R.A., Goss-Custard, J.D., West, A.D., Durell, S.E.A.L.V.D., Caldow, R. W.G., Mcgrorty, S., *et al.* 2000a. Predicting mortality in novel environments: tests and sensitivity of a behaviour-based model. *Journal of Applied Ecology*, 37:564–588.

Stillman, R.A., West, A.D., Goss-Custard, J.D., Caldow, R.W.G., Mcgrorty, S., Durell, S.E.A.L.V.D., *et al.* 2003. An individual behaviour-based model can predict shorebird mortality using routinely collected shellfishery data. *Journal of Applied Ecology*, 40:1090–1101.

Stillman, R.A., Caldow, R.W.G. and Alexander, M.J. 2000b. Individual variation in intake rate: the relative importance of foraging efficiency and dominance. *Journal of Animal Ecology*, 69:484–493.

Stillman, R.A., Poole, A.E., Goss-Custard, J.D., Caldow, R.W.G., Yates, M.G., Triplet, P., *et al.* 2002. Predicting the strength of interference more quickly using behaviour-based models. *Journal of Animal Ecology*, 71:532–541.

Stringham, S.F. 1983. Roles of adult males in grizzly bear population biology. *Bears Their Biology and Management*, 5:140–151.

Sutherland, W.J. 1996. *From Individual Behaviour to Population Ecology*. Oxford: Oxford University Press.

Sutherland, W.J. 1998. The importance of behavioural studies in conservation biology. *Animal Behaviour*, 56:801–809.

Sutherland, W.J. 2006. Predicting the ecological consequences of environmental change: a review of the methods. *Journal of Applied Ecology*, 43:599–616.

Sutherland, W.J. and Norris, K. 2002. Behavioural models of population growth rates: implications for conservation and prediction. *Philosophical Transactions of the Royal Society B: Biological Sciences*, 357:1273–84.

Swenson, J.E. 2003. Implications of sexually selected infanticide for the hunting of large carnivores. In Festa-Bianchet, M. and Apollonio, M. (eds.) *Animal Behavior and Wildlife Conservation*. Washington DC: Island Press.

Tavecchia, G., Pradel, R., Genovart, M. and Oro, D. 2007. Density-dependent parameters and demographic equilibrium in open populations. *Oikos*, 116:1481–1492.

West, A.D., Goss-Custard, J.D., Mcgrorty, S., Stillman, R.A., Sarah, E.A., Durell, V., *et al.* 2003. The Burry shellfishery and oystercatchers: using a behaviour-based model to advise on shellfishery management policy. *Marine Ecology Progress Series*, 248:279–292.

West, A.D., Goss-Custard, J.D., Stillman, R.A., Caldow, R.W.G., Durell, S.E.A.L.V. D. and McGrorty, S. 2002. Predicting the impacts of disturbance on shorebird mortality using a behaviour-based model. *Biological Conservation*, 106:319–328.

Whitman, K., Starfield, A.M., Quadling, H.S. and Packer, C. 2004. Sustainable trophy hunting of African lions. *Nature*, 428:175–178.

Wielgus, R., Sarrazin, F., Ferriere, R. and Clobert, J. 2001. Estimating effects of adult male mortality on grizzly bear population growth and persistence using matrix models. *Biological Conservation*, 98:293–303.

Williams, B.K., Nichols, J.D. and Conroy, M.J. 2001. *Analysis and Management of Animal Populations. Modeling, Estimation and Decision Making*. Academic Press.

Wolf, M. and Weissing, F.J. 2012. Animal personalities: Consequences for ecology and evolution. *Trends in Ecology & Evolution*, 27:452–461.

Wright, J.P. and Jones, C.G. 2006. The concept of organisms as ecosystem engineers ten years on: progress, limitations, and challenges. *Bioscience*, 56:203.

# Manipulating animal behavior to ensure reintroduction success

DEBRA SHIER

## 10.1 INTRODUCTION

A key pathway through which animals interact with their surroundings is behavior. In a sense, the behavior of an animal is a mediator between the animal and its surroundings, allowing the animal the opportunity to adjust and improve its performance in an ever-changing environment. This is what makes behavior such an important part of conservation and management (Chapter 1). Since behavior is central for the adaptation of animals to their environment, ineffective or inappropriate behavior can greatly reduce animals' fitness, cause management programs to fail and exacerbate human–wildlife conflict. It is in these cases that managers need to intervene and modify the problematic behavior in order to mitigate its conservation implications.

Natural selection has shaped animal behavior to ensure that organisms respond adaptively to their environment (Chapter 2). For animals that are flexible in their behavior, much of this adaptation is modified by learning through experience. If the environment is altered abruptly (often as a result of anthropogenic activity) the existing behavioral patterns may reduce fitness and even lead species toward extinction (e.g. ecological traps – see Chapter 4). This is especially true for captive-raised individuals. Because the environment in which an animal is raised will strongly affect its behavior as an adult, animals raised in captivity and later released into the wild may, therefore, show maladaptive behaviors if the conditions in which they were raised did not prepare them for life in the wild.

In order to successfully modify detrimental behaviors in wildlife, we must have a firm understanding of how the animal in question perceives its environment and of the learning processes underlying the behaviors it

*Conservation Behavior: Applying Behavioral Ecology to Wildlife Conservation and Management*, eds. O. Berger-Tal and D. Saltz. Published by Cambridge University Press. © Cambridge University Press 2016.

displays. Chapter 3 of this volume is dedicated to learning, while Chapter 6 centers on animals' sensory systems and on how to use our knowledge of these systems in order to successfully manipulate animal behavior. In this chapter I will focus on the pivotal role of behavioral modifications of captive-bred animals targeted for reintroduction into the wild.

In response to rapid loss of species, worldwide captive breeding followed by reintroduction to the wild has become an increasingly common approach to species conservation (Chapter 8). In theory, captive breeding of wild animals allows for the propagation of a large number of animals because they are protected from environmental elements such as weather extremes, low food availability, predators, diseases and negative impacts from humans. However, evaluations of reintroduction programs show that most of these programs fail to produce viable populations (Griffith et al. 1989, Beck et al. 1994, Miller et al. 1994, Maynard et al. 1995, Wolf et al. 1998), undermining the very purpose of captive propagation for conservation. In response, over the last three decades there has been a surge of interest in improving captive breeding and reintroduction methodology via conversion of this management practice into a scientific discipline, which requires scientific rigor, reintroduction outcome assessment and clear definitions of success (Seddon et al. 2007). During this time captive breeding protocols have become more sophisticated and reintroduction methodologies more rigorous (Griffiths & Pavajeau 2008).

Early evaluations show that captive-born and reared individuals are less likely to survive following release compared to those captured from and released to the wild (Griffith et al. 1989, Beck et al. 1994, Miller et al. 1994, Maynard et al. 1995) and it has been long thought that the differential success is due to the ineffective behavior of released animals (Kleiman 1989). In fact, scientists as far back as Darwin have recognized that captivity can radically alter animal behavior (Darwin 1868, Price 1984, Lickliter & Ness 1990, Carlstead 1996, McPhee 2003). Captive environments often fail to replicate key features of the wild (Yoerg & Shier 1997), are predictable or unchanging, and thus, captive-born individuals may fail to develop effective behavioral skills for reproduction in captivity and/or survival following release. Therefore, captive-born animals are typically inexperienced and are exposed to elevated risks upon release and, consequently, must go through a behavioral acclimatization period. Likewise, in this stable benign environment wild-caught individuals may lose the behavioral flexibility or the range of behaviors that enable response to a variable and unpredictable environment. This loss can occur within the lifetime of the individuals as they habituate to captive conditions (Yoerg & Shier 1997) or across genera-tions (McPhee 2003). For example, wild-caught kangaroo rats show degraded behavior after 4 years in captivity (Yoerg & Shier 2000). By contrast, Mallorcan midwife toads, *Alytes muletensis*, captive-reared in a predator-free environment begin to lose anti-predator behavior after nine to twelve generations in captivity (Kraaijeveld-Smit et al. 2006). Thus, managers should strive to modify behavior and/or provide opportunities for wild-type behavioral maintenance prior to release to expedite the animals' acclimation to the wild and free them from potential evolutionary traps (see Chapter 4).

In the wild, dynamic environmental selective pressures shape beha-vioral evolution toward improved survival and reproductive success (Reed 1985). In captivity, however, the environmental conditions are stable and benign and thus vastly different from those in which the species evolved. (Hediger 1964, Price 1970, Soule 1986, Frankham & Loebel 1992, Seidensticker & Forthman 1998, McPhee & McPhee 2012). The captive environment can relax or increase selective pressures, change the direc-tion of selection or exert completely novel pressures (Price 1970, Endler 1986). In response, animals adjust their behavior to cope with their captive environment. This process can cause genetic and phenotypic divergence between captive and wild populations and result in effects on behavior that are detrimental to reintroduction (Carlstead 1996, McPhee & McPhee 2012).

In particular, research has revealed that captive-born and-raised animals are often behaviorally deficient compared to their wild coun-terparts (Kleiman 1989, Miller et al. 1994, Biggins et al. 1999). Skills associated with fitness, such as anti-predator behavior, food finding ability, locomotory skills, social behavior including breeding and nest-ing and/or refuge use, may be suboptimal in captive-reared animals (Rabin 2003). For example, captive-born and-reared golden-lion tamar-ins (*Leontopithecus rosalia rosalia*) were deficient in locomotor and foraging skills when compared with wild-born conspecifics (Stoinski et al. 2003). In fish, hatchery-raised individuals tend to seek refuge less (Brown & Laland 2001) and allocate more time to freezing in the presence of a predator than wild-raised fish (El Balaa & Blouin-Demers 2011), making them more vulnerable to predators and redu-cing the time available for other behaviors such as foraging. This behavioral deficiency puts individuals originating from an established captive population at a disadvantage when reintroduced to the wild. In black-tailed prairie dogs, *Cynomus ludovicianus*, juveniles raised in captivity were less wary in the presence of predators when compared to juveniles raised in the wild, and these same individuals were less likely to survive reintroduction to the wild than their wild-reared counterparts (Shier & Owings 2007). If conservation benefits of captive breeding are to be realized, animals reared in captivity must be behaviorally competent in order to be reintroduced (Alberts 2007). At a minimum, captive-reared animals need to be able to select appropriate habitat for settlement, find shelter, find, acquire and process suitable food, interact appropriately with conspecifics, avoid predators and navigate and locomote successfully in natural habitats (Kleiman 1989, Chizar et al. 1993, Alberts 2007).

For some species, the development of effective behavioral skills for survival following release to the wild requires modifications to the captive environment and/or behavioral modifications through direct training. Captive housing enrichment not only provides a cognitive challenge and may reduce abnormal behavior, if present (Shyne 2006), it can also facilitate the development of effective locomotory skills. Enrichment for primates held in zoos often includes enclosures that allow for movement through complex three-dimensional arboreal habitat (Reading et al. 2013). This type of enrichment has been shown to affect the locomotory behavior of released animals. For example, the golden-lion tamarin reintroduction program allows animals slated for release to limited free ranging in a complex three-dimensional space (Beck et al. 1994, Stoinski et al. 2003, Stoinski & Beck 2004). Tamarins allowed to free range fell from trees less, spent more time on natural substrates and traveled at greater heights following release compared to tamarins released without being permitted to free range (Stoinski & Beck 2004). While there have been no direct fitness assessments of tamarins allowed to free range compared to those that are not allowed to free range, adults that survived longer than 6 months following release used more natural substrates and locomoted more compared to adults that survived less than 6 months (Stoinski & Beck 2004).

Behavioral modification training can be used in a variety of captive contexts: to facilitate the development of effective behavior by naïve animals, to help wild-born animals to maintain their wild-type behavior (Kleiman 1989) and/or to provide cognitive stimulation and challenge (Meehan & Mench 2007). Many captive breeding programs that are implemented for the purpose of reintroduction have begun to institute some form of behavioral modification training prior to release (Kleiman 1989, McLean 1997).

## 10.2 PREDATOR TRAINING – A KEY COMPONENT OF SURVIVAL COMPETENCY

One of the critical components of survival following reintroduction is avoiding predators. The ability to recognize and detect predators through attention to sensory cues, remaining vigilant, seeking and effectively utilizing cover, and responding to predator presence when appropriate are all important aspects of behavioral competency (Chizar *et al.* 1993). Because ineffective anti-predator behavior often results in death, historically, it was thought to develop in the absence of experience. Certainly, it is the case that many young animals perform species-typical anti-predator behaviors without prior experience. The C-start or fast-start startle response in fish (Eaton *et al.* 1977) or freezing behavior exhibited by three-spined sticklebacks, *Gasterosteus aculeatus*, in response to a looming stimulus are classic examples of a comparatively innate antipredator behavior (Giles 1984). Yet, a growing literature shows that anti-predator behavior is complex, the development of effective anti-predator skills can require some form of experience and animals that initially exhibit ineffective anti-predator skills can be trained to recognize and respond to live or model predators (see reviews in Griffin *et al.* 2000, Brown & Laland 2001) or predator cues (Brown & Smith 1998, Mirza & Chivers 2000, Wisenden *et al.* 2004).

Predator training typically involves some form of associative learning (Chapter 3), in which animals are purposely exposed to predators or cues of predator presence (CS; Conditioned Stimulus, predator or predator model) paired with an aversive stimulus (US; Unconditioned Stimulus, e.g. spray of water). The outcome of which is a learned association between the predator and a negative experience. This type of predator training teaches animals to recognize and avoid predators. Some may consider anti-predator training through exposure to predators to contradict welfare of captive animals because predator training often requires subjecting them to stressful stimuli (Teixeira *et al.* 2007). However, exposure to predators in a training context is a short-term acute stressor and thus not likely to lead to distress (defined as a biological cost of stress that causes the animal to divert energy away from normal biological functions such as the immune system; Moberg 2000). By contrast, nature contains stressors that can enhance cognitive processes and hone animals' adaptive behavioral responses (Teixeira *et al.* 2007, Reading *et al.* 2013), and without exposure animals slated for release will likely experience distress arising from release into a novel wild habitat for which the animal has no adaptive response (Teixeira *et al.* 2007, Zidon *et al.* 2009, Swaisgood 2010). Thus, predator training may reduce the negative effects of distress experienced by releases and improve welfare.

Predator avoidance training is increasingly being implemented in recovery programs that require captive breeding and reintroduction (e.g. captive breeding programs for: small eutherian mammals, Yoerg 1996, McLean *et al*. 1999; marsupials, McLean *et al*. 2000; birds, Heatley 2002, Shier 2013).

Developing a training protocol to modify a target species' behavior requires an understanding of several aspects of the target species: the species' natural history, in particular, its habitat type, lifestyle, the length of time isolated from predators (generations or duration in captivity), social behavior, the range and type of predators (e.g. native vs. invasive, mammalian vs. raptorial) that the target species encounters in the wild, the variation and form of anti-predator behavior wild-born individuals exhibit (e.g. freeze, flee, concealment, return to refuge, evasive maneuvers, vigilance), and the extent to which the target species interacts with potential competitors. Understanding these aspects of the target species will allow the researcher to determine: (1) the number of predator types with which to train the target species; (2) the type of training that may be most effective; (3) the number of training events required to modify behavior; and (4) appropriate competency goals.

### 10.2.1    Identifying predator stimuli

Most prey species have multiple predators and their anti-predator defenses have coevolved with various aspects of these predators such as morphology and hunting style. Recognizing and avoiding different predators may require different predator-specific strategies (Owings & Coss 1977, Cheney & Seyfarth 1990, Caro 2005). Consider black-tailed prairie dogs that must avoid being killed by raptorial predators, terrestrial mammals and reptiles. In the presence of mammals and avian predators, adult prairie dogs bark repetitiously, warning offspring and non-descendant kin of impending danger (Hoogland 1995). Upon hearing a bark alarm call, prairie dogs scan for predators and, if one is detected, run to a burrow mound and either enter or begin calling while facing the predator (Hoogland 1995). Behavior toward snakes is very different, because of both the distinct threat posed by snakes and the apparent low vulnerability of adult prairie dogs to rattlesnakes (Owings & Loughry 1985). During interactions with snakes, adults typically orient toward and approach the snake in an elongated posture, sniff, head-bob, jump away and give distinct jump-yip calls and/or footdrum (Owings & Owings 1979, Halpin 1983,

Owings & Loughry 1985). With multiple predators and so much variation in anti-predator defenses, it would be impractical to train captive-born animals with every predator that they will encounter in the wild. So how do you determine which predators to use for anti-predator training? One way is to determine which predator is expected to exert the most pressure on the target species at the release site and start there. However, for many species, this data is not available. Another strategy is to classify multiple predators by hunting style and/or body morphology and train the target species with one archetype predator that represents each category. This approach is promising because several species exhibit anti-predator responses to broad categories of predators (prairie dogs, Hoogland 1995; titi monkeys, Caesar et al. 2012). Further, some predator training research has shown that training may be generalizable across different predator types (e.g. fox to cat or domesticated dog to fox; McLean et al. 2000, Griffin & Evans 2003), especially if they have similar body morphology (McLean et al. 2000). While this approach may elicit anti-predator responses to nonpredatory species and thus be energetically costly in the short term, if the trained individuals survive, there will be opportunities to refine predator recognition post release to reduce energetic costs.

### 10.2.2 Selecting the CS and US

For each predator type, a predator (CS) is paired with a (US) to teach naïve animals to recognize, detect and avoid these animals when they counter them in the wild. While exposure to live predators may provide important information regarding variation in cues, such as morphology, body posture, speed and gait, in most situations risk of injury is too great to allow direct interactions. Thus, for training, most researchers have used either live predators behind some type of barrier or in a separate but adjacent enclosure (Chivers & Smith 1994, Yoerg & Shier 1997, Shier & Owings 2006, Shier & Owings 2007), or some form of predator cue(s) (e.g. a taxidermic predator model: visual cue, Griffin et al. 2000; or predator odor: olfactory cue, Brown & Smith 1998, Mirza & Chivers 2000, Wisenden et al. 2004) as the CS.

For a US to be effective, it must elicit the same motivational state in the subject as a natural predation event (Griffin et al. 2000). In the terrestrial environment, the most common aversive stimulus used is a simulated capture (Griffin & Evans 2003, de Azevedo & Young 2006b, Mesquita & Young 2007, Moseby et al. 2012). Simulated captures are thought to elicit fear similar to a natural predation event. Other effective negative experiences used include pairing predator exposure with a live dog trained to chase but not capture individuals to be trained (McLean *et al.* 2003), water spray (McLean *et al.* 1996), loud noises (McLean *et al.* 1996), visual cues such as presentation of a dead individual of the target species in the mouth of the predator (Maloney & McLean 1995) or alarm vocalizations (McLean *et al.* 1996, Shier & Owings 2006, Campbell & Snowdon 2009). While presentations of heterospecific alarm calls have been used, results indicate that they may not be as effective as conspecific calls. New Zealand robins, *Petroica australis*, exposed to blackbird alarm calls paired with the simulated mobbing by a model predator elicited lower rates of reactive behavior than controls (Maloney & McLean 1995). In the aquatic environment, chemical cues such as alarm pheromone from the skin of fish are most commonly employed to train hatchery-reared fish to recognize and avoid predators (Brown & Smith 1998, Berejikian *et al.* 2003, Wisenden *et al.* 2004, Olson *et al.* 2012).

### 10.2.3 The use of sensory cues

Classification of predators into categories that represent different hunting styles may also help determine the ideal form of training. Detection of predatory raptors is primarily through visual, motion and/or auditory cues (Carlile *et al.* 2006). Thus, training a species to detect a raptor predator may be most effective if a model raptor, say a barn owl, mounted in a hunting pose with wings extended and talons out, is presented above the training arena (CS). Because raptors are looming stimuli that elicit an innate overhead fright response (Giles 1984, Maier *et al.* 2004, Shier & Owings 2006), overhead motion toward the subject can be used as the US. Thus, if the model raptor is placed on a cable, the visual cue can be paired with a motion cue and the target species will experience a simulated attack of a raptor (CS and US). Including a playback of the raptor vocalizations prior to and during predator presentation has been shown to enhance vigilance following training (Shier & Owings 2006). Pairing the visual and motion cues with the auditory cue creates a multimodal cue that the subject can use for early predatory detection and may provide a survival benefit following release. Avoidance strategies such as early detection are predicted to have the greatest influence on the outcome of a potential predation interaction (Endler 1991).

By contrast, to facilitate recognition and avoidance of reptile predators, such as snakes that utilize a sit and wait ambush hunting strategy, visual presentation of a live or model predator (CS) is required for learned predator recognition. Yet, scent cues are likely to be key to detection post release as visual detection ability may be limited if, say, the snake is hidden within a burrow entrance or is cryptically colored. Thus, if a model rather than a live predator is used, scent cues may enhance training benefits. For training a species to recognize terrestrial mammalian predators such as cougars, wolves, coyotes, stoats and weasels, which actively hunt, stalk and pounce, visual, olfactory, auditory and/or motion cues may be necessary, but may be most effective if combined with other stimuli to create a multimodal cue. Multimodal cues improve detection and location information, reduce ambiguity and are thought to amplify the information context in a signal (Narins et al. 2005, Partan & Marler 2005), and some species have been shown to exhibit anti-predator behavior when presented with multimodal cues but not with the same cues presented in isolation (e.g. tadpoles, *Oophaga pumilio*; Stynoski & Noble 2012).

Much research on life skills training has been conducted in aquatic systems (Brown & Laland 2001). For fish and amphibians, both visual and chemical predator recognition cues (CS) have been effectively used during training (Chivers & Smith 1998, Smith 1999, Manassa et al. 2013b). In the aquatic environment, chemical alarm cues released by injured or disturbed prey animals are thought to be a reliable indicator of predation risk (Ferrari et al. 2010, Crane & Mathis 2011) because they are normally released when either the skin is damaged during a predatory event (Chivers & Smith 1998) or when the animal exhibits a fright response (Crane & Mathis 2011). Much research has shown that these chemical alarm cues can be used as a US and have wide-reaching effects in the aquatic environment, but are most effective when paired with a predator model.

### 10.2.4 The number of training events

The most effective number of training events will likely depend on the length of time in isolation (the number of generations in captivity), the target species, the style of predatory behavior and/or the retention and competency goals. Large-scale loss of behavioral responses are more likely the result of separation from selection over several generations (relaxed selection), while lack of experience with predators during an animal's lifetime is more likely to cause suboptimal anti-predator behavior (Brown & Laland 2001, Blumstein 2006). Thus, it follows that animals sequestered in captivity from their predators for several generations may have more impoverished survival skills and require more training than those isolated for only a few generations (Brown & Laland 2001). For species in which isolation from predators is relatively recent, the number of training events required to elicit behavioral competency in anti-predator skills may vary. In some species, predator training has been shown to be effective in only one to a few exposures (Griffin et al. 2000, Brown & Laland 2001). In fact, it has been thought that too many exposures may result in habituation of captive animals to the predator and negate any beneficial effects of training (Griffin et al. 2000, Teixeira & Young 2014). Yet, evidence from some species suggests that extended training may not cause habituation and can, in fact, be beneficial. In Nile tilapia (Oreochromis niloticus), only three training trials were required for conditioned fish to exhibit species-typical anti-predator behavior in which individuals swam quickly away from the stimulated predator in multiple directions: the "scatter-effect" (Mesquita & Young 2007). After twelve training trials, however, tilapia were still exhibiting the erratic swimming behavior during a simulated capture, and a few of the tilapia expressed a new anti-predator response, rising to the surface of the training tank and freezing (Mesquita & Young 2007). This behavior was considered adaptive because the simulated predator was presented from the bottom of the tank. By the twenty-seventh trial, all trained fish exhibited both behaviors. Thus, extending the number of training events in this case was beneficial. Whether habituation to behavioral modification training occurs will depend, in part, on the type of predator that the target species is being trained to recognize and avoid. For example, for prey species with aerial raptorial predators, predator recognition is visual and/or auditory. Because raptors are looming stimuli that elicit an innate overhead fright response (Giles 1984), the number of training trails necessary will likely be minimal (Shier & Owings 2006, Shier & Owings 2007).

### 10.2.5 Timing of training – effects of ontogeny and retention

To be most effective, anti-predator training should mimic the critical features of ontogenetic processes in the wild; the stimuli used, the developmental timing of the training and the social and physical context in which it occurs must all be appropriate. While it has been posited that it would be maladaptive for animals to be capable of learning anti-predator skills only at specific developmental stages (Griffin et al. 2000), most animals learn better earlier in life (Vargas & Anderson 1998, Vargas & Anderson 1999), and some species have sensitive periods in which they imprint (Bateson 1981). Consider salmon, which have a sensitive period for olfactory imprinting. Research shows that only coho salmon (Oncorhynchus kisutch) exposed to an odorant during the smolt stage demonstrate an increased attraction for this odorant (Dittman et al. 1996). Taking this further, a study on Atlantic salmon (Salmo salar) indicates that the sensitive period is critical for appropriate predator response. Ten-to 15-week-old salmon respond more strongly to predator odor than younger or older juvenile fish (Hawkins *et al.* 2008). Thus, providing training early in an animal's life, but not too early, promises to improve training outcomes and post-release survival (Yoerg & Shier 1997, Shier & Owings 2006, Reading *et al.* 2013), and targeting sensitive periods, if they exist, may maximize the effectiveness. No study has yet directly compared the anti-predator behavior and post-release survival of captive-reared adults that were trained as juveniles or adults. Thus, it is not yet clear whether captive-reared animals trained as adults could, with practice, perform tasks as efficiently as animals trained as juveniles. However, research shows that maturation and experience can interact to shape anti-predator behavior. When Heermann's kangaroo rats (*Dipodomys heermanni*) were exposed to a snake, wild-caught adults were more wary compared to both captive-born juvenile and adult kangaroo rats. Captive-born juveniles were more active than either group of adults and showed no differences in behavior in the presence or absence of a live snake (Yoerg & Shier 1997). Kangaroo rats are solitary as adults but young are philopatric, providing a short window of time for social learning from their mothers. Young kangaroo rats shadow their mother in the presence of snakes and may learn anti-predator skills from her (Yoerg & Shier 1997). While animals that miss critical experiencing during their imprinting period may still learn to perform tasks, they would likely do so less efficiently than animals that received training during the appropriate time period (Reading *et al.* 2013).

Mortality following reintroduction is highest during the establishment phase, in the days to weeks following release (Shier & Swaisgood 2012), and the cause of death is often predation (Short & Smith 1994, Fischer & Lindenmayer 2000, Letty *et al.* 2000). The elevated risk of predation is not only the result of inexperience, but likely is also impacted by the distress caused by the transfer and release into a novel environment (Teixeira *et al.* 2007). This, in turn, dictates taking in and analyzing large amounts of spatial data and modifying behavior accordingly (Berger-Tal & Saltz 2014). For anti-predator training to translate to influence post-release survival, the trained animals need to exhibit sustained behavioral modification during this establishment period. Once settled, experience with predators will reinforce the training. Retention time will likely reflect the target species' memory capacity for learning, the species' lifespan and the effectiveness of training. Effectiveness of training and retention can be examined by retesting trained subjects at various points in time following training to compare behavioral responses. Fish studies show anti-predator training retention of as long as 75 days (Chivers & Smith 1994, Brown & Smith 1998, Mesquita

& Young 2007), while a study on flightless greater rheas (*Rhea americana*) showed birds retained behavioral modification for a minimum of 3 months (de Azevedo & Young 2006a). Thus, depending on the life history of the target species, training programs should begin while captive individuals are young and be reinforced in the months to weeks prior to release.

### 10.2.6 Effects of social context on anti-predator training

It is well established that social interactions play a major role in the acquisition and development of learned behaviors across taxa (Galef 1988). Interactions with experienced conspecifics can facilitate learning in inexperienced young; reducing the time, energy and fitness costs associated with exploring the environment and/or learning survival skills alone (Brown & Laland 2001, Shier & Owings 2007). For most group-living species and those with extended parental care, the development of predator-avoidance skills is socially learned (reviewed in Griffin 2004). Social training regimes are likely to be more effective for these species because they mimic natural processes. Captive-reared juvenile black-tailed prairie dogs trained with predators in the presence of experience conspecifics (demonstrators) were more wary than those trained without an experienced adult and were nearly twice as likely to survive reintroduction. While wild-reared juveniles originating from the same family groups were more vigilant than demonstrator-trained juveniles prior to release, one year following reintroduction there were no differences in survival between captive-reared juveniles trained with demonstrators and their wild-reared counterparts (Shier & Owings 2007). Importantly, these results suggest that predator avoidance training can emulate experience in the wild if the social context of the training regime used is species appropriate. Social demonstrators may also be critical for species that use mobbing as a form of anti-predator defense. For these species, predator recognition may be learned via context-appropriate playbacks (i.e. playbacks of conspecific mobbing calls), but observing experienced conspecifics may be required for learning to mob. For example, captive-reared cotton top tamarins (*Saguinus oedipus*) trained with snakes paired with conspecific mobbing vocalizations alone did not learn to mob the snakes (Campbell & Snowdon 2006). The authors surmised that observational learning from demonstrators may enhance learning to mob in tamarins.

There is abundant evidence from fish that incorporating the social context into anti-predator behavioral modification has the potential to improve training outcomes. Hatchery-reared fish are produced in mass to support the fishery industry and for the conservation and management of endangered species (Brown & Laland 2001). Due to the excessive mortality rates post release, two decades of research have been dedicated to elucidating the mechanisms that underlie anti-predator recognition and avoidance in fish. Anti-predator training can inculcate and/or improve anti-predator responses in fish (Brown & Laland 2001) and basic research has shown that members of shoals monitor and utilize behavior and cues of group members (Elgar 1989, Brown & Laland 2001, Griffin 2004, Ferrari et al. 2010). The social transmission of predator recognition and avoidance has been reported in a variety of fish species (Brown & Laland 2001) and can spread rapidly through a predator-naïve population after introduction of a predatory species (Chivers & Smith 1995).

Developing a social training regime is expected to improve the efficacy of training through the incorporation of appropriate social cues and contexts and to reduce the cost (Brown & Laland 2001, Olson et al. 2012). Furthermore, because shoals are composed of mixed species groups, which show social transmission of fright responses (Krause 1993) and predator recognition (e.g. coral reef fish; Manassa et al. 2013a), social training regimes need not be limited to intraspecific learning opportunities.

### 10.2.7 Future directions in anti-predator training and research
A new method of group training was developed using fathead minnows (*Pimephales promelas*) as a model. Researchers exposed a group of thirty naïve minnows to conspecific skin extract and predator odor cues for 2 hours while a control group was not exposed to any odors. Twenty-four hours later fish from both groups were tested for a behavioral response to predator odor. Conditioned fish reduced activity during exposure to predator odor but the control fish did not (Olson et al. 2012). While more research is needed to verify the efficacy of this approach, results for large-scale training are promising. Overall, more controlled studies are needed in which trained and non-trained animals are released into the wild simultaneously and monitored to examine the fitness outcomes of predator training programs. Such studies, while costly in the short run (both monetarily and in terms of success), may greatly enhance our understanding of the relationship between training and post-release survival.

### 10.3 FORAGING SKILLS

The ability to identify, locate, manipulate, ingest and process natural food and water resources is critical for survival following release to the wild (Alberts 2007). Akin to behavioral deficiencies seen in anti-predator behavior, captive-born and/or reared animals often demonstrate foraging deficiencies (Kleiman 1989, Beck *et al.* 1994, Brown & Laland 2001, but see Alberts 2007 for a review of iguana species that show no deficiency). Captive-born golden-lion tamarins show two recognition problems. They make inappropriate food choices and they fail to recognize high-quality foraging sites (Kleiman *et al.* 1990, Stoinski *et al.* 2003). Not surprisingly, these foraging deficiencies are caused by the depauperate rearing environment. For example, foods offered in captivity often differ from those that the target species utilizes in the wild. The artificial diet of hatchery-reared turbot (*Scophthalmus maximus*) was found to impair foraging behavior of released individuals (Ellis *et al.* 2002).

For herbivorous prey species, simply offering animals the foods that they will encounter following release may help develop food recognition, foraging skills and an effective gut flora (Ortega-Reyes & Provenza 1993, Young 1997, Villalba *et al.* 2004). For many large herbivores, their tenure in a habituation enclosure with representative vegetation may also be sufficient to develop their foraging skills (Dolev *et al.* 2002). Yet, similar to antipredator behavior, there may be a sensitive period during which animals learn these skills most efficiently (Ortega-Reyes & Provenza 1993, Stoinski *et al.* 2003) and experienced social group members or, in solitary species, mothers, may be important for identifying appropriate plant species for consumption or avoidance

Predator species need experience to efficiently recognize, locate, hunt, chase and kill prey. While regulations in some countries restrict foraging training activities, this type of experience can be critical for post-release survival in animals slated for reintroduction (Reading *et al.* 2013). In fish, spatial heterogeneity and variable cue presentation can significantly improve the transition of hatchery-reared cod (*Gadus morhua*) from a captive pellet diet to live fish prey (Braithwaite & Salvanes 2005). But for other predator species, experience with live prey may be required as interactions with live prey are dynamic and multi-modal signals that include motion cues may be critical for development. Consider the development of hunting behavior in the black-footed ferrets (*Mutela nigripes*). Juvenile ferrets raised in pens with live prairie dogs displayed higher predator proficiency than cage-reared ferrets (Vargas & Anderson 1998, Biggins *et al.* 1999, Vargas & Anderson 1999). Prairie dogs constitute 99% of a black-footed ferret's diet in the wild and have evolved substantial antipredator defenses (Hoogland 1995), thus in the absence of foraging skill experience, released ferrets will likely die. Similarly, young felids learn to recognize prey species by observing the prey items that their mother brings back from her hunting trips (see review in Kitchener 1999). As young grow and accompany their mother on hunting trips, they learn to locate preferred prey (Kitchener 1999). If experience with prey items is limited to species that the target species is unlikely to encounter following release, the captive-reared animals may have the skills to hunt and take down prey but their ability to select appropriate prey may be lacking. In an experimental reintroduction of cheetahs (*Acinonyx jubatus*) that only had experience of killing Barbary sheep (*Ammotrugus Eewia*) in captivity, the cheetahs attempted to hunt giraffe calves (*Girafsa camelopardulis*), African buffalo (*Syncerus casfer*), zebra (*Equus burchelli*) and wildebeest (*Connochaets taurinus*), despite the presence of their typical natural prey – impala (*Aepyceros Melampus*) (Pettifer 1981). Whereas, when cheetahs were provided with a variety of dead and live injured prey species during rehabilitation, including impala, they were able to effectively hunt impala following release (Houser et al. 2011).

## 10.4 HABITAT SELECTION BEHAVIOR

Reintroductions are plagued by immediate release site rejection and long-distance movements following release (Griffith *et al.* 1989, Kleiman 1989, Short *et al.* 1992, Miller *et al.* 1999). In fact, it is not uncommon for newly released animals to travel much farther than species typical dispersal distances (Stamps & Swaisgood 2007). Delayed settlement has been shown to reduce survival following release (Bright & Morris 1994, Moehrenschlager & Macdonald 2003, Shier & Swaisgood 2012). Thus, determining how to expedite settlement promises to improve reintroduction outcomes.

Several factors involving the pre-release environment are likely to influence the probability of successful settlement and survival following release. While it is common to consider some measure of overall habitat quality when selecting a release site habitat (Griffith *et al.* 1989, Fischer & Lindenmayer 2000, Moorhouse *et al.* 2009), the habitat selection behavior of releasees may prove to have an even larger impact on reintroduction success.

In the wild, animals have been shown to imprint on cues in their natal habitat and experience in natal habitat can increase an animal's preference for post-dispersal habitat that contain cues comparable to those in their natal habitat (Natal Habitat Preference Induction, NHPI; Davis & Stamps 2004). Thus, matching captive and release habitat cues may expedite settlement (Stamps & Swaisgood 2007) and reduce risky exposure time during settlement.

If an animal's ability to survive and thrive in a new habitat requires life-skills training that is habitat specific, then pre-release skill training must include appropriate habitat cues for satisfactory performance in the post-release habitat. This can be illustrated with fathead minnows. In the presence of cover (weeds) there was no difference in the survival of minnows conditioned with conspecific alarm cues compared to those conditioned with control odor. Yet, when no cover was present during training, minnows conditioned with conspecific alarm cues survived longer than those conditioned with control odor (Gazdewich & Chivers 2002). In contrast, if experience in the original habitat is primarily used as a shortcut for assessing habitat quality, then managers need to ensure that cues in the captive habitat match those in the release habitat, or confine and protect animals at the release site long enough for them to learn that it is suitable for long-term occupancy (Stamps & Swaisgood 2007).

Important cues could include the presence of conspecifics and/or their odors or vocalizations, familiar refuge types, vegetation type or structure, substrate type, and food (Stamps & Swaisgood 2007). In some cases, animals might learn to use man-made structures while in captivity. These structures can then be used in the post-release environment to encourage settlement. For instance, latency to find an acclimation cage burrow composed of black plastic corrugated tubing was shorter for Stephens' kangaroo rats held in captivity with the same type of artificial burrows compared to those held with cardboard artificial burrows (Shier, unpublished data). Similarly, mallard ducks, *Anas platyrnchos*, hatched in elevated nestboxes, reared in captivity and subsequently released, returned to use the elevated nestboxes as adults. Likewise, wild-hatched ducks reared in the elevated nestboxes chose to use the elevated nestboxes for laying once they became sexually mature. By contrast, mallards hatched on the ground and reared in a ground nest made their nests on the ground as adults (Hess 1972). These types of man-made cues are likely to have only short-term benefits. Ideally, if the release site has been selected far enough in advance, habitat characteristics from the site could be replicated in captivity and provide a survival benefit following release.

## 10.5 SOCIAL BEHAVIOR

Among wildlife, social living varies from solitary to colonial and hence the experiences required for instilling effective social skills will vary. Despite living alone, solitary species require some basic social skills to thrive in the wild, including: production and reception of intraspecific signals, assessment of intent and adaptive response to social situations and location, and attraction to/of suitable partners with which to mate (Alberts 2007). Providing opportunities for the development of these behaviors will require exposure to conspecifics during development. For solitary species with some parental care, such as during a period of philopatry, a sensitive period during which social skills are learned may include social learning from mothers. Solitary animals are typically aggressive with conspecifics. Thus, effective captive housing will limit direct interactions while allowing for regular communication. The captive housing regime for the solitary Morro Bay kangaroo rat (*Dipodomys heermanni morroensis*) is illustrative of this point. In the captive facility, "social" cages were employed in which animals were housed in individual units but adjacent to opposite-sex neighbors. Clear, perforated barriers separated individuals to allow for visual, auditory and scent communication and facilitated estrous cycling (Yoerg 1994). When clear barriers were replaced with opaque ones, females stopped cycling (Yoerg 1995). Hence, designing species-specific housing is one important component of effective captive rearing. Because of the social limitations required in housing solitary animals, exposure to conspecifics outside of home cages may be required for the development of social skills. Yet, caution must be taken to ensure sufficient space and refuges during interactions to minimize the probability of injury. Ideally, experience with experienced conspecifics would be given at several time points during development to allow captive-reared animals to hone social skills.

For individuals that live in stable groups, a social group, rather than the individual or a single additional social partner, forms the backdrop for social learning (Heyes & Galef 1996), and social training regimes are likely to be more effective because they mimic natural processes (Shier & Owings 2007). While the social context is important for behavioral modification during training in each of the behavioral categories previously discussed (anti-predator, foraging and habitat selection), the development of effective mating behavior is critical because it is directly related to population dynamics and therefore is important for both captive propagation and fitness following release. Effective mating behavior can depend on species recognition, selection of high-quality mates and communication as a form of courtship. Maladaptive responses by inexperienced young in any of these ways can prevent reproduction.

To begin, captive young must imprint on the correct cues for species and mate recognition. Historically, puppet rearing has been used to reduce filial and sexual imprinting on humans (Valutis & Marzluff 1999). For captive-reared animals slated for reintroduction, both filial and sexual imprinting on conspecifics is important. Filial imprinting on caretakers can reduce stress in captive animals but cause released animals to seek out human contact, which may translate to reduced survival following release. Crows reared without puppets were less fearful of human caretakers and had lower survival following release compared with crows reared with puppets (Valutis & Marzluff 1999). Sexual imprinting on conspecifics rather than humans is important both for captive propagation and for fitness following release. For species in which social housing is an option, social integration during development will likely provide the appropriate species-specific cues and conditions for sexual imprinting.

Historically, breeding considerations in conservation breeding programs have been primarily focused on genetics – minimizing inbreeding by mating nonrelatives and maximizing effective population size ($N_e$) by minimizing variance in reproduction among individuals (Grahn et al. 1998). Yet, there has been a call for integration of species-appropriate female mate choice into captive propagation (Grahn et al. 1998, Quader 2005). In the absence of mate choice, problems such as social and/or genetic incompatibility, lowered offspring quality, and/or reduction in female fecundity and investment in offspring may manifest. To illustrate this point consider female zebra finches (*Taenopygia guttata*). Female finches prefer males with leg bands of certain colors. Females mated to preferred males invest more and show a bias in the sex ratio compared to females mated to nonpreferred males (Burley 1981). An example of how mate choice directly influences fitness can be found in house mice (*Mus musculus*). Female mice mated to preferred males had larger litters than females mated to nonpreferred males. When mated pairs preferred each other, pre-weaning pup mortality was substantially lower (8.1% compared to 23.9% for pairs who did not prefer each other; Drickamer et al. 2000). Further, if the release is planned as a reinforcement of a low-density resident population and captive-born animal behavior is suboptimal, wild conspecifics may reject them as mates. Hence, captive programs for the purpose of reintroduction may need to provide animals with the opportunity to develop and refine courtship and mating behavioral skills.

## 10.6 BENEFITS OF A COMPLEX CAPTIVE ENVIRONMENT

Care must be taken to ensure that the captive environment does not encourage the development of maladaptive preferences and/or behaviors (see Chapter 8). Rearing bank voles in small impoverished cages has been shown to inculcate preferences for similarly impoverished captive environments over enriched ones (Cooper & Nicol 1991). This preference for a barren environment was strongly associated with the development of stereotypic coping behavior, which, once established, persisted even with cage enrichment training (Cooper et al. 1996). This can be illustrated by captive-reared Guam rails. Captive-reared Guam rails showed a strong preference for traveling along roads, apparently because they were familiar with more open habitat from their pre-release zoo enclosures, seeking more open habitat similar to their former zoo enclosures in the United States, and were unwilling to enter the dense closed forests that comprise their native habitat (Witteman et al. 1990, cited in Stamps & Swaisgood 2007).

In addition to enhancing an animal's survival skills, exposure to predators, variation in forage, complex habitat features, a selection of mates, etc. may provide a necessary challenge for the development of cognitive ability in captive-reared animals (Meehan & Mench 2007). Animals exposed to greater environmental complexity show increased neural synaptic connections (Rosenzweig 1979) and improved cognitive function (Young 2003). There is ample evidence that environmental enrichment reduces frustration and stereotypes in captive animals (Young 2003, Swaisgood & Shepherdson 2006), but the incorporation of opportunities for problem solving into captive environments has only recently begun to be evaluated. To date, research has shown that cognitive challenges reduce aggression (pigs; Rauterberg et al. 2013), increase social association and play (crested macaques; Whitehouse et al. 2013), improve learning (African cichlids; Kotrschal & Taborsky 2010), can alter time budgets of captive animals to match those of their wild counterparts (chimpanzees; Yamanashi & Hayashi 2011) and enhance captive survival (Bell et al. 2009). Integrating problem-solving protocols in captivity promises not only to improve captive welfare and fitness, but is likely to increase post-release survival for animals slated for reintroduction. Improved cognitive abilities and the resultant reduction in stress and improved behavioral flexibility will help animals respond quickly to environmental dynamics found in the wild.

## 10.7  ASSESSMENT OF BEHAVIORAL COMPETENCY

As stated in the previous section, the goal of captive conditioning of naïve animals is to improve release outcomes by inculcating effective species-specific behavior into captive-reared animals. The task of selecting animals for release, then, involves setting competency criteria for fitness-related behaviors. Ideally, predator avoidance, foraging, habitat selection, and social and mating behavior of the target species have been studied in the wild prior to species decline. However, this is rarely the case. Thus, one approach would be to use controlled experiments to assess behavioral competency. For example, one could test wild-caught founders at the onset of the program to establish competency goals. In this way, captive-reared animals can be tested in the same tests at various time points during their tenure in captivity to determine if they meet the established criteria and can be included in the release group. While this approach relies on population means, individual variation in behavioral responses may be important for population viability following release (Watters & Meehan 2007). In a study with swift foxes, bold animals were more likely to perish following release (Bremner-Harrison et al. 2004). Because highly bold/ aggressive animals may not tolerate captive conditions, they may be artificially selected out of the captive population. However, bold animals may provide a critical role in the social group. For example, bold captive-reared black-tailed prairie dog juveniles spent more time digging in trials with predators compared to shy juveniles. Whereas, juveniles judged as shy spent more time vigilant and alarm calling (Shier, unpublished). Thus, groups composed of a range of behavioral types, released together may be more likely to survive reintroduction and persist following release.

Following release, behavioral competency can be reassessed by determining if released individuals choose to settle in a stable area that will allow them to survive and reproduce, are able to locate, identify, ingest and process food resources, are able to recognize and detect predators, and are capable of integrating into the natural breeding population in the wild.

## 10.8 CONSIDERATIONS FOR CAPTIVE BEHAVIORAL MANAGEMENT IN AN ANTHROPOGENIC WORLD

In the face of human-induced rapid environmental change (HIREC; Sih 2013), habitat alternations and invasive species abound, posing unique challenges for captive-reared releasees. While addressing the issue of reintroduction in an anthropogenic world is beyond the scope of this chapter, modifying captive animal behavior in response to novel predators or prey species in the release habitat or in response to critical habitat change is increasingly becoming a necessary component of captive behavioral modification.

One of the greatest threats to global biodiversity is the human-mediated movement of plants and animals beyond their natural distributions (Williamson 1999). Invasive species can substantially modify ecosystems and once established can be impossible to eradicate. When novel predators or toxic prey species are introduced, captive-reared naïve animals may require training to recognize and detect the novel species, and behave appropriately. In the case of captive-reared prey species and novel predators, such as house cats, predator training may be effective (greater bilbies; Moseby *et al.* 2012). Yet, for situations in which a novel toxic prey species is invasive, conditioned taste aversion (CTA) has been shown to be effective. Conditioned taste aversion (CTA) is a training strategy in which an animal is trained to associate the taste of a food item with post-consumption illness and thereafter avoids that food (Garcia *et al.* 1974). CTA has been used to mitigate the impacts of invasion by the highly toxic cane toad (*Bufo marinus*) for two native predators in Australia, the northern quoll (*Dasyurus hallucatus*) and the blue-tongued skink (*Tiliqua scincoides intermedia*). Both species were trained with CTA to avoid cane toads following release and in controlled experiments, trained subjects survived at significantly higher rates when compared to toad-naïve subjects (quolls: O'Donnell *et al.* 2010; skinks: Price-Rees *et al.* 2013). Thus, CTA appears to be an effective strategy for mitigating the impacts of invasive toxic prey species when eradication is not an option. It is possible that this same type of training could be used to reduce human/wildlife conflict following release of listed predator species that may come into contact with livestock.

A second area in which behavioral modification has been shown to benefit captive-reared animals in an anthropogenic world is in conditioning species to avoid artificial structures such as roads or powerlines. As large scavengers, California condors (*Gymnogyps californianus*) typically perch in the tree tops in the wild. The structure of those trees matches that of power line poles. Early in the species' reintroduction program, electrocution by powerlines was a primary source of mortality for released condors. Thus, every bird that is released today is trained to associate power poles with a mild shock prior to release, which has significantly improved release outcomes (Wallace M., pers. comm.).

## 10.9 SUMMARY

While the priority for species conservation is to keep animals *in-situ* whenever possible, in the face of human population growth and its associated consequences on biodiversity loss, captive breeding is increasingly implemented to protect imperiled species. Yet, impoverished captive environments do little to prepare animals for release back into the wild and ineffective behavioral skills are the primary cause of mortality following reintroduction. Behavioral modification training paired with captive environmental enrichment can help to make captive propagation and reintroduction an effective recovery strategy by facilitating the development of species-specific wild-type behaviors, stimulating cognitive abilities to improve animal welfare while in captivity and outcomes following release, and transitioning captive animals back into an anthropogenic world.

## REFERENCES

Alberts, A.C. 2007. Behavioral considerations of headstarting as a conservation strategy for endangered Caribbean rock iguanas. *Applied Animal Behaviour Science*, 102: 380–391.

Bateson, P.P.G. 1981. Ontogeny. In McFarland, D. (ed.) *The Oxford Companion to Animal Behavior*, pp 414–426. Oxford: Oxford University Press.

Beck, B.B., Rapaport, L.G. and Wilson, A.C. 1994. Reintroduction of captive-born animals. In Feistner, A. (ed.) *Creative Conservation*, pp 265–286. London: Chapman and Hall.

Bell, J.A., Livesey, P.J. and Meyer, J.F. 2009. Environmental enrichment influences survival rate and enhances exploration and learning but produces variable responses to the radial maze in old rats. *Developmental Psychobiology*, 51: 564–578.

Berejikian, B.A., Tezak, E.P. and LaRae, A.L. 2003. Innate and enhanced predator recognition in hatchery-reared chinook salmon. *Environmental Biology of Fishes*, 67: 241–251.

Berger-Tal, O. and Saltz., D. 2014. Using the movement patterns of reintroduced animals to improve reintroduction success. *Current Zoology*, 60: 515–526.

Biggins, D., Vargas, A., Godbey, J.L. and Anderson, S.H. 1999. Influences on pre-release experience on reintroduced black-footed ferrets (*Mustela nigripes*). *Biological Conservation*, 89: 121–129.

Blumstein, D.T. 2006. The multipredator hypothesis and the evolutionary persistence of antipredator behavior. *Ethology*, 112: 209–217.

Braithwaite, V.A. and Salvanes, A.G.V. 2005. Environmental variability in the early rearing environment generates behaviourally flexible cod: implications for rehabilitating wild populations. *Proceedings of the Royal Society Biological Sciences Series B*, 272: 1107–1113.

Bremner-Harrison, S., Prodohl, P.A. and Elwood, R.W. 2004. Behavioural trait assessment as a release criterion: boldness predicts early death in a reintroduction programme of captive-bred swift fox (*Vulpes velox*). *Animal Conservation*, 7: 313–320.

Bright, P.W. and Morris, P.A. 1994. Animal translocation for conservation: performance of dormice in relation to release. *Journal of Applied Ecology*, 31: 699–708.

Brown, C. and Laland, K. 2001. Social learning and life skills training for hatchery reared fish. *Journal of Fish Biology*, 59: 471–493.

Brown, G.E. and Smith, R.J.F. 1998. Acquired predator recognition in juvenile rainbow trout (*Oncorhynchus mykiss*): conditioning hatchery-reared fish to recognize chemical cues of a predator. *Canadian Journal of Fisheries and Aquatic Sciences*, **55**: 611–617.

Burley, N. 1981. Sex ratio manipulation and selection for attractiveness. *Science (Washington, DC)*, **211**: 721–722.

Caesar, C., Byrne, R., Young, R.J. and Zuberbuehler, K. 2012. The alarm call system of wild black-fronted titi monkeys, *Callicebus nigrifrons*. *Behavioral Ecology and Sociobiology*, **66**: 653–667.

Campbell, M.W. and Snowdon, C.T. 2006. Auditory playback alone is insufficient to condition captive-reared cotton-top tamarins to mob a predator. *American Journal of Primatology*, **68**: 56.

Campbell, M.W. and Snowdon, C.T. 2009. Can auditory playback condition predator mobbing in captive-reared *Saguinus oedipus*? *International Journal of Primatology*, **30**: 93–102.

Carlile, P.A., Peters, R.A. and Evans, C.S. 2006. Detection of a looming stimulus by the Jacky dragon: selective sensitivity to characteristics of an aerial predator. *Animal Behaviour*, **72**: 553–562.

Carlstead, K. 1996. Effects of captivity on the behavior of wild mammals. In Kleiman, D.G., Allen, M.E., Thompson, K.V. and Lumpkin, S. (eds.) *Wild Mammals in Captivity*, pp. 317–333. Chicago: University of Chicago Press.

Caro, T.M. 2005. *Antipredator Defenses in Birds and Mammals*. Chicago: University of Chicago Press.

Cheney, D.L. and Seyfarth, R.M. 1990. *How Monkeys See the World: Inside the Mind of Another Species*. Chicago: University of Chicago Press.

Chivers, D.P. and Smith, R.J.F. 1994. Fathead minnows, *Pimephales promelas*, acquire predator recognition when alarm substance is associated with the sight of unfamiliar fish. *Animal Behaviour*, **48**: 597–605.

Chivers, D.P. and Smith, R.J.F. 1995. Free-living fathead minnows rapidly learn to recognize predators. *Journal of Fish Biology*, **46**: 949–954.

Chivers, D.P. and Smith, R.J.F. 1998. Chemical alarm signalling in aquatic predator–prey systems: a review and prospectus. *Ecoscience*, **5**: 338–352.

Chizar, D., Smith, H.M. and Racliffe, C.W. 1993. Zoo and laboratory experiments on the behavior of snakes: assessments of competence in captive-raised animals. *American Journal of Zoology*, **33**: 109–116.

Crane, A.L. and Mathis, A. 2011. Predator-recognition training: a conservation strategy to increase postrelease survival of hellbenders in head-starting programs. *Zoo Biology*, **30**: 611–622.

Darwin, C.R. 1868. *The Variation of Animals and Plants under Domestication*. Baltimore, MD: Johns Hopkins University Press.

Davis, J.M. and Stamps, J.A. 2004. The effect of natal experience on habitat preferences. *Trends in Ecology & Evolution*, **19**: 411–416.

de Azevedo, C.S. and Young, R.J. 2006a. Do captive-born greater rheas *Rhea americana* Linnaeus (Rheiformes, Rheidae) remember antipredator training? *Revista Brasileira de Zoologia*, **23**: 194–201.

de Azevedo, C.S. and Young, R.J. 2006b. Shyness and boldness in greater rheas *Rhea americana* Linnaeus (Rheiformes, Rheidae): the effects of antipredator training on the personality of the birds. *Revista Brasileira de Zoologia*, **23**: 202–210.

Dittman, A.H., Quinn, T.P. and Nevitt, G.A. 1996. Timing of imprinting to natural and artificial odors by coho salmon (*Oncorhynchus kisutch*). *Canadian Journal of Fisheries and Aquatic Sciences*, 53: 434–442.

Drickamer, L.C., Gowaty, P.A. and Holmes, C.M. 2000. Free female mate choice in house mice affects reproductive success and offspring viability and performance. *Animal Behaviour*, 59: 371–378.

Eaton, R.C., Bombardieri, R.A. and Meyer, D.L. 1977. The Mauthner initiated startle response in Teleost fish. *Journal of Experimental Biology*, 66: 65–82.

El Balaa, R. and Blouin-Demers, G. 2011. Anti-predatory behaviour of wild-caught vs captive-bred freshwater angelfish, *Pterophyllum scalare*. *Journal of Applied Ichthyology*, 27: 1052–1056.

Elgar, M.A. 1989. Predator vigilance and group size in mammals and birds: a critical review of the empirical evidence. *Biological Reviews*, 64: 13–33.

Ellis, T., Hughes, R.N. and Howell, B.R. 2002. Artificial dietary regime may impair subsequent foraging behaviour of hatchery-reared turbot released into the natural environment. *Journal of Fish Biology*, 61: 252–264.

Endler, J.A. 1986. The newer synthesis some conceptual problems in evolutionary biology. *Oxford Surveys in Evolutionary Biology*, 224–244.

Endler, J.A. 1991. Interactions between predators and prey. In Krebs, J.R. and Davies, N.B. (eds.) *Behavioural Ecology: An Evolutionary Approach*, 3rd edn.Oxford: Blackwell Scientific Publication.

Ferrari, M.C.O., Wisenden, B.D. and Chivers, D.P. 2010. Chemical ecology of predator–prey interactions in aquatic ecosystems: a review and prospectus. *Canadian Journal of Zoology*, 88: 698–724.

Fischer, J. and Lindenmayer, D.B. 2000. An assessment of the published results of animal relocations. *Biological Conservation*, 96: 1–11.

Frankham, R. and Loebel, D.A. 1992. Modeling problems in conservation genetics using captive *Drosophila* populations: rapid genetic adaptation to captivity. *Zoo Biology*, 11: 333–342.

Galef, B.G. 1988. Imitation in animals: history, definition, and interpretation of data from the psychological laboratory. In Zentall, T.R. and Galef, B.G. (eds.) *Social Learning: Psychological and Biological Perspectives*, pp. 3–28. New Jersey: L. Erlbaum.

Garcia, J., Hankins, W.G. and Rusiniak, K.W. 1974. Behavioral regulation of the milieu interne in man and rat. *Science (Washington DC)*, 185: 824–831.

Gazdewich, K.J. and Chivers, D.P. 2002. Acquired predator recognition by fathead minnows: influence of habitat characteristics on survival. *Journal of Chemical Ecology*, 28: 439–445.

Giles, N. 1984. Development of the overhead fright response in wild and predator-naive three-spined sticklebacks, *Gasterosteus aculeatus*. *Animal Behaviour*, 32: 276–279.

Grahn, M., Landefors, A. and von Schantz, T. 1998. The importance of mate choice in improving viability in captive populations. In Caro, T. (ed.) *Behavioral Ecology and Conservation Biology*, pp. 341–363. Oxford: Oxford University Press.

Griffin, A.S. 2004. Social learning about predators: A review and prospectus. *Learning & Behavior*, 32: 131–140.

Griffin, A.S., Blumstein, D.T. and Evans, C.S. 2000. Training captive-bred or translocated animals to avoid predators. *Conservation Biology*, **14**: 1317–1326.

Griffin, A.S. and Evans, C.S. 2003. The role of differential reinforcement in predator avoidance learning. *Behavioural Processes*, **61**: 87–94.

Griffith, B., Scott, J.M., Carpenter, J.W. and Reed, C. 1989. Translocation as a species conservation tool: status and strategy. *Science*, **245**: 477–480.

Griffiths, R.A. and Pavajeau, L. 2008. Captive breeding, reintroduction, and the conservation of amphibians. *Conservation Biology*, **22**: 852–861.

Halpin, Z.T. 1983. Naturally occurring encounters between black-tailed prairie dogs (*Cynomys ludovicianus*) and snakes. *American Midland Naturalist*, **109**: 50–54.

Hawkins, L.A., Magurran, A.E. and Armstrong, J.D. 2008. Ontogenetic learning of predator recognition in hatchery-reared Atlantic salmon, *Salmo salar. Animal Behaviour*, **75**: 1663–1671.

Heatley, J.J. 2002. Antipredator conditioning in Mississippi sandhill cranes (*Grus canadensis pulla*). M.S. Thesis. Louisiana State University, pp. 1–96.

Hediger, H. 1964. *Wild Animals in Captivity* (Translated by G. Sircom; Foreword by Dr. Edward Hindle.). London: Butterworths Scientific Publications.

Hess, E.H. 1972. The natural history of imprinting. *Annals of the New York Academy of Sciences*, **193**: 1972.

Heyes, C.M. and Galef, B.G. 1996. *Social Learning in Animals: The Roots of Culture*. London: Academic Press.

Hoogland, J.L. 1995. *The Black-Tailed Prairie Dog: Social Life of a Burrowing Mammal*. Chicago: University of Chicago Press.

Houser, A., Gusset, M., Bragg, C.J., Boast, L.K. and Somers, M.J. 2011. Pre-release hunting training and post-release monitoring are key components in the rehabilitation of orphaned large felids. *South African Journal of Wildlife Research*, **41**: 11–20.

Kitchener, A.C. 1999. Watch with mother: a review of social learning in the Felidae. *Mammalian Social Learning: Comparative and Ecological Perspectives*: 236–258.

Kleiman, D.G. 1989. Reintroduction of captive mammals for conservation. *Bioscience*, **39**: 152–161.

Kleiman, D.G., Beck, B.B., Baker, A. *et al*. 1990. The conservation program for the golden lion tamarin, *Leontopithecus rosalia. Endangered Species Update*, **8**:82–85.

Kotrschal, A. and Taborsky, B. 2010. Environmental change enhances cognitive abilities in fish. *PLoS Biology*, **8**: e1000351.

Kraaijeveld-Smit, F.J.L., Griffiths, R.A., Moore, R.D. and Beebee, T.J.C. 2006. Captive breeding and the fitness of reintroduced species: a test of the responses to predators in a threatened amphibian. *Journal of Applied Ecology*, **43**: 360–365.

Krause, J. 1993. Transmission of fright reaction between different species of fish. *Behaviour*, **127**: 37–48.

Letty, J., Marchandeau, S., Clobert, J. and Aubineau, J. 2000. Improving translocation success: an experimental study of anti-stress treatment and release method for wild rabbits. *Animal Conservation*, **3**: 211–219.

Lickliter, R. and Ness, J.W. 1990. Domestication and comparative psychology: status and strategy. *Journal of Comparative Psychology*, **104**: 211–218.

Maier, J.X., Neuhoff, J.G., Logothetis, N.K. and Ghanzanfar, A.A. 2004. Multisensory integration of looming signals by rhesus monkeys. *Neuron*, **43**: 177–181.

Maloney, R.F. and McLean, I.G. 1995. Historical and experimental learned predator recognition in free-living New Zealand robins. *Animal Behaviour*, **50**: 1193–1201.

Manassa, R.P., Dixson, D.L., McCormick, M.I. and Chivers, D.P. 2013a. Coral reef fish incorporate multiple sources of visual and chemical information to mediate predation risk. *Animal Behaviour*, **86**: 717–722.

Manassa, R.P., McCormick, M.I., Chivers, D.P. and Ferrari, M.C.O. 2013b. Social learning of predators in the dark: understanding the role of visual, chemical and mechanical information. *Proceedings of the Royal Society Biological Sciences Series B*, **280**: 20130720.

Maynard, D.J., Flagg, T.A. and Mahnken, C.V.M. 1995. A review of seminatural culture strategies for enhancing the postrelease survival of anadromous salmonids. In Schramm, H.L. and Piper, R.G. (eds.) *American Fisheries Society Symposium; Uses and effects of cultured fishes in aquatic ecosystems*, pp. 307–314.

McLean, I., Lundie-Jenkins, G. and Jarman, P.J. 1996. Teaching an endangered mammal to recognise predators. *Biological Conservation*, **75**: 51–62.

McLean, I., Holzer, C. and Studholme, B. 1999. Teaching predator recognition to a naive bird. *Biological Conservation*, **87**: 123–130.

McLean, I., Schmitt, N.T., Jarman, P.J., Duncan, C. and Wynne, C.D.I. 2000. Learning for life: training marsupials to recognise introduced predators. *Behaviour*, **137**: 1361–1376.

McLean, I.G. 1997. Conservation and the development of behavior. In Clemmons J. R. and Buchholz R. (eds.) *Behavioral Approaches to Conservation in the Wild*, pp. 132–156. Cambridge: Cambridge University Press.

McPhee, M.E. 2003. Generations in captivity increases behavioral variance: considerations for captive breeding and reintroduction programs. *Biological Conservation*, **115**: 71–77.

McPhee, M.E. and McPhee, N.F. 2012. Relaxed selection and environmental change decrease reintroduction success in simulated populations. *Animal Conservation*, **15**: 274–282.

Meehan, C.L. and Mench, J.A. 2007. The challenge of challenge: can problem solving opportunities enhance animal welfare? *Applied Animal Behaviour Science*, **102**: 246–261.

Mesquita, F.d.O. and Young, R.J. 2007. The behavioural responses of Nile tilapia (*Oreochromis niloticus*) to anti-predator training. *Applied Animal Behaviour Science*, **106**: 144–154.

Miller, B., Ralls, K., Reading, R., Scott, J. and Estes, J. 1999. Biological and technical considerations of carnivore translocation: a review. *Animal Conservation*, **2**: 59–68.

Miller, B.D., Biggins, D., Hanebury, L. and Vargas, A. 1994. Reintroduction of the black-footed ferret (*Mustela nigripes*). In Olney, P.J.S., Mace, G.M. and Feistner, A.

T.C. (eds.) *Creative Conservation: Interactive Management of Wild and Captive Animals*, pp. 455–464. London: Chapman and Hall.

Mirza, R.S. and Chivers, D.P. 2000. Predator-recognition training enhances survival of brook trout: evidence from laboratory and field enclosure studies. *Canadian Journal of Zoology*, **78**: 2198–2208.

Moehrenschlager, A. and Macdonald, D.W. 2003. Movement and survival parameters of translocated and resident swift foxes *Vulpes velox*. *Animal Conservation*, **6**: 199–206.

Moorhouse, T.P., Gelling, M. and Macdonald, D.W. 2009. Effects of habitat quality upon reintroduction success in water voles: evidence from a replicated experiment. *Biological Conservation*, **142**: 53–60.

Moseby, K.E., Cameron, A. and Crisp, H.A. 2012. Can predator avoidance training improve reintroduction outcomes for the greater bilby in arid Australia? *Animal Behaviour*, **83**: 1011–1021.

Narins, P.M., Grabul, D.S., Soma, K.K., Gaucher, P. and Hodl, W. 2005. Cross-modal integration in a dart-poison frog. *Proceedings of the National Academy of Sciences of the United States of America*, **102**: 2425–2429.

O'Donnell, S., Webb, J.K. and Shine, R. 2010. Conditioned taste aversion enhances the survival of an endangered predator imperilled by a toxic invader. *Journal of Applied Ecology*, **47**: 558–565.

Olson, J.A., Olson, J.M., Walsh, R.E. and Wisenden, B.D. 2012. A method to train groups of predator-naive fish to recognize and respond to predators when released into the natural environment. *North American Journal of Fisheries Management*, **32**: 77–81.

Ortega-Reyes, L. and Provenza, F.D. 1993. Amount of experience and age affect the development of foraging skills of goats browsing blackbrush (*Coleogyne ramosissima*). *Applied Animal Behaviour Science*, **36**: 169–183.

Owings, D.H. and Coss, R.G. 1977. Snake mobbing by California ground squirrels: adaptive variation and ontogeny. *Behaviour*, **62**: 50–69.

Owings, D.H. and Loughry, W.J. 1985. Variation in snake-elicited jump-yipping by black-tailed prairie dogs: ontogeny and snake-specificity. *Zeitschrift fuer Tierpsychologie*, **70**: 177–200.

Owings, D.H. and Owings, S.C. 1979. Snake-directed behavior by black-tailed prairie dogs (*Cynomys ludovicianus*). *Zeitschrift fuer Tierpsychologie*, **49**: 35–54.

Partan, S.R. and Marler, P. 2005. Issues in the classification of multimodal communication signals. *American Naturalist*, **166**: 231–245.

Pettifer, H.L. 1981. The experimental release of captive-bred cheetah (*Acinonyn jubatus*) into the natural environment. Chapman, J.A. and Pursley, D. (eds.) *Worldwide Furbearer Conference Proceedings*, pp. 1001–1024.

Price-Rees, S.J., Webb, J.K. and Shine, R. 2013. Reducing the impact of a toxic invader by inducing taste aversion in an imperilled native reptile predator. *Animal Conservation*, **16**: 386–394.

Price, E.O. 1970. The effect of early outdoor experience on the behavior of a genetically heterogeneous strain of the laboratory rat. *American Zoologist*, **10**: 291.

Price, E.O. 1984. Behavioral aspects of animal domestication. *Quarterly Review of Biology*, **59**: 1–32.

Quader, S. 2005. Mate choice and its implications for conservation and management. *Current Science (Bangalore)*, **89**: 1220–1229.

Rabin, L.A. 2003. Maintaining behavioural diversity in captivity for conservation: Natural behaviour management. *Animal Welfare*, 12: 85–94.

Rauterberg, S., Sonoda, L.T., Fels, M. *et al.* 2013. Cognitive enrichment in the farrowing pen – a first approach to use early behavioural conditioning of suckling piglets to reduce aggressive behaviour during rearing. *Zuchtungskunde*, 85: 376–387.

Reading, R.P., Miller, B. and Shepherdson, D. 2013. The value of enrichment to reintroduction success. *Zoo Biology*, 32: 332–341.

Seddon, P.J., Armstrong, D.P. and Maloney, R.F. 2007. Developing the science of reintroduction biology. *Conservation Biology*, 21: 303–312.

Seidensticker, J. and Forthman, D.L. 1998. Evolution, ecology, and enrichment. Basic considerations for wild animals in zoos. In: Shepherdson, D.J., Mellen, J.D. and Hutchins, M. (eds.) *Second Nature: Environmental Enrichment for Captive Animals*. Washington DC: Smithsonian Institution Press.

Shier, D.M. 2013. Captive breeding, anti-predator behavior and reintroduction of the Pacific pocket mouse (*Perognathus longimembris pacificus*), annual report to United States Fish and Wildlife Service. San Diego Zoo Institute for Conservation Research, Escondido, CA, p. 41.

Shier, D.M. and Owings, D.H. 2006. Effects of predator training on behavior and post-release survival of captive prairie dogs (*Cynomys ludovicianus*). *Biological Conservation*, 132: 126–135.

Shier, D.M. and Owings, D.H. 2007. Effects of social learning on predator training and post-release survival in juvenile black-tailed prairie dogs (*Cynomys ludovicianus*). *Animal Behaviour*, 73: 567–577.

Shier, D.M. and Swaisgood, R.R. 2012. Fitness costs of neighborhood disruption in translocations of a solitary mammal. *Conservation Biology*, 26: 116–123.

Short, J., Bradshaw, S.D., Giles, J., Prince, R.I.T. and Wilson, G.R. 1992. Reintroduction of macropods (Marsupialia: Macropodoidea) in Australia – a review. *Biological Conservation*, 62: 189–204.

Short, J. and Smith, A. 1994. Mammal decline and recovery in Australia. *Journal of Mammalogy*, 75: 288–297.

Shyne, A. 2006. Meta-analytic review of the effects of enrichment on stereotypic behavior in zoo mammals. *Zoo Biology*, 25: 317–337.

Sih, A. 2013. Understanding variation in behavioural responses to human-induced rapid environmental change: a conceptual overview. *Animal Behaviour*, 85: 1077–1088.

Smith, R.J.F. 1999. What good is smelly stuff in the skin? Cross function and cross taxa effects in fish "alarm substances." In: Johnson, R.E., Muller-Schwarze, D. and Sorensen, P.W. (eds.) *Advances in Chemical Signals in Vertebrates*. New York: Springer.

Soule, M.E. 1986. *Conservation Biology: The Science of Scarcity and Diversity*. Sunderland: Sinauer.

Stamps, J.A. and Swaisgood, R.R. 2007. Someplace like home: experience, habitat selection and conservation biology. *Applied Animal Behaviour Science*, 102: 392–409.

Stoinski, T.S. and Beck, B.B. 2004. Changes in locomotor and foraging skills in captive-born, reintroduced golden lion tamarins (*Leontopithecus rosalia rosalia*). *American Journal of Primatology*, 62: 1–13.

Stoinski, T.S., Beck, B.B., Bloomsmith, M.A. and Maple, T.L. 2003. A behavioral comparison of captive-born, reintroduced golden lion tamarins and their wild-born offspring. *Behaviour*, **140**: 137–160.

Stynoski, J.L. and Noble, V.R. 2012. To beg or to freeze: multimodal sensory integration directs behavior in a tadpole. *Behavioral Ecology and Sociobiology*, **66**: 191–199.

Swaisgood, R. and Shepherdson, D. 2006. *Environmental Enrichment as a Strategy for Mitigating Stereotypies in Zoo Animals: A Literature Review and Meta-analysis*. Oxon: Cabi Publishing.

Swaisgood, R.R. 2010. The conservation–welfare nexus in reintroduction programs: a role for sensory ecology. *Animal Welfare*, **19**: 125–137.

Teixeira, B. and Young, R.J. 2014. Can captive-bred American bullfrogs learn to avoid a model avian predator? *Acta Ethologica*, **17**: 15–22.

Teixeira, C.P., De Azevedo, C.S., Mendl, M., Cipreste, C.F. and Young, R.J. 2007. Revisiting translocation and reintroduction programmes: the importance of considering stress. *Animal Behaviour*, **73**: 1–13.

Valutis, L.L. and Marzluff, J.M. 1999. The appropriateness of puppet-rearing birds for reintroduction. *Conservation Biology*, **13**: 584–591.

Vargas, A. and Anderson, S.H. 1998. Ontogeny of black-footed ferret predatory behavior towards prairie dogs. *Canadian Journal of Zoology*, **76**: 1696–1704.

Vargas, A. and Anderson, S.H. 1999. Effects of experience and cage enrichment on predatory skills of black-footed ferrets (*Mustela nigripes*). *Journal of Mammalogy*, **80**: 263–269.

Villalba, J.J., Provenza, F.D. and Han, G.-d. 2004. Experience influences diet mixing by herbivores: implications for plant biochemical diversity. *Oikos*, **107**: 100–109.

Watters, J.V. and Meehan, C.L. 2007. Different strokes: can managing behavioral types increase post-release success? *Applied Animal Behaviour Science*, **102**: 364–379.

Whitehouse, J., Micheletta, J., Powell, L.E., Bordier, C. and Waller, B.M. 2013. The impact of cognitive testing on the welfare of group housed primates. *Plos One*, **8**: e70308.

Williamson, M. 1999. Invasions. *Ecography*, **22**: 5–12.

Wisenden, B.D., Klitzke, J., Nelson, R., Friedl, D. and Jacobson, P.C. 2004. Predator-recognition training of hatchery-reared walleye (*Stizostedion vitreum*) and a field test of a training method using yellow perch (*Perca flavescens*). *Canadian Journal of Fisheries and Aquatic Sciences*, **61**: 2144–2150.

Wolf, C.M., Garland, T., Jr. and Griffith, B.J. 1998. Predictors of avian and mammalian translocation success: reanalysis and phylogenetically independent contrasts. *Biological Conservation*, **86**: 243–255.

Yamanashi, Y. and Hayashi, M. 2011. Assessing the effects of cognitive experiments on the welfare of captive chimpanzees (*Pan troglodytes*) by direct comparison of activity budget between wild and captive chimpanzees. *American Journal of Primatology*, **73**: 1231–1238.

Yoerg, S.I. 1994. Captive breeding and anti-predator behavior of the Heermann's kangaroo rat (*Dipodomys heermanni*). 1993 Annual Report. California Department of Fish and Game, Sacramento, CA, pp. 1–50.

Yoerg, S.I. 1995. Captive breeding and anti-predator behavior of the Heermann's kangaroo rat (*Dipodomys heermanni*). 1994 Annual Report. California Department of Fish and Game, Sacramento, CA, pp 1–49.

Yoerg, S.I. 1996. Captive breeding and anti-predator behavior of the Heermann's kangaroo rat (*Dipodomys heermanni*). 1995 Annual Report. California Department of Fish and Game, Sacramento, CA, pp. 1–48.

Yoerg, S.I. and Shier, D.M. 1997. Maternal presence and rearing condition affect responses to a live predator in kangaroo rats, *Dipodomys heermanni arenae*. *Journal of Comparative Psychology*, 111: 362–369.

Yoerg, S.I. and Shier, D.M. 2000. Captive breeding and anti-predator behavior of the Heermann's kangaroo rat (*Dipodomys heermanni*). Final Report. California Department of Fish and Game, Sacramento, pp. 1–56.

Young, R.J. 1997. The importance of food presentation for animal welfare and conservation. *Proceedings of the Nutrition Society*, 56: 1095–1104.

Young, R.J. 2003. *Environmental Enrichment for Captive Animals*. Oxford: Blackwell Scientific.

Zidon, R.D., Salta, D., Shore, L.S. and Motro, U. 2009. Behavioral changes, stress, and survival following reintroduction of Persian fallow deer from two breeding facilities. *Conservation Biology*, 23: 1026–1035.

# Part IV

# Behavioral indicators

Just as criminal profilers study the behavior of their suspects in order to understand their state of mind and predict their actions, wildlife managers can use the behavior of animals to get insights into their state and the state of their environments.

The conservation behavior framework identifies two types of behavioral indicators – indicators that can reveal the effects of anthropogenic activities before a numerical response is evident, and indicators used to evaluate the effectiveness of management programs at their early stages. While the context in which the two types of indicators are used is very different, their mechanism is identical. In fact, the same behavioral indicator can be used to identify the detrimental effects of an anthropogenic disturbance, and later on to indicate whether the management program aimed at mitigating the disturbance is working effectively. We therefore chose to not divide the two chapters in this section according to the designation of the behavioral indicators, but rather to do it according to their scale of reference.

Behavioral indicators can operate on staggeringly different scales: From assessing the health and well-being of a single individual to helping predict climate change across the globe. However, in this case, the differences in scale of reference usually involve very different behaviors and call for a diversity of management approaches. We refer to behavioral indicators that can inform on the individual state of the animals observed and on the state of their respective populations as "direct behavioral indicators." When behavioral indicators tell us about other populations within the community, the state of the ecosystem or about environmental changes on a global scale, we term them "indirect behavioral indicators." Chapters 11 and 12 of this section give an overview and discuss the usefulness of direct and indirect behavioral indicators, respectively.

# Direct behavioral indicators as a conservation and management tool

BURT P. KOTLER, DOUGLAS W. MORRIS
AND JOEL S. BROWN

## 11.1 INTRODUCTION

Here is an exercise to try with your students or colleagues regarding wildlife conservation and management. Tell them they are managing an area containing a population of an endangered, charismatic, flagship wildlife species, say mountain nyala in Bale Mountains National Park, Ethiopia. Invite them to write down the one or two things they would most want to know in order to best manage the population. The answers will vary. Some may inquire into the population size or density; others may want to know what the nyala are eating; others may want to know about the nyalas' levels of genetic heterozygosity. But what we really want to know is "what is the state of the population in terms of growth rate and relationship to resource density?" "what are the threats to the population?" and "what are the population's prospects for the future?" Are these questions we can answer? Will knowledge of population size or genetics or diet allow us to answer these? Or can answers best be obtained from other information? If so, how can such information be acquired? What are the best indicators?

Ideally, indicators of population well-being must be reliable. Further, they should be easy to measure, respond quickly to environmental change and forecast the future. Measurements of population sizes are frequently used in management decisions and may excel in identifying when small population issues are of concern, but are woefully inadequate as indicators of population processes. Such metrics do not necessarily respond quickly to environmental change. Most populations experience time-lagged dynamics. But time lags mean that density is a trailing indicator of current conditions. We must search elsewhere for leading indicators – indicators that predict the future rather than simply recapitulating the

*Conservation Behavior: Applying Behavioral Ecology to Wildlife Conservation and Management*, eds. O. Berger-Tal and D. Saltz. Published by Cambridge University Press. © Cambridge University Press 2016.

past. Perhaps we can find our indicators in the traits of organisms that have been shaped by evolution (Grafen 1982, Lucas & Grafen 1985, Mitchell & Valone 1990). One attractive class of characteristics comes from foraging theory and measures of behavior (Stephens & Krebs 1986). These can be classified into behavioral indicators based on **diet, patch use** or **habitat selection.**

Consider indicators of population well-being further. An example involving the Baltic tellin (*Macoma balthica*) illustrates this well. Baltic tellins, benthic bivalves from the Dutch Wadden Sea, suffer predation from red knots (*Calidris canutus*) (van Gils *et al.* 2009). They live in the muddy substrate of the intertidal zone and extend their siphons into the water at high tide to filter feed. Red knots probe the mud with their bills in search of tellin. The deeper the tellin burrow, the safer they are, but this comes at the cost of reduced feeding rates. Consequently, individuals in excellent condition seek safety and burrow more deeply than those in poor condition. Hence, a shift in bivalves toward the top layer indicates a deteriorating environment. As a successful leading indicator, there is a strong negative correlation between burrowing depth in one year and the direction and magnitude of change in population size into the next (population growth rate). By tracking how deeply tellin bury themselves, one can know the state of the population and whether populations are increasing or decreasing (Figure 11.1).

Why do behavioral indicators such as burrowing depth work? When an organism forages, several things can happen that affect its fitness. First, it may find and consume food. The intake of energy and nutrients can then be used for maintenance, and under the appropriate conditions, reproduction. These are directly connected to survivorship and fecundity, the two components of fitness. Furthermore, instead of finding food an organism may become food for others. This, too, contributes to survivorship, and animals that make decisions that reduce predation risk will be at an advantage. Thus the decisions of foragers should be under natural selection, favoring animals that feed quickly, efficiently and safely. Furthermore, when faced with options concerning different types of foods, resource patches or habitats organisms should choose the ones that contribute toward highest fitness. Because fitness reflects the per capita population growth rate (Mitchell & Valone 1990), there should be strong, fundamental connections between foraging behaviors and future population dynamics (Sutherland 1996). Finally, decisions affecting foraging behavior are often based on costs that include the marginal value of energy and the forager's future prospects (e.g. Brown 1988, 1992). Insofar as the behaviors reveal present

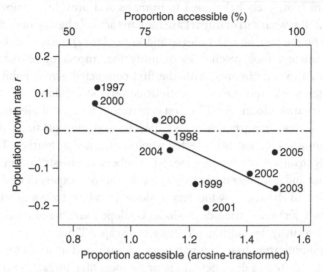

**Figure 11.1:** Baltic tellin (*Macoma balthica*) population growth rate between year t and t + 1 regressed against the proportion of bivalves accessible to predators (red knots [*Calidris canutus*]) in the top 4 cm of sediment in year t. Proportion accessible provides a behavioral indicator of population growth rate. After van Gils *et al.* 2009.

circumstances and perceptions of the future, foraging behaviors provide reliable indicators (Kotler *et al.* 2007, Morris *et al.* 2009). Because foraging occurs daily, behavioral indicators can be measured as needed and should respond quickly to environmental change, providing leading indicators of changes in population state.

The conservation behavior framework (Berger-Tal *et al.* 2011) proposes two functions for behavioral indicators: one as indicators of conservation threats and the other as indicators of management success (Vickery *et al.* 2010). In regards to conservation threats, behavioral indicators can monitor population well-being and identify and provide detailed knowledge of potential detrimental impacts that need addressing, say from changes in environmental quality or human disturbance. The Baltic tellin provides an example of monitoring a population's well-being wherein behavioral indicators reveal both the current state of the population and its prospects for the coming year.

Nubian ibex (*Capra nubiana*) in the Negev Desert of Israel demonstrate how behavioral indicators can identify and monitor impacts detrimental to a population's state. These spectacular and threatened wild goats form a large population at the equally spectacular desert canyon of Ein Avdat. There, they seem completely habituated to humans and provide a major tourist attraction. But what sort of impact are humans actually having on individual ibex and on the population? Tadesse and Kotler (2012) measured patch use in experimental food patches to quantify the impact of tourists. They conducted two experiments, with the first conducted across habitats and days of the week, and the other conducted on a steep slope with tourists present at various locations. The first experiment measured the amount of food left in resource patches across habitats and through time. Ibex left greater amounts of food unharvested when confronted by tourists. This was especially so on weekends when greater numbers of tourists further disrupt habitat use and force ibex onto steep cliffs. The second experiment also used patch use and showed how tourists upslope from ibex have a much larger disruptive effect than when downslope. Upslope tourists reduce the ibex's sense of security by cutting off lines of escape. Thus in this example, behavioral indicators based on patch use identify and quantify the impact of tourist activity as a detrimental factor for ibex. Managers can act on this information when planning hiking trails and the timing and intensity of tourist access to the preserve.

## 11.2 BEHAVIORAL INDICATORS FOR CONSERVATION: GUIDING CONCEPTS, THEORIES AND PRESCRIPTIONS

Behavioral indicators can take many forms, and be categorized in several ways. Here, we address indicators for three different scales at which animals make foraging decisions: diet choice, patch use and habitat selection. We focus on foraging metrics since the need to seek and harvest food is a universal property of all organisms. Foraging and feeding generally occupies much of an animal's time and attention. Finally, the richness, frequency and diversity of feeding behaviors provide varied opportunities for animals to reveal aspects of their well-being, their environmental quality and even their future prospects (see Box 11.1 for other possible indicators based on movement). For each, we review the underlying theory that gives meaning to the indicators, present relevant examples and discuss prescriptions for applying each indicator, including the questions that each can address.

### 11.2.1 Diet choice

The population of mule deer of the Kaibab Plateau in Arizona famously increased in response to the removal of predators and livestock in the early part of the twentieth century. They grew from an estimated population of

Box 11.1:    Animal movement as a behavioral indicator
ODED BERGER-TAL AND DAVID SALTZ

Movement is a fundamental element in the ontogeny of all organisms. While for some organisms movement is strictly a passive phenomenon (Nathan *et al.* 2008), in animals most movement is active and involves decision-making. As such, it facilitates the execution of most other behaviors. An animal may decide to explore its surroundings, or to forage in a particular patch or to escape a predator. All these actions require the animal to move. According to theory, changes to an animal's internal state or to its environment should trigger subsequent changes to the animal's movement behavior (Fryxell *et al.* 2008, Nathan *et al.* 2008), which will be manifested in all of the above-mentioned behaviors and others.

The interaction between the animal's internal state and the environment occurs simultaneously at various scales and combined with the interaction between the various internal states (e.g. hunger and fear) will produce complex movement trajectories (Getz & Saltz 2008) that may be difficult to interpret. Nevertheless, the importance of understanding movement in conservation-related issues has become well established in recent years (Jeltsch *et al.* 2013). As the collection of fine-scale movement data becomes increasingly simple and affordable (Johnson *et al.* 2002, Cagnacci *et al.* 2010), we are better able to recognize and identify changes in behavior, making movement behavior an ideal behavioral indicator reflecting the state of the moving individual and its surroundings. In this box we will briefly describe two important applications of movement behavioral indicators: monitoring the success of translocation programs, and mapping the landscape of fear for animals in the wild. Large-scale movement patterns, such as dispersal or migration, can also give us indispensable information regarding the quality of the environment or even regarding global climate changes, and these are described in more detail in Box 12.1.

## POST-RELEASE MOVEMENT BEHAVIOR AS AN INDICATOR OF TRANSLOCATION SUCCESS

Translocated animals find themselves in a completely unfamiliar environment and need to rapidly gain information on their surroundings in order to survive (see Chapter 8). Therefore, following their release, translocated animals should perform exploratory movements aimed at increasing their knowledge of their environment in order to allow for efficient foraging, anti-predatory behavior, mate search and so on. However, exploratory movement is costly (e.g. energetic demands, higher predation risk, missed opportunities), and so animals face a trade-off between their need to explore a novel environment and their need to exploit the already familiar resources to ensure their current subsistence (Eliassen *et al.* 2007; Berger-Tal *et al.* 2014). Optimal performance will depend on the level of

**Box 11.1:** (cont.)

knowledge the animal possesses – as the animal becomes more familiar with its environment it should change its movement behavior accordingly, reducing its exploratory movements and increasing its exploitative movement behavior. We term this change in movement "post-release behavioral modification" (PRBM; Berger-Tal & Saltz 2014).

The precise nature of the PRBM process will depend on the species in question and the spatial characteristics of the environment. In many cases, animals in exploratory mode will cover larger distances and move in straighter lines (e.g. beavers, Nolet & Rosell 1994; fish, Crook 2004; elk, Fryxell et al. 2008; black rhinos, Gottert et al. 2010; Arabian oryx, Berger-Tal & Saltz 2014), but in densely vegetated areas, exploration movement may be represented by slow and tortuous movements (e.g. Persian fallow deer, Berger-Tal & Saltz 2014). In most cases, once the animal becomes familiar with its environment it will establish a home range and move within the boundaries of its home range in a bi-modal pattern – slow and tortuous within known resource patches, fast and direct between resource patches (Fryxell et al. 2008, Berger-Tal & Saltz 2014). Monitoring the PRBM process can inform managers whether the released animals have acclimatized to their new surroundings. If the movements of released individuals do not reflect a gradual accumulation of knowledge (indicated by a gradual change in movement behavior), this may serve as a warning signal, alerting managers to a potential problem.

## ESCAPE BEHAVIOR AS AN INDICATOR OF DISTURBANCES

Fleeing is a common response of animals to approaching threats. "Flight-initiation distance" (FID) is the simple measurement of the distance at which the animal begins to flee from that threat (Tarlow & Blumstein 2007) and has also been sometimes termed "flush distance" or "escape flight distance" (Taylor & Knight 2003). FID provides managers with an accurate index of fear in animals (Miller et al. 2006), and is therefore an extremely useful tool for quantifying the effects of human disturbance on wildlife (e.g. Manor & Saltz 2005, Stankowich 2008, Weston et al. 2012) and to assess wild animals' welfare state (Dwyer 2004). FID is usually measured by approaching individual animals (by foot or using a potentially disturbing vehicle) until they move away (Tarlow & Blumstein 2007).

Animals should flee an approaching threat when the costs of staying exceed the benefits (Ydenberg & Dill 1986, Stankowich & Blumstein 2005). This means that FID is affected by many factors, both internal and external, including the angle of approach (Burger & Gochfled 1991), the season (Richardson & Miller 1997), the reproductive state of the animal (Bauwens & Thoen 1981) and the size of the group the animal is in (Burger & Gochfeld 1991). Being such an easy to implement method, FID is a common

**Box 11.1:  (cont.)**

behavioral measurement in mammals (see Stankowich 2008 and references within), birds (Weston *et al.* 2012), and has also been used with reptiles (Li *et al.* 2014) and even insects (Bateman & Fleming 2014). It is important to note that an animal's "range of disturbance" is likely to be much greater than the FID (Ward & Cupal 1979, Stankowich 2008). Animals can detect disturbances (Detection Distance) and respond physiologically to this disturbance (Physiological Initiation Distance) at larger distances than the FID, and these physiological effects may have negative consequences to animals' populations (Weston *et al.* 2012).

FID has many useful management applications. It is most commonly used to measure the distances beyond which people can be said to minimally disturb wildlife populations (Tarlow & Blumstein 2007). As such, FID has been used to measure the effects of hikers, cyclists, cars, boats, aircraft and more. Such studies have yielded important insights, such as the fact that humans on foot usually elicit the strongest flight response across ungulate species, and that the context matters, i.e. hikers walking off trails elicit stronger responses than hikers on trails (Stankowich 2008). The measured distance in which various anthropogenic disturbances impact wildlife species can then be used to help design buffer zones between protected and disturbed areas (Blumstein & Fernández-Juricic 2010), calculate the width of corridors and wildlife overpasses, and the true impact zone of anthropogenic activities. FID can also be used as an indicator as to which species are more vulnerable to anthropogenic disturbances. For example, Møller (2008) looked at 56 species of European birds and found that species with longer FIDs are more sensitive to disturbances and have declining population trends across Europe. Larger FIDs can serve as an indicator of hunting and other anthropogenic activity throughout the landscape (Manor & Saltz 2005, Stankowich 2008). Another potential application of FID is assessing the effectiveness of management efforts, such as limiting the speed of vehicles driving in coastal habitats, by measuring the response of animals to these efforts (Schlacher *et al.* 2013), or assessing the effectiveness of captive-rearing methods by looking at how released individuals respond to anthropogenic threats (Zidon *et al.* 2009).

## REFERENCES

Bateman, P.W. and Fleming, P.A. 2014. Switching to plan B: changes in the escape tactics of two grasshopper species (Acrididae: Orthoptera) in response to repeated predatory approaches. *Behavioral Ecology and Sociobiology*, 68:457–465.

Bauwens, D. and Thoen, C. 1981. Escape tactics and vulnerability to predation associated with reproduction in the lizard *Lacerta vivipara*. *Journal of Animal Ecology*, 50:733–743.

**Box 11.1:** (cont.)

Berger-Tal, O., Nathan, J., Meron, E. and Saltz, D. 2014. The exploration-exploitation dilemma: a multidisciplinary framework. *PLoS One*, 9:e95693.

Berger-Tal, O. and Saltz, D. 2014. Using the movement patterns of reintroduced animals to improve reintroduction success. *Current Zoology*, 60:515–526.

Blumstein, D.T. and Fernández-Juricic, E. 2010. *A Primer of Conservation Behavior*. Sunderland: Sinauer Associates.

Burger, J. and Gochfeld, M. 1991. Human distance and birds: tolerance and response distances of resident and migrant species in India. *Environmental Conservation*, 18:158–165.

Cagnacci, F., Boitani, L., Powell, R.A. and Boyce, M.S. 2010. Animal ecology meets GPS-based radio telemetry: a perfect storm of opportunities and challenges. *Philosophical Transactions of the Royal Society B-Biological Sciences*, 365:2157–2162.

Crook, D. 2004. Movements associated with home-range establishment by two species of lowland river fish. *Canadian Journal of Fisheries and Aquatic Science*, 61:2183–2193.

Dwyer, C.M. 2004. How has the risk of predation shaped the behavioural responses of sheep to fear and distress? *Animal Welfare*, 13:269–281.

Eliassen, S., Jorgensen, C., Mangel, M. and Giske, J. 2007. Exploration or exploitation: life expectancy changes the value of learning in foraging strategies. *Oikos*, 116:513–523.

Fryxell, J.M., Hazell, M., Borger, L. *et al.* 2008. Multiple movement modes by large herbivores at multiple spatiotemporal scales. *Proceedings of the National Academy of Sciences of the United States of America*, 105:19114–19119.

Getz, W.M. and Saltz, D. 2008. A framework for generating and analyzing movement paths on ecological landscapes. *Proceedings of the National Academy of Sciences of the United States of America*, 105:19066–19071.

Goettert, T., Schoene, J., Zinner, D., Hodges, J.K. & Boeer, M. 2010. Habitat use and spatial organisation of relocated black rhinos in Namibia. *Mammalia*, 74:35–42.

Jeltsch, F., Bonte, D., Pe'er, G. *et al.* 2013. Integrating movement ecology with biodiversity research – exploring new avenues to address spatiotemporal biodiversity dynamics. *Movement Ecology*, 1:6

Johnson, C.J., Parker, K.L., Heard, D.C. and Gillingham, M.P. 2002. Movement parameters of ungulates and scale-specific responses to the environment. *Journal of Animal Ecology*, 71: 225–235.

Li, B.B., Belasen, A., Pafilis, P., Bednekoff, P. and Foufopoulos, J. 2014. Effects of feral cats on the evolution of anti-predatory behaviours in island reptiles: insights from an ancient introduction. *Proceeding of the Royal Society B*, 281: 20140339

Box 11.1: (cont.)

Manor, R. and Saltz, D. 2005. Human impacts on gazelle habitat use patterns and flight distance in a heavily disturbed area in Israel. *Journal of Wildlife Management*, **69**:1683–1690.

Miller, K.A., Garner, J.P. and Mench, J.A. 2006. Is fearfulness a trait that can be measured with behavioural tests? A validation of four tests for Japanese quail. *Animal Behaviour*, **71**:1323–1334.

Møller, A.P. 2008. Flight distance and population trends in European breeding birds. *Behavioral Ecology*, **19**:1095–1102.

Nathan, R., Getz, W.M., Revilla, E. *et al.* 2008. A movement ecology paradigm for unifying organismal movement research. *Proceedings of the National Academy of Sciences of the United States of America*, **105**:19052–19059.

Nolet, B.A. and Rosell, F. 1994. Territoriality and time budgets in beavers during sequential settlement. *Canadian Journal of Zoology*, **72**:1227–1237.

Richardson, C.T. and Miller, C.K. 1997. Recommendations for protecting raptors from human disturbance: a review. *Wildlife Society Bulletin*, **25**:634–638.

Schlacher, T.A., Weston, M.A., Lynn, D. and Connolly R.M. 2013. Setback distances as a conservation tool in wildlife–human interactions: testing their efficacy for birds affected by vehicles on open-coast sandy beaches. *PLoS ONE*, **8**:e71200.

Stankowich, T. and Blumstien, D.T. 2005. Fear in animals: a meta-analysis and review of risk assessment. *Proceedings of the Royal Society B*, **272**:2627–2634.

Stankowich, T. 2008. Ungulate flight responses to human disturbance: a review and meta-analysis. *Biological Conservation*, **141**:2159–2173.

Tarlow, E.M. and Blumstein, D.T. 2007. Evaluating methods to quantify anthropogenic stressors on wild animals. *Applied Animal Behaviour Science*, **102**:429–451.

Taylor, A.C. and Knight, R.L. 2003. Behavioral responses of wildlife to human activity: terminology and methods. *Wildlife Society Bulletin*, **31**:1263–1271.

Ward, A.L.W. and Cupal, J.J. 1979. Telemetered heart rate of three elk as affected by activity and human disturbance. In: Ward A.L.W. and Cupal, J. J. (eds.), *Proceedings of the Symposium on Dispersed Recreation and Natural Resource Management: A Focus on Issues, Opportunities and Priorities*. Logan: Utah State University. pp. 47–55.

Weston, M.A., McLeod, E.M., Blumstein, D.T. and Guay, P.-J. 2012. A review of flight initiation distances and their application to managing disturbance to Australian birds. *Emu*, **112**:269–286.

Ydenberg, R.C. and Dill, L.M. 1986. The economics of fleeing from predators. *Advances in the Study of Behavior*, **16**:229–249.

Box 11.1: (cont.)

Zidon, R., Saltz, D., Shore, L.S. and Motro, U. 2009. Behavioral changes, stress, and survival following reintroduction of Persian fallow deer from two breeding facilities. *Conservation Biology*, 23:1026–1035.

4000 to as many as 100,000 animals by 1926. The deer population crashed to less than 9000 by 1940 – decreasing by as much as 60% in 1927 alone. During the increase phase the deer fed mostly on aspen, but by the population peak their diet expanded to include many conifers (Binkley *et al.* 2006). The dramatic ups and downs of the deer could have been forecasted by their diet breadths.

A healthy population may routinely exhibit seasonal fluctuations in diet. Departures from typical diets provide indicators of environmental quality. For instance, cottontail rabbits typically feed on grasses and forbs during the summer season and fall back on stripping bark from shrubs during the winter when higher-quality foods are scarce. For such animals, the presence of bark-feeding during the summer would indicate a food-stressed population, and the absence of bark-feeding in the winter would indicate an unusually mild or food-rich winter (Dalke & Sime 1941, Hoffmeister 1989). In applying such an indicator, some Canadian trappers take all beavers from any lodge where beavers cache conifers because they believe that the animals occupying the lodge cannot survive until spring (Morris, pers. comm.)

As these examples show, diet choice or diet breadth can provide valuable insights of environmental quality. In general, foragers should rank food items from highest to lowest according to energy reward per unit handling time ($e/h$). In accord with Pulliam's (1974) classic diet model, increasing the availability of higher-quality foods should cause the animal to cease harvesting lower-quality food items. More generally, if the environment offers the feeding animal an average harvest rate of $f$, then the forager should include in its diet all foods with $e/h > f$; accepting food items with lower $e/h$ would reduce its long-term average harvest rate.

Consider a food of intermediate quality that is sufficient by itself to permit positive population growth when abundant. If animals avoid this food, then current availability of better food is sufficient to generate positive fitness. If avoidance occurs only briefly, it indicates only a short spike in food availability, but if persistent then better food continues to be abundant, the population should grow, and habitat quality is excellent.

Now consider a food that is insufficient to maintain positive population growth regardless of availability. If a manager observes animals feeding on this food occasionally or for short durations, then it indicates either temporary food shortage, or the need for the special nutrients contained in this food. If feeding occurs more regularly, it likely indicates famine and eventual population decline.

Thus environmental richness can be tracked through diet selection and diet breadth. With no knowledge of the value of any particular food item, a manager can track food availability simply by tracking diet breadth through direct observations of animals, scat and pellet analysis or stomach content analysis, with more selective diets indicating a higher-quality environment, (e.g. Kotler 1985, Kjellander & Nordström 2003), as in the case of feral cats in Australia (Yip *et al.* 2015).

Even better, when food items can be ranked in preference according to $e/h$, then environmental quality can be tracked simply by knowing the least-preferred item consumed. The lower this cut-off of $e/h$ is, the lower the environment's quality. Preference, $e/h$, can be estimated by measuring the energetic content of food items in the laboratory in conjunction with digestibility studies, and handling times can be measured through direct observations or other suitable means (e.g. Fryxell *et al.* 1994, Kotler *et al.* 2010). Alternatively, relative preferences can be revealed by creating artificial resource patches provisioned with one food type or the other and measuring the remaining food in each after exploitation (see below; Kotler *et al.* 1994).

### 11.2.2 Patch use

> "I have watched the face of many a newly wolfless mountain, and seen ... a maze of new deer trails ... every edible bush and seedling browsed ... to anemic desuetude ... as if someone had given God a new pruning shears. ..." **Aldo Leopold**, 1944, *Thinking Like a Mountain*

The thoroughness with which a forager depletes a patch of discrete food items such as squirrels seeking acorns under an oak or moose browsing a willow tree provides a treasure trove of insights into the animal's state, its fears, its prospects and even its impact on resources and prey. Both pikas in Colorado (Huntly 1987) and hyraxes in South Africa (Druce *et al.* 2006) inhabit rocky refugia from which they venture into the surrounding landscapes consuming forbs, grasses and foliage. In both cases, these species forage more thoroughly near than away from shelter. Consequently, the standing crop of vegetation is lowest near and highest away. This central place effect (Olsson *et al.* 2008) reveals that predation risk increases with a pika's or hyrax's distance from a refuge. These foragers' pruning shears are controlled by fear and hunger.

Patch use may also be seen in predators via the partial consumption of their prey. Predators should always consume the yummy bits, but may eschew the less nutritious or digestible portions of a prey. Backswimmers (*Notonecta spp.*) are voracious insect predators found in many freshwater ecosystems. They feed by piercing their victims and extracting the contents. Interestingly, backswimmers often eat only part of each victim, turning each into a depletable resource patch (Sih 1980). Charnov's (1976) Marginal Value Theorem predicts that foragers should exploit resource patches until their instantaneous harvest rate falls to the environmental average. A consequence is that foragers should exploit patches less thoroughly in richer environments. Backswimmers when preying on mosquito larvae consume less of each when the larvae are more abundant (Sih, 1980). Thus, partial consumption provides an indicator of environmental quality.

Patch use derives from the heterogeneous distribution and abundance of resources resulting from spatial variation in production, dispersal, redistribution and exploitation. That is, resources occur in patches, and foragers respond to and contribute toward this patchiness. The most common way in which they respond is by partial consumption of a resource clump, prey item or plant. Key here is that the forager's harvest rate declines with time spent exploiting the patch; there are diminishing returns.

Patch use can reveal a forager's estimate of environmental richness, as well as how its foraging costs vary in time and space. Examples of patch use can range from kangaroo rats extracting seeds from sandy soils (Brown 1988), to browsing mammals cropping stems of woody food plants (Tadesse & Kotler 2013), hummingbirds drinking nectar from flowers (e.g. Baum & Grant 2001), Andean condors feeding on sheep carcasses (Speziale *et al.* 2008) and even mountain lions hunting mule deer and moving from wood lot to wood lot as the probability of prey capture drops (Brown *et al.* 1999, Laundre 2010).

**Giving-up densities (GUDs):** Patch use can provide a versatile and powerful behavioral indicator by measuring the giving-up density

(GUD): the amount or density of resources in a food patch after a forager leaves (Brown 1988). The theory sees fitness as influenced by multiple inputs, including energy, survivorship (i.e. avoiding predation) and benefits accrued from activities other than foraging. An optimal forager should exploit a resource patch until its harvest rate (the quitting harvest rate, $QHR$) falls to equal its energetic ($C$), predation ($P$) and missed opportunity ($MOC$) costs of foraging (Brown 1988, 1992).

$$QHR = C + P + MOC$$

See the next equation below for a more explicit version that applies when reproduction is seasonal.

When the harvest rate is a function of the density of food in a patch, the giving-up density provides an estimation of the QHR (Brown 1988, Kotler & Brown 1990). The GUD relates in many ways to Tilman's (1988) R* concept, where the GUD measures the species' foraging efficiency on its resources. A more efficient forager can profitably harvest its resources to a lower GUD. The GUD can measure both foraging costs and competitive abilities. As such, GUDs can be used to reveal characteristics of the forager, characteristics of the environment, and aspects of the community.

Measuring GUDs often requires creating artificial food patches in which foragers experience diminishing returns (Figure 11.2). This can be accomplished by filling a tray with food that has been mixed into an inedible substrate. In such patches, foragers at first experience a high harvest rate as they consume the easy to find food items close to the surface. As they continue, subsequent food items become harder and harder to find. The harvest rate drops lower and lower until it no longer exceeds foraging costs, and the forager should leave the resource patch. A forager should exploit a patch until its marginal benefits of patch exploitation equal its marginal costs.

Differences in GUDs among patches available to the same foragers for the same foraging bout should reflect differences in perceived predation risk, metabolic costs (if some patches are more exposed to the weather or other physical factors influencing physiological costs or require more energy to harvest, for instance because of different substrate types or foraging modes; Nolet et al. 2006), or food preferences and characteristics (if patches offer different foods or combinations of food). Day-to-day or week-to-week variation in GUDs may reflect weather, background food availability and perhaps the movement of predators. Seasonal variation in GUDs can reflect climate, food scarcity

**Figure 11.2:** Examples of feeding trays that provide diminishing returns of harvest rates to foragers and in which giving-up densities can be measured. Clockwise from the upper left are trays for fox squirrels, Egyptian fruit bats, Nubian ibex, quail, klipspringers, striped mice, and gerbils. In the middle are trays for goats and springboks.

or surplus, breeding versus non-breeding conditions and seasonal changes in predator activity or abundance. Finally, year-to-year differences in GUDs may indicate overall habitat quality, climate change and the population's health or prospects.

Food patches can take a variety of forms, so long as they offer diminishing returns. For example, Sanchez (2006) created a food patch for lesser Egyptian fruit bats by placing fruit juice or sugar water into a small cup that also contained twenty pieces of rubber hose strung together and anchored to the inside of the cup (Figure 11.2). Bats feeding from these cups had to push down harder and harder on the tubes to continue draining liquid from the cup. This led to diminishing returns and informative GUDs. Using these food patches,

Sanchez and colleagues demonstrated that alcohol is a toxin negatively affecting food quality, that energetic state of the forager affects patch use and that ethanol and sugar are complementary resources for the bats (Sanchez et al. 2008a,b).

Sometimes it is not possible or desirable to create food patches for a target animal. Perhaps the animals do not acclimate easily to food trays, perhaps there are legal or ethical restrictions due to the animal's conservation status, or maybe their natural history precludes creating a successful food patch. When this happens, it may be possible to substitute natural GUDs (see below). The key requirement for natural GUDs is a correlation between the metric and actual harvest rates. There must be diminishing returns to the foragers as they feed.

Managers can use GUDs to obtain useful information on the characteristics or ecologies of individuals. Provisioning different food patches with different food types allows one to reveal food preferences simply by comparing GUDs. Nubian ibex prefer alfalfa to rye grass: they had lower GUDs on standardized pellets made from each (Kotler et al. 1994). More than mere preferences, GUDs allow one to assess the actual value of the food. Effects of plant toxins can be similarly assessed. For example, both tannins and oxalates adversely affect food quality for fox squirrels, but GUDs reveal that oxalates have twice the impact as tannins (Schmidt et al. 1998). In contrast, free-ranging goats in a semi-arid savannah experienced equally detrimental effects from consuming tannins or oxalates (Shrader et al. 2008).

Using feeding trays, the importance of drinking water can be assessed by testing the complementarity of food and water. How do GUDs in feeding trays on days when bowls full of water are immediately adjacent to trays compare to days when water is absent? Organisms such as Australian ravens (Kotler et al. 1998), Nubian ibex (Hochman & Kotler 2006), springbok (Landeman et al., unpublished data) and free ranging goats (Shrader et al. 2008) show strong complementarity, with GUDs in the presence of water being half that of GUDs in the absence of water. At the same time, organisms such as klipspringers (Druce et al. 2009) show no complementarity, GUDs with and without water being the same. Management of the former might include water provisioning while that of the latter would not.

Giving-up densities in food patches can be used to map the "landscape of fear" (Laundre et al. 2001). Doing so connects the foraging cost of predation with physical features of the landscape. The concept is simple. Just place an array of feeding trays across a physical landscape, measure landscape features

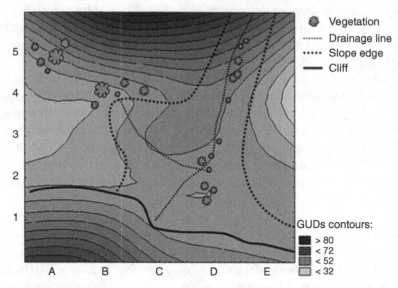

**Figure 11.3:** Landscape of fear for Nubian ibex (*Capra nubiana*) in the Negev Desert, Israel, given by contour lines of equal giving-up densities measured in feeding trays. Higher values denote higher giving-up densities and greater predation costs of foraging. After Iribarren and Kotler 2012b.

at each station, and collect GUDs. Since the microclimatic conditions for each tray should be similar, and providing any given forager has access to most or all trays (i.e. the foraging or home range scale), energetic costs of foraging are similar and missed opportunity costs are identical across the landscape. The landscape here is at a scale equal to or smaller than an individual's or colony's home range. Variation in GUDs from tray to tray therefore represents spatial variation in the predation cost of foraging. Least squares or other smoothing algorithms can be used to create a topographical map of fear, where the contour lines are lines of equal GUD, and changes in "elevation" represent changes in the cost of predation. Multiple regression analyses can then establish relationships between features of the physical landscape and predation risk. For example, for Nubian ibex major determinates of "elevation" (predation cost of foraging, or fear) include distance to the nearest cliff – representing escape terrain – and vegetation cover – representing poor sight lines (Iribarren & Kotler 2012a; Figure 11.3). Similarly, visibility contributed greatly to the landscape of fear in both Cape ground squirrels (van der Merwe & Brown 2008) and free-ranging goats (Shrader *et al.* 2008). The landscape of fear can be used to detect the effects of anthropogenic disturbances (roads, tourist routes, noise sources, etc.) on the way animals perceive their environment (Ciuti *et al.* 2012)

GUDs can be used to reveal habitat preferences (Tadesse & Kotler 2012), the value of escape substrate (Brown *et al.* 1992, Kotler *et al.* 2001), the importance of sight lines (Iribarren & Kotler 2012b), and the consequences of moon phase, cloud cover and temperature (Kotler *et al.* 1993). Such studies need not be limited to GUDs obtained from artificial feeding trays. Tadesse and Kotler (2013) used natural giving-up densities of mountain nyala at Bale Mountains National Park in the Ethiopian highlands. They examined habitat use, foraging and vigilance by mountain nyala in grasslands versus closed canopy woodlands. For six shrub species common to both habitats, Tadesse and Kotler measured the stem diameter at point of browse as a measure of GUD. Higher values correspond to more woody and less digestible bites and therefore lower harvest rates and GUDs. In the open grassland, mountain nyalas had lower GUDs, lower vigilance rates and took more bites per minute than in the woodlands. This reveals that grasslands provide a better habitat based on superior sight lines and lower predation costs of foraging. Conserving at least some open habitats that provide nyalas with safety from predators such as spotted hyenas becomes critical for the management of this endangered species. Other outstanding examples of natural GUDs include lesser spotted woodpeckers (*Dendrocopus minor*) foraging on insect larvae under tree bark as revealed by x-rays of tree branches (Olsson *et al.* 1999), and Bewick's swans (*Cygnus columbianus bewickii*) foraging on fennel pondweed tubers in shallow marine sediment (Nolet *et al.* 2006). For species of conservation concern, natural GUDs can provide a non-invasive and non-disruptive indicator for identifying preferred habitats. GUD analyses can also be applied to reserve design (Chapter 7).

Like diet selection, patch use and GUDs can be used to gauge habitat quality. Here scale is extremely important. Within the foraging scale of an individual, foragers will leave lower GUDs at better locations because of the landscape of fear (see above). But between sites or across seasons, food availability and missed opportunity costs play major roles (Olsson *et al.* 2007, Vickery *et al.* 2010), and higher GUDs reveal richer environments and times. Seeing where and when GUDs begin to rise with food augmentations can assist in calibrating the amounts of resources required to improve habitat quality, reduce famine, or induce positive population growth (Rieuceu *et al.* 2009). And, as habitat quality improves, GUDs generally rise fastest in risky microhabitats and least in the safest spots.

Implementing programs that include measuring patch use with GUDs can be challenging. Bedoya-Perez *et al.* (2013) provide a valuable overview of technique and application. Managers must first identify an acceptable food. One way to do this is to place a number of candidate food types in the environment and see which ones are taken. Often the type of food that is most taken is the best. Next, a suitable foraging context with diminishing returns needs to be created. Often, placing the food in a tray and mixing it with an inedible substrate will work, but sometimes more creativity is necessary (e.g. Sanchez 2006). Once the appropriate context has been found, it can be adjusted to provide a suitable range of GUDs. If animals too often finish all of the food, GUDs can be raised by adding substrate, using a deeper tray, switching to a less desirable food, making access to the tray more challenging (e.g. Berger-Tal et al. 2009) or using a smaller food particle size. Sometimes, trays will need to be left in the natural environment for a long time to overcome neophobia. It can be valuable to pair food patches with remote sensing cameras or RFID systems to record forager identity and to record the time and duration of visits by foragers (Emerson & Brown 2012, 2013, McArthur et al. 2012).

**Measuring K:** Carrying capacity is a central, fundamental concept in ecology and reflects the density-dependent population consequences for food, energy, nutrients, shelter and safety. Carrying capacity is reached when per capita population growth rate is 0. On average, at K, individuals by using all of their profitable opportunities just replace themselves over their lifetimes. Carrying capacity can be difficult to quantify. Yet because it is inextricably tied to population growth rate and therefore fitness, adaptive traits including patch use and other foraging behaviors of individuals can reveal this parameter (van Gils et al. 2004).

Morris and Mukherjee (2007) provide an innovative approach for measuring carrying capacity. Rather than directly attempting to measure K, they used an abrupt switch in foraging behavior to identify when K had been reached. It works as follows. Given that a forager's state depends on population density and demography (Houston & McNamara 1999), and that foraging behavior also depends on state (e.g. Brown 1988), it should be possible to identify the transition between population growth and population decline (= $K$; Morris & Mukherjee 2007).

When populations are growing, the quitting-harvest rate ($QHR$) equation from above can be rewritten as

$$QHR = C + \frac{\mu F}{\frac{\partial F}{\partial E}} + \frac{\phi_t}{p\left(\frac{\partial F}{\partial E}\right)}$$

where $C$ is the metabolic cost of foraging in the patch, and the remaining two terms are just P and MOC from before. The foraging cost of predation, P, and missed opportunity cost, MOC, have been rewritten in terms of $\mu$ (the instantaneous rate of being killed by a predator while in the patch), $F$ (survivor's fitness), $\phi_t$ (the marginal fitness-value of time that could be allocated to alternative activities), $p$ (the probability of surviving the foraging interval) and $\frac{\partial F}{\partial E}$ (the marginal fitness-value of energy) (Brown 1992, Brown & Kotler 2004). Here, P consists of the risk of predation times the marginal rate of substitution of energy for survivorship, given by the ratio of survivor's fitness divided by the marginal value of energy. MOC consists of the marginal value of time multiplied by the marginal rate of substitution of energy for alternative activities, given by the reciprocal of the probability of surviving times the marginal value of energy. If the population exceeds $K$, then individuals should reduce reproduction and maximize their own survival subject to maintaining an adequate energetic state

$$QHR = C + \frac{\mu p + \phi_t}{\phi_F\left(\frac{\partial F}{\partial E}\right)}$$

where $\phi_F$ is the marginal value of survival (Gilliam & Fraser 1987, Brown 1992, Brown & Kotler 2004). This equation derives from the previous one, but for individuals that do not have the option of using energy for reproducing, i.e. those at or above K.

The important difference between these equations is that the value of energy is lower for individuals maintaining their state than it is for individuals that are converting energy directly into descendants. Thus, if one measures $QHR$ across a range of densities both below and above $K$, then there should be a discontinuity at $K$ in the relationship between $QHR$ and density.

Morris and Mukherjee (2007) performed an experiment to test this prediction. They sequentially moved red-backed voles (*Myodes gapperi*) from one identical field enclosure to another while estimating $QHR$ from giving-up densities in foraging patches. The experiment yielded a remarkable fit with theory (Figure 11.4). Not only did the data reveal the predicted foraging discontinuity, the estimated carrying capacity of approximately forty animals ha$^{-1}$ compared favorably with field estimates ($\sim$ thirty animals ha$^{-1}$; Morris & Mukherjee 2007). It may often be impractical for conservation biologists to conduct similar experiments, but measuring giving-up densities is often simple enough to include within existing monitoring programs. A sharp discontinuity in $GUD$s during either a population increase or decline represents a leading indicator of changing population

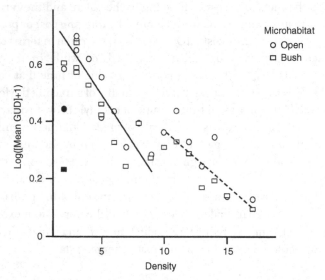

**Figure 11.4:** An illustration of a discontinuity in density-dependent foraging behavior by red-backed voles that was used to infer their carrying capacity in two 0.25 ha enclosures containing natural vegetation in northern Ontario, Canada. The lines represent least-squares fits to data on opposite sides of the switch point (ten animals). The solid circle and square are outliers excluded from the analysis. After Morris and Mukherjee (2007).

dynamics. Monitoring for such breaks could be particularly valuable during a population's recovery following a catastrophe, some habitat modification or a species reintroduction.

Measuring K can take at least two contrasting approaches. One is referred to above (Morris & Mukherjee 2007). This approach requires nothing more than a way to measure GUDs and a regular population monitoring program. Another requires more knowledge of the organism and the environment (van Gils et al. 2004). It requires site-independent quantification of the foragers' functional response (Piersma et al. 1995), measurements of rates of energy expenditure in all patch types in the study area (Wiersma & Piersma 1994, Nolet et al. 2001), plus an estimate of the relative proportion of patch types. From this, net intake rates can be predicted from prey density estimates (e.g. Gyimesi et al. 2012) and carrying capacity predicted from optimal patch use models (Southerland 1996, Gill et al. 2001, van Gils et al. 2004).

**Vigilance and apprehension:** Snow leopards are elusive and hard to study, even with camera traps and GPS equipment. Yet their prey seem highly aware of where and when snow leopards are present. Ale and Brown (2009) took advantage of this by quantifying vigilance behavior of Himalayan tahr in Sagarmatha National Park in Nepal. The reflection of the snow leopard in the eyes of the tahr revealed the numbers, whereabouts and habitat preferences of the snow leopards. Observing the vigilance of the tahr allowed Ale and Brown to actually spot leopards. Yet, the best knowledge on the snow leopards was gained by "asking" the tahr.

Vigilance can be defined as when a forager stops its food harvesting activities and directs its attention toward its surroundings. A closely related and more general concept is apprehension (Brown *et al.* 1999, Dall *et al.* 2001). Apprehension refers to when a forager redirects some portion of its attention toward predator detection and away from harvesting tasks. Via the penalty of multi-tasking, a more apprehensive forager makes more mistakes in its food harvesting tasks and thus suffers reduced harvest rate (Dall *et al.* 2001, Kotler *et al.* 2002). Foragers can remain apprehensive and yet still harvest food. Oftentimes, researchers will lump vigilance and apprehension together and call it all vigilance. We will follow that convention here.

Vigilance can provide a powerful behavioral indicator, especially in combination with others. In particular, vigilance provides foragers a critical tool for risk management. When foragers need to exploit depletable resource patches and when they can also use time allocation to manage risk, optimal vigilance, $u^*$, is given by:

$$u^* = \sqrt{\frac{mF}{bf_{max}\left(\frac{\partial F}{\partial e}\right)} - \frac{k}{b}}$$

where $m$ is the encounter rate with predators, $F$ is survivor's fitness, $b$ is the effectiveness of vigilance in reducing mortality, $f_{max}$ is the forager's harvest rate when no vigilance is used and $1/k$ is the lethality of the predator in the absence of vigilance. Rates of vigilance can be used to detect changes in predator numbers, as well as changes in the prospects of the foragers themselves, as shown by the example of tahr and snow leopard. Furthermore, foragers will combine their vigilance and time allocation in a manner that minimizes giving-up density (= maximizes food harvested from the patch; Brown 1999). When feasible, time allocation and vigilance should be used as joint indicators.

A surprising example of vigilance and time allocation as joint behavioral indicators comes from gerbils. Most studies on vigilance focus on diurnal social foragers. In contrast, gerbils are nocturnal and solitary. Yet, they

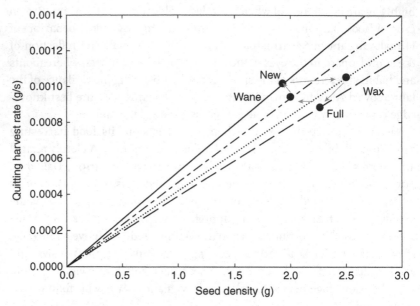

**Figure 11.5:** Harvest rate curves for Negev Desert gerbils foraging during four moon phases. Estimates of quitting harvest rates appear as functions of seed density in the resource patch. Curves are derived by estimating attack rates and handling times from giving-up density and time allocation data fitted to a Holling disc equation. The estimated quitting harvest rate for each moon phase comes from the mean giving-up density and the disc equation. Shallower slopes correspond to higher levels of vigilance; giving-up densities lying closer to the origin correspond to greater time allocation. Arrows track moon phases through time. After Kotler *et al.* 2010.

manage risk of predation using both vigilance and time allocation (Kotler *et al.* 1991, Kotler *et al.* 2010). We know this by the way the gerbils use resource patches. Vigilant animals necessarily harvest resources from a patch more slowly since they are splitting their attention between foraging tasks and predator detection. So, harvest rate curves showing the relationship between harvest rate and food remaining in a patch (Figure 11.5) reveal a forager's level of vigilance via their slopes (Brown 1999). Shallower curves (slower harvest rates) correspond to higher levels of vigilance and slower, more cautious patch exploitation. Similarly, differences in time allocation appear as points along the harvest curve. The point reflecting initial food abundance in the patch appears on the curve in the upper right-hand corner, and patch depletion proceeds along the curve toward the origin. Points farther along the curve and closer to the origin represent a greater allocation of time (Brown 1999).

Risk management for any given context can be deconstructed into time allocation and vigilance by knowing initial food abundance, final food abundance (GUD) and time spent in the resource patch. By integrating Holling's disc equation from initial to final density (i.e. harvest rate = $aR/(1+ahR)$, a type II functional response where $R$ is current food abundance), one obtains an equation in terms of these variables and whose coefficients are $h$, handling time, and $1/a$, the inverse of attack rate (Kotler & Brown 1990).

$$t = (1/a)\left[\ln\left(R_0/R_f\right)\right] + h(R_0 - R_f)$$

The subscripts refer to initial and final density of resources in the patch. The disc equation can thus be parameterized and GUDs can be converted into quitting harvest rates.

Kotler et al. (2010) did this for Allenby's gerbil over the four phases of a lunar cycle (Figure 11.5). Here, risk management affected gerbil state, which fed back to further shape risk management behavior. In particular, the new moon represents a time of the month when dark, moonless hours allow gerbils to forage safely and achieve a high energy state. At this time, they use little vigilance and allocate a great deal of time to patch exploitation. As the moon waxes, moonlight hours increase, and gerbils respond by increasing vigilance and decreasing time allocation. They allow their state to drop. As the moon waxes into full moon their state continues to drop. Even as the gerbils continue to increase their vigilance, they must increase time allocation to protect their state. Eventually, the moon starts to wane, and dark hours appear and become more abundant. Consequently, gerbils reduce vigilance, increase time allocation further and begin to rebuild their energy state. Finally, the moon wanes back to new, gerbils continue to decrease vigilance, but ease up on time allocation as they finish rebuilding their energy state (Kotler et al. 2010). In this manner, the feedback between risk, state and behavior is necessary for understanding risk management.

The manager can take two different approaches to using vigilance behaviors. If focal observations on the target animal are desirable, vigilance rates can be quantified by percent of time spent vigilant or by the number of times per focal observation that an animal stops foraging to observe its surroundings. The advantage of this approach is its simplicity. But, it may miss aspects of vigilance connected to broader forms of apprehension. It also requires ethograms to partition time between feeding and vigilance activities. If the species of interest will utilize feeding patches, then apprehension/vigilance can be calculated by knowing time allocated to patches, the initial food abundances of the patches and the GUDs. Time allocated to each food patch can be measured using direct observations, video cameras or RFID technologies (e.g. Kotler *et al.* 2010).

### 11.2.3 Habitat selection

Muskrats (*Ondatra zibethicus*) are semi-aquatic rodents of North America and invasive to parts of Europe and South America. They feed on wetland plants and invertebrates and seek safety from diverse predators such as mink, coyotes, bobcats and birds of prey. Work in North Central USA (Errington 1963) and in Saskatchewan, Canada (Messier *et al.* 1990), reveal striking changes in their overall habitat choices at high and low densities. At low densities, dens occur as burrows along the shorelines of ponds or as lodges in vegetation such as common bulrush (deeper water). At high densities, the muskrats spill-over into vegetation such as prairie bulrush (shallower water). During a high-population year, Messier *et al.* (1990) found 57% and 35% of muskrat lodges in common bulrush and prairie bulrush habitat, respectively. This changed to 73% and 12% during a low-population year. Access to food and safety dictate habitat quality, and in a small but growing population one should see an expansion of habitat use by the muskrats. Space use can be a behavioral indicator.

During population declines when animals are experiencing deprivation, they should expand their diets to include less-favorable foods and deplete food patches to lower giving-up densities. Yet they may actually contract their use of available space and habitats. This contraction of habitat use may come about because those in less suitable habitats are more likely to die, less likely to reproduce, or via behavior; the remaining individuals contract their use of available habitats toward places with more food or where risk from predation is least likely. Conversely, a growing population may manifest through the use and occupation of a broader range of habitats. This is the essence of density-dependent habitat selection theory (Fretwell & Lucas 1969, Morris 1987, Rosenzweig 1979, 1981). By animals voting with their hooves, paws, fins, wings or feet, density-dependent habitat selection provides several valuable indicators for wildlife conservation.

**Resource selection functions:** Bison in Prince Albert National Park in Canada make somewhat surprising residents within the forested habitats of the southwest section of the park (Fortin *et al.* 2009). There, they move through a landscape comprising forests, meadows, lakes, streams and roads. As they seek food they can fall victim to wolves. Resource selection functions (RSFs) for bison fitted with GPS collars revealed strong affinities for areas abundant with meadows. Seasonally the bison avoid areas with thick riparian vegetation that limits sight lines, especially in the winter. They avoid bodies of water in the summer, but seek them out in the winter, and avoid areas with deep snow. Habitat selection appears to be driven by factors associated with not just food, but also risk of predation.

Arguably, RSFs that estimate the probability of use of different resources (or components of habitat) relative to their availability provide the most straightforward means of studying habitat selection (e.g. Boyce and McDonald 1999, Boyce et al. 2002, Manly et al. 2002, McLoughlin et al. 2010, and others). RSFs, often generated for animals from global positioning systems (e.g. Cagnacci et al. 2010), can be converted into maps illustrating favored versus unfavored habitats. Patterns of space use can be used to motivate management and recovery strategies. Rather than directly dealing with resources, RSFs provide measures of space use that can be correlated with habitat attributes (resources). Insofar as an animal's selectivity for habitats indicates their importance and contribution to fitness, RSFs provide an effective means of scoring habitat suitability from vegetation, topographical or geologic maps.

A common way to generate the RSF is to obtain a series of animal locations across the landscape of interest. These locations receive scores of 1. In addition, random points are collected by calculating the mean distance that animals move and then choosing points at random within that radius for each actual location. These random points receive scores of 0. Then, every location is matched with scores from a set of environmental variables. Such variables may include elevation, vegetation type, slope aspect, snow depth and even group size. The RSF is then generated with logistic regression (e.g. Fortin et al. 2009).

Although RSFs can reliably measure current space use and habitat values, most do not incorporate underlying density-dependent habitat selection (Fortin et al. 2008, Morris et al. 2009, McLoughlin et al. 2010). Incorporating density is both valuable and crucial for connecting the current state of the population with its future prospects. RSFs carried to the next level of density-dependence connect resource use with changing conditions and underlying evolutionary mechanisms responsible for habitat preference. As in the case of the muskrats, habitat use and selectivity change dramatically with density, and are highly responsive to population trends and trajectories.

**Isodars:** Springbok in Augrabies Falls National Park, South Africa, forage over a complex mosaic of habitats comprising twelve major

**Table 11.1.** *Combination of giving-up densities and activity densities allows the classification of habitats into core, refuge, rich but risky, and unsuitable.*

| | Giving-up density | |
| --- | --- | --- |
| | Low | High |
| **Abundance** | | |
| Low | Refuge habitat for prey; low risk and poor opportunities | Unsuitable; risky habitat with poor opportunities |
| High | Core/suitable habitat; safe and rich in opportunities | Highest-quality habitat for predators; risky habitat but rich in opportunities |

vegetation types (Reid 2004). They concentrate their activities on four types of *Acacia mellifera* (black-thorn acacia) scrubland defined by the co-dominant vegetation: (1) *Euphorbia spp.*, (2) *Stipagrostis hochstetteriana*, (3) *Zygophyllum dregeanum*, and (4) *Monechma spartioides*. The springbok are especially active in habitats (1), (2) and (3), with less of a presence in (4). This corresponds nicely with GUDs being ranked according to (1) < (2) < (3) = (4). Yet more information can be gleaned by taking advantage of how the individual springbok distribute their activity and numbers among habitats over periods when populations have grown or shrunk. By voting with their "hooves" and by the use of activity densities and isodar analysis, we can conclude that habitats (2) and (3) are similarly productive and habitat (4) is the least productive (Table 11.1). By combining GUDs, density-dependent habitat use, and isodars we can surmise that habitat (1) is safe and bountiful, as is (2). Habitat (3) is bountiful, but risky – likely valuable habitat for the leopards that prey on the springbok. Habitat (4) has two strikes against it. It offers little food and high predation risk to springbok (Figure 11.6). But, what is an isodar and how is it measured, analyzed and interpreted?

Theories of density-dependent habitat selection (Fretwell & Lucas 1969, Rosenzweig 1981) become operational in the form of isodars (Morris 1988). An isodar is all combinations of a species' densities between two habitats such that the fitness gained from the habitats is equal. They represent the dynamic solution to an evolutionary game of habitat selection (Cressman & Křivan 2010). Isodars are found through the regression analysis of a species' densities in two or more habitats (Morris 1987, 1988; Figure 11.7). Human examples include trying to pick the fastest line among many at a supermarket or the fastest driving lane on a freeway.

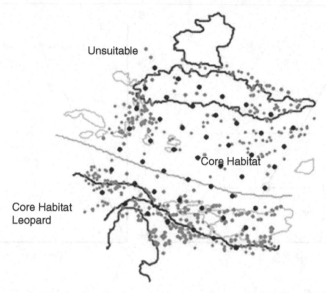

**Figure 11.6:** The landscape of springbok at Augrabies Falls National Park, South Africa. Giving-up densities and activity densities allow the habitats to be divided into core habitat for springbok, core habitat for leopards and unsuitable habitat for springbok. Grey dots correspond to trees, dark grey outlines are rocks and the light grey line is a dry river bed.

Imagine a species whose population grows logistically within two separate habitats such that $dN_i/dt = r_iN_i[(K_i-N_i)/K_i]$ for $i$ = habitats A and B, where $r_i$ is the intrinsic rate of population increase, $N_i$ is population size and $K_i$ is carrying capacity for populations in habitat i. In its simplest guise, density-dependent habitat selection predicts that individuals will distribute themselves between habitats so that fitness is equal. Thus:

$$r_A[(K_A-N_A)/K_A] = r_B[(K_B-N_B)/K_B]$$

This can be re-arranged to give:

$$N_B = (r_B - r_A)K_B/r_B + r_AK_BN_A/(r_BK_A)$$

Morris (1987, 1988) referred to this as an "isodar."

When individuals can assess and choose their use of habitats, isodars can inform us through their slopes and intercepts when we plot the abundance of a species in habitat B (y-axis) versus its abundance in A (x-axis). For instance, if both habitats offer the same opportunities and are identical in all but appearance ($r_A=r_B$ and $K_A=K_B$), then the isodar will have a slope of 1 and a y-intercept of 0. If one habitat is intrinsically better or safer than another ($r_A\neq r_B$), then the intercept will be positive if B > A. The slope of the isodar

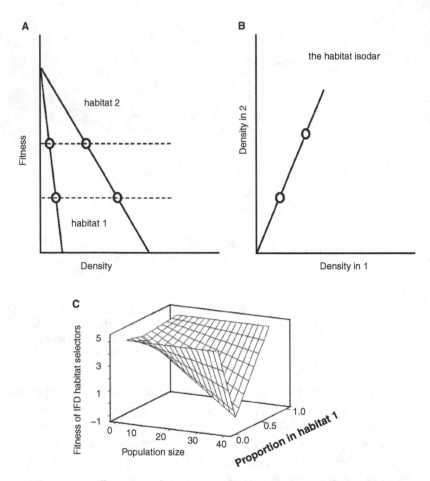

**Figure 11.7:** Illustrations demonstrating how one can convert fitness-density functions (A) into isodars (B), and invader strategy landscapes (Apaloo *et al.* 2009; C) of habitat selection. Points along the isodar correspond to different densities where the expectation of fitness is equal in both habitats (dashed horizontal lines in A). The invader strategy landscape is calculated by comparing the fitness an individual would achieve if it used habitats in the ratio specified by the isodar when all others use the two habitats at all other possible ratios. After Morris (2011).

reveals the strength of density-dependence. For instance, if the resources of a habitat are divided evenly among the individuals within the habitat (resource matching), the slope of the isodar will be 2 if habitat B has twice the value as A. In general, a slope < 1 reveals that fitness declines with density more slowly in habitat A, while vice-versa for a slope > 1. All four

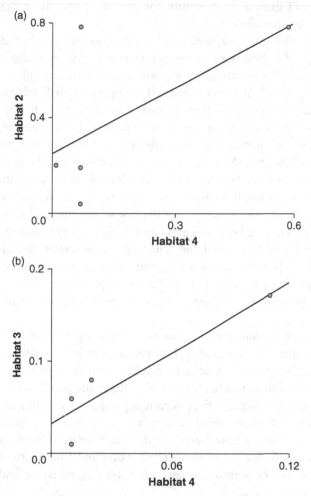

**Figure 11.8:** The relationship (isodars) between springbok activity density (population size) in habitats 2 and 4 and habitats 3 and 4 for springbok at Augrabies Falls National Park, South Africa. Both habitats 2 and 3 are superior to habitat 4. After Reid 2004.

combinations of intercepts and slopes are possible and revealing. The magnitudes of these slopes and intercepts permit an actual quantification of habitat differences to the species. For the springbok, the slopes and intercepts of isodars among the different habitats provided habitat rankings. Both habitats 2 (positive intercept) and 3 (slope > 1) are superior to 4 (Figure 11.8). When conjoined with GUDs we know the contribution of predation risk to habitat preferences, and when conjoined with the overall abundance of these habitats within the park we know the park's overall suitability for springbok.

At the community level, isodar theory can be extended to include multiple species of interacting competitors (Morris 1999, 2003a,b). Multiple species applications converge on Rosenzweig's (1981) concept of "isolegs" where the presence of a competitor will strongly and predictably influence the habitat selection of another. In fact, extension of the theory permits managers and researchers to use temporal sequences (or spatial replicates) of habitat-specific population sizes to determine each species' fitness functions in both habitats (Morris et al. 2011), and determine the potential for another species or ecotype to invade (Morris et al. 2012). Once the fitness functions are known, it is easy to build the species' fitness invasion landscape. A species can invade at any density where the landscape reveals positive fitness. And, because fitness and foraging are so intimately intertwined, one can even build the landscape with GUDs (Morris 2014). Conversely, when the landscape reveals negative population growth, a species is en route to extinction. We foresee opportunities for ecologists to combine space use with foraging behavior to restrict and eliminate invasive species.

For species of conservation concern, habitat degradation (or improvement) changes the relationship between a population's fitness and its density. Such a change should be revealed by the diet or patch use activities of individuals within that habitat and by shifts in the population's distribution over other habitats. It thus becomes possible to monitor changes in habitat quality through overall patterns of habitat selection. Furthermore, managers can forecast how changes in the quality or quantity of one habitat will influence future population distributions and the ability of habitat selection to buffer populations against habitat degradation and habitat destruction.

An abrupt shift in the population size or activity of a species in one habitat without corresponding changes in other habitats could indicate success or failure at maintaining the integrity of that habitat. Watching whether species flee from or flood into a modified habitat may provide the best initial indicator of management impacts. These movements and shifts in habitats can occur much more rapidly than actual changes in overall population sizes from births and deaths. This becomes most dramatic as individuals of a species whose habitat is declining in quality will eventually congregate for a "last stand" in the last available patches of suitable habitat. The last great flocks and aggregations of passenger pigeons moving into the Midwest United States may have been just such an event as huge tracts of masting beech and chestnut forests were dwindling in the East (Greenberg 2014).

Finally, isodar analysis can be useful for detecting when species may be caught in an ecological trap – a situation where human habitat modification gives the impression of suitability to a species when in fact the habitat has deteriorated (Chapter 4). Here, an independent assessment of survivorship or reproductive success is necessary, along with the density estimates needed to calculate the isodars. Negative isodar intercepts reveal traps, as do isodars running through the origin coupled with differences in survivorship or reproductive success across habitats. Shochat *et al.* (2005) provide an instructive application of isodars to understand the decline of five shrub- and ground-nesting birds from a tall-grass prairie reserve in Oklahoma. Isodar analysis permitted the researchers to see the habitat through the eyes of ground-nesting songbirds. Measures of nesting success permitted an objective evaluation of the birds' assessments. Sadly the birds seemed blithely unaware that human management techniques for prairie restoration favored the very reptiles that depredated their nests in the most heavily managed habitats.

### 11.2.4  Allee effects

A curious phenomenon involving black-crowned night herons occurred in Chicago, Illinois, USA. These animals are of conservation concern. During the summer of 2007, nesting pairs for the first time were noted in treelines of Lincoln Park, a large, yet very urban park. The numbers have grown rapidly since then, resulting in c. 400 adults by summer 2011 (Hunt, V., pers. comm.). Concurrent with this rise was an equally dramatic decline of the night herons from a long-standing roost near Lake Calumet on the southern boundary of Chicago. It seems the new colony siphoned birds from the old, although no direct proof is yet available. Vicky Hunt, in conjunction with the Lincoln Park Zoo, suggests that while negative density-dependence may occur over the wider range of feeding habitats, positive density-dependence occurs at roosting sites as safety in numbers protects eggs, nestlings and perhaps even adults from mammalian and avian nest predators. Here, habitat choice seems to involve an Allee effect (Chapter 4).

Most studies of habitat selection assume negative relationships between fitness and density (e.g. Green & Stamps 2001). But at low population sizes fitness often increases with density. Increasing abundance can overcome constraints on mating opportunities, social activities, safety through numbers and other density-limited effects (e.g. Courchamp *et al.* 2008, Gregory & Courchamp 2010). Two outcomes are particularly relevant to conservation. (1) Net population growth at low density may be negative (a pseudo-sink, Watkinson & Sutherland 1995; often referred to as a "strong" Allee effect). (2) Low-density populations are at risk of stochastic extinction (Gregory & Courchamp 2010, Kramer & Drake 2010). Despite their importance, detecting Allee effects can be difficult. Crucial data connecting fitness with density can be hard to collect, and by the time sufficient data have been collected on population sizes versus fitness it may be too late for meaningful conservation measures. Habitat selection behavior can provide an indicator and a solution.

Imagine two habitats in which fitness increases with density at low population size (Allee effect) before declining at higher abundances (Figure 11.9A). Imagine further that we have census data across a broad range of densities, including those in the region of the Allee effect. At the lowest population sizes individuals should occupy only the richer of the two habitats. But, as density increases it eventually reaches an unstable point where individuals maximize fitness by also settling in the poorer habitat, drawing others from the richer habitat with them (Fretwell & Lucas 1969, Greene & Stamps 2001, Morris 2002). The instability of the Allee effect is captured by a characteristic J-shaped isodar (Figure 11.9B). Even if the J-shape is not obvious, there will often be a hiatus in the isodar at low population size (Morris 2002). Consistent with the theory, habitat selection by small mammals using xeric and mesic forest habitats in Alberta, Canada, revealed just such an effect (Morris 2002). With extreme Allee effects, the movement of some individuals from the occupied habitat to the new habitat can create a positive feedback prompting the siphoning effect possibly occurring with the black-crowned night herons of Chicago.

## 11.3 BEHAVIORAL INTERACTIONS AND MECHANISMS OF SPECIES COEXISTENCE

Maintenance of biodiversity requires an understanding of the mechanisms by which species coexist with their close competitors (e.g. Brown 1989a). While some species may disappear due to the elimination of their fundamental niche, it is likely that far more species are threatened by a disappearance of their realized niche and a disruption of their mechanisms of coexistence. What comprises a mechanism of coexistence? Theory tells us that frequency-dependent selection is necessary to support any sort of diversification. In regards to species

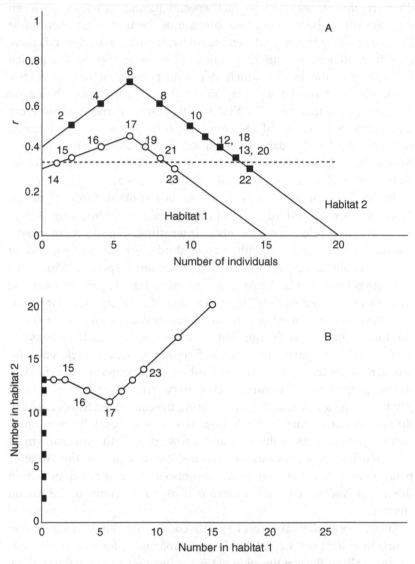

**Figure 11.9:** An example demonstrating how density-dependent habitat selection can reveal an Allee effect. Here, populations were censused sixteen times at regular time intervals. Numbers appearing by data points correspond to the total population size in both habitats, and the scale on the abscissa corresponds to population size in habitat i. For the first seven census periods, all individuals were in habitat 2. Thereafter, both habitats were occupied. A. Fitness-density relationships in two habitats with Allee effects. B. The data in "A" plotted as an isodar. The horizontal dashed line in "A" represents the unstable threshold where individuals flow directionally from habitat 2 toward habitat 1. After Morris (2002).

diversity, this comes down to each species having an advantage when rare. For this to be the case, two things must be true. First, there must be an axis of environmental heterogeneity or niche axis. Second, there also needs to be an evolutionary trade-off for the species such that each has a part of the axis at which it profits more than its competitor. Mechanisms of coexistence may range from familiar ones based on diet selection, patch use or habitat selection in time and space to more unusual ones based on life history trade-offs or temporal and spatial variability in the abundance of food resources (Kotler & Brown 1988, Brown 1989a,b, Kotler *et al.* 1993b, Brown *et al.* 1994, Perrin & Kotler 2005, Brown & Kotler 2007, Kotler & Brown 2007).

Within 6000 or so hectares of the Arabuko-Sokoke forest of coastal Kenya, the Amani sunbird is one of the most abundant bird species. Yet, in Kenya it resides nowhere else. Interestingly, understanding the mechanisms of coexistence with other sunbird species may be paramount to successfully managing this highly endangered species. Within the *Brachystegia* woodlands, the Amani sunbird appears to compete with the widespread collared sunbird. Investigations of foraging behaviors show how the collared sunbird is both the behavioral dominant and a cream skimmer. It moves swiftly from leaf to leaf seeking insects. It harvests few insects per leafy patch, but moves frequently, with a preference for broader, larger leaves. The Amani sunbird is the opposite and displays the foraging tactics of a crumb-picker, harvesting more insects per leaf patch, moving less frequently and favoring the compound leaves found in the canopies of the mature *Brachystegia* (Oyugi *et al.* 2012). Without direct observations of these behaviors and knowledge of the mechanism of coexistence, the significance of mature *Brachystegia* for the Amani's persistence would have remained unnoticed. Lose these trees within this forest, and the collared sunbird is likely to out-compete the Amani sunbird.

Knowledge of mechanisms of species coexistence can aid conservation efforts in at least two ways. First, if one is managing for biodiversity, this approach allows the identification of the key heterogeneities in the environment that promote species coexistence. With this knowledge, management can embrace the entire ecological community and tailor efforts toward preserving precisely the necessary variability. Second, in cases involving a flagship species, preservation efforts can focus on essential factors within the environment rather than just on habitat correlations. Consider for example two coexisting squirrel species. Measurements of GUDs, habitat distributions and direct interactions of fox squirrels (*Sciurus niger*) and gray squirrels (*S. carolinensis*) show that these species coexist by habitat selection and a trade-off between food and safety (Van der Merwe *et al.* 2007). Gray squirrels are superior to fox squirrels at interference and resource competition, but fox squirrels are better at dealing with predators. Fox squirrels predominate where predators are prevalent, but are rare where predators are rare or ineffective. Thus the key to maintaining fox squirrels is to promote predators and the habitats the predators favor. This should be the focus of efforts to conserve the endangered Delmarva fox squirrel (Nelson *et al.* 2005).

Implementation of a "mechanisms of species coexistence" approach may differ widely depending on the communities and species of interest. Many will be based on patch use behavior and measurements of giving-up densities, along with complementary data on activity densities and estimates of population densities. In these terms, the mechanisms generate unique sets of predictions that can be tested experimentally (e.g. Brown 1989b, Kotler *et al.* 1993b, Ziv *et al.* 1993, Brown *et al.* 1994). Using behaviors to test among possible mechanisms of coexistence can provide a quick and efficient means for managers to identify the key features of the environment necessary for maintaining biodiversity.

## 11.4 WHAT SHOULD WE DO WHEN INDICATORS FAIL?

Adaptive evolution rewards those individuals that match behaviors, such as habitat selection and patch use, with fitness. Although adaptive behavior may be a valid indicator of habitat quality, density need not be (e.g. van Horne 1983). As seen with ecological traps and the prairie birds in Oklahoma (Shochat *et al.* 2005), mismatches can occur when organisms are constrained from making otherwise adaptive decisions, or when individuals misinterpret cues of habitat quality. At one extreme of the continuum, formerly reliable cues become disconnected from fitness, and individuals become trapped in habitats of low quality (Dwernychuk & Boag 1972, Schlaepfer *et al.* 2002, Hawlena *et al.* 2010). At the other extreme, individuals misinterpret reliable cues and fail to take advantage of high-quality habitat (a perceptual trap; Patten & Kelly 2010). Somewhere in the middle, habitat choice may be constrained by phenotype, learning or development (e.g. Wecker 1963, Stamps 2001, Davis & Stamps 2004, Mabry & Stamps 2008) such that individuals occupy habitats yielding lower than maximal fitness. In each instance, it is likely that habitat patterns in density will unfailingly correspond to behavioral habitat choice, but not necessarily to innate habitat quality. A manager acting too quickly might allocate conservation efforts toward the wrong habitat.

Astute conservation managers, on the other hand, will look first to additional indicators. Quitting-harvest rates of individuals "caught" in perceptual traps should reveal the habitat's superior quality, and vigilance behavior can often be used to indicate predation risk. Quitting-harvest rates and vigilance may not be reliable indicators, however, when prey are caught in an ecological trap with novel predators or other survival risks that they fail to recognize. Comparisons of resource-selection functions before and after habitat change may not be able to document which habitat conditions are better, but should document changes in habitat use. Regardless, the lesson is to avoid making rash decisions on habitat management that can hinder evolutionary adjustments to habitat quality. The simple strategy of philopatry, for example, will often be an efficient and effective mechanism by which species can evolutionarily wiggle free of an ecological trap (Kokko & Sutherland 2001). The challenge for the conservation manager is to ensure that the population, and its habitats, persist long enough for behavioral adjustments and/or adaptive evolution to solve the problem.

## 11.5  MANAGEMENT IMPLICATIONS

Conservation biologists and managers need to know the population states of vulnerable and endangered species, the identity and impacts of various threats and the future prospects of these populations under various management strategies. Furthermore, managers need to evaluate the efficacy of their actions. We have seen here that indicators based on adaptive behaviors can provide reliable, responsive and easy to use metrics. Behavioral indicators accomplish these goals by providing leading indicators of habitat quality and population well-being. Behavioral indicators often are based on diet, patch use or habitat selection. These indicators work when an animal's behavioral decisions reflect its circumstances and its fitness consequences.

Their uses are manifold:

(1) Diet selection and diet breadth provide indicators of *habitat quality*. Animals that avoid food items of intermediate quality are experiencing excellent habitat quality, while those consuming poor food items are experiencing poor habitat quality;

(2) Giving-up densities in natural or artificial resource patches reveal various *foraging costs* and aspects of *food and habitat quality*. Because of trade-offs of food and safety, they provide *indicators of population*

*well-being* and *forecast the future* by indicating whether environmental quality is improving or deteriorating;

(3) Variance in giving-up densities can reveal characteristics of individuals, characteristics of the environment and characteristics of the community. Day-to-day and week-to-week variation reveals changes in weather and food availability; seasonal variation reveals changes in climate, food availability, breeding condition and predator abundance; year-to-year variation reveals changes in overall environmental quality and prospects for the future;

(4) Many characteristics of individuals can be revealed using GUDs. These include food preferences, attack rates and handling times on various food types, harvest rates, the complementarity of different food or nutrient inputs such as food and water, effects of toxins, effects of predation risk, the energetic cost of predation and more;

(5) Abrupt changes in giving-up densities with changing population densities can pinpoint the population's *carrying capacity*;

(6) Giving-up densities excel at measuring *human impacts in time and space*;

(7) Spatial variance in giving-up densities reveals the *landscape of fear*;

(8) Giving-up densities in combination with measures of habitat use can identify a species' core, refuge and unsuitable habitats and those most beneficial to the species' predators;

(9) Data on giving-up densities, population densities and activity densities allow for the testing of multiple *mechanisms of species coexistence* and assist in *community-level management of biodiversity*;

(10) Vigilance behavior can reveal *information about elusive predators* as viewed through the eyes of their prey;

(11) Density-dependent habitat selection behavior and isodars can reveal *differences between habitats due to productivity, predation, the intensity of density-dependence and interactions with competing species*;

(12) Activity densities reveal details of *habitat selection* and *resource selection functions*;

(13) *Isodars* can be calculated from the numbers and activities of a species across habitats. Isodars identify habitat differences in productivity or safety and differences in density-dependence or carrying capacity. Isodars along with other behavioral indicators can reveal the

existence of disruptive ecological features including Allee effects and ecological traps;

(14) Range collapse to only the very best of habitat types can reveal *populations in critical condition;*

(15) Applying behavioral indicators can allow *the implementation of novel management schemes* that promote species coexistence and biodiversity or stabilize predator–prey interactions.

## REFERENCES

Ale, S.B. and Brown, J.S. 2009. Prey behavior leads to predator: a case study of the himalayan tahr and the snow leopard in Sagarmatha (Mt. Everest) National Park, Nepal. *Israel Journal of Ecology and Evolution*, 55:315–327.

Apaloo, J., Brown, J.S. and Vincent, T.L. 2009. Evolutionary game theory: ESS, convergence stability, and NIS. *Evolutionary Ecology Research*, 11:489–515.

Baum, K.A. and Grant, W.E. 2001. Hummingbird foraging behavior in different patch types: simulation of alternative strategies. *Ecological Modelling*, 137:201–209.

Bedoya-Perez, M.A., Carthy, A.J.R., Mella, V.S.A., McArthur, C. and Banks, P.B. 2013. A practical guide to avoid giving up on giving-up densities. *Behavioral Ecology and Sociobiology*, 67:1541–1553.

Berger-Tal, O., Mukherjee, S., Kotler, B.P. and Brown, J.S. 2009. Look before you leap: is risk of injury a foraging cost? *Behavioral Ecology and Sociobiology*, 63:1821–1827.

Berger-Tal, O., Polak, T., Oron, A., Lubin, Y., Kotler, B.P. and Saltz, D. 2011. Integrating animal behavior and conservation biology: a conceptual framework. *Behavioral Ecology*, 22:236–239.

Binkley, D., Moore, M.M., Romme, W.H. and Brown, P.M. 2006. Was Aldo Leopold right about the Kaibab deer herd? *Ecosystems*, 9:227–241.

Boyce, M.S. and McDonald, L.L. 1999. Relating populations to habitats using resource selection functions. *Trends in Ecology & Evolution*, 14:268–272.

Boyce, M.S., Vernier, P.R., Nielsen, S.E. and Schmiegelow, F.K.A. 2002. Evaluating resource selection functions. *Ecological Modelling*, 157:281–300.

Brown, J.S. 1988. Patch use as an indicator of habitat preference, predation risk, and competition. *Behavioral Ecology and Sociobiology*, 22:37–47.

Brown, J.S. 1989a. Desert rodent community structure: a test of 4 mechanisms of coexistence. *Ecological Monographs*, 59:1–20.

Brown, J.S. 1989b. Coexistence on a seasonal resource. *American Naturalist*, 133:168–182.

Brown, J.S. 1992. Patch use under predation risk. I. Models and predictions. *Annales Zoologici Finnici*, 29:301–309.

Brown, J.S. 1999. Vigilance, patch use and habitat selection: foraging under predation risk. *Evolutionary Ecology Research*, 1:49–71.

Brown, J.S., Arel, Y., Abramsky, Z. and Kotler, B.P. 1992. Patch use by gerbils (*Gerbillus allenbyi*) in sandy and rocky habitats. *Journal of Mammalogy*, 73:821–829.

Brown, J.S., Kotler, B.P. and Mitchell, W.A. 1994. Foraging theory, patch use, and the structure of a Negev Desert granivore community. *Ecology*, 75:2286–2300.

Brown, J.S., Laundré, J.W. and Gurung, M. 1999. The ecology of fear: optimal foraging, game theory, and trophic interactions. *Journal of Mammalogy*, 80:385–399.

Brown, J.S. and Kotler, B.P. 2004. Hazardous duty pay and the foraging cost of predation. *Ecology Letters*, 7:999–1014.

Brown, J.S. and Kotler, B.P. 2007. Foraging and the ecology of fear. In Stephens D. W., Brown, J.S. and Ydenberg, R.C. (eds.), *Foraging: Behavior and Ecology*, pp. 437–482. Chicago: University of Chicago Press.

Cagnacci, F., Boitani, L., Powell, R.A. and Boyce, M.S. 2010. Challenges and opportunities of using GPS-based location data in animal ecology. *Philosophical Transactions of the Royal Society B*, 365:2157–2312.

Charnov, E.L. 1976. Optimal foraging, marginal value theorem. *Theoretical Population Biology*, 9:129–136.

Ciuti, S., Northrup, J.M., Muhly, T.B., Simi, S., Musiani, M., Pitt, J.A. and Boyce M. S. 2012. Effects of humans on behaviour of wildlife exceed those of natural predators in a landscape of fear. *PLoS ONE*, 7:e50611.

Courchamp, F., Berec, L. and Gascoigne, J. 2008. *Allee Effects in Ecology and Conservation*. Oxford: Oxford University Press.

Cressman, R. and Křivan, V. 2010. The ideal free distribution as an evolutionarily stable state in density-dependent population games. *Oikos*, 119:1231–1242.

Dalke, P.D. and Sime, P.R. 1941. Food habits of the eastern and New England cottontails. *Journal of Wildlife Management*, 5:216–228.

Dall, S.R.X., Kotler, B.P. and Bouskila, A. 2001. Attention, apprehension, and gerbils searching in patches. *Annales Zoologica Fennica*, 38:15–23.

Davis, J.M. and Stamps, J.A. 2004. The effect of natal experience on habitat preferences. *Trends in Ecology & Evolution*, 19:411–416.

Druce, D.J., Brown, J.S., Castley, J.C., Kerley, G.I.H., Kotler, B.P. and Slotow, R. 2006. Scale-dependent foraging costs: habitat use by rock hyraxes (*Procavia capensis*) determined by giving-up densities. *Oikos*, 115:513–525.

Druce, D.J., Brown, J.S., Kerley, G.I.H., Kotler, B.P., Mackey, R.A. and Slotow, R. 2009. Spatial and temporal scaling in habitat utilization by klipspringers (*Oreotragus oreotragus*) determined using giving-up densities. *Austral Ecology*, 34:577–587.

Dwernychuk, L.W. & Boag, D.A. 1972. Ducks nesting in association with gulls – an ecological trap? *Canadian Journal of Zoology*, 50:559–563.

Emerson, S.E. and Brown, J.S. 2012. Using giving up densities to test for dietary preferences in primates: an example with samango monkeys (*Cercopithecus (nictitans) mitis erythrarchus*). *International Journal of Primatology*, 33:1420–1438.

Errington, P.L. 1963. *Muskrat Populations*. Iowa State University Press, Ames, Iowa, USA. 664 pp.

Fortin, D., Morris, D.W. and McLoughlin, P.D. 2008. Habitat selection and the evolution of specialists in heterogeneous environments. *Israel Journal of Ecology & Evolution*, 54:311–328.

Fortin, D., Fortin, M.E., Beyer, H.L., Duchesne, T., Courant, S. and Dancose, K. 2009. Group-size-mediated habitat selection and group fusion-fission dynamics of bison under predation risk. *Ecology*, 90:2480–2490.

Fretwell, S.D. and Lucas, H.L. Jr. 1969. On territorial behavior and other factors influencing habitat distribution in birds. *Acta Biotheoretica*, 19:16–36.

Fryxell, J.M, Vamosi, S.M., Walton, R.A. and Doucet, C.M. 1994. Retention time and the functional response of beavers. *Oikos*, 71:207–214.

Gill, J.A., Sutherland, W.J. and Norris, K. 2001. Depletion models can predict shorebird distribution at different spatial scales. *Proceedings of the Royal Society B-Biological Sciences*, 268:369–376.

Gilliam, J.F. and Fraser, D.F. 1987. Habitat selection under predation hazard: a test of a model with foraging minnows. *Ecology*, 68:1856–1862.

Grafen, A. 1982. How not to measure inclusive fitness. *Nature*, 298:425–426.

Gregory, S.D. and Courchamp, F. 2010. Safety in numbers: extinction arising from predator-drive Allee effects. *Journal of Animal Ecology*, 79:511–514.

Green, C.M. and Stamps, J.A. 2001. Habitat selection at low population densities. *Ecology*, 82:2091–2100.

Greenberg, J. 2014. *A Feathered River across the Sky: The Passenger Pigeon's Flight to Extinction*. New York: Bloomsbury Publishing.

Gyimesi, A., Varghese, S., De Leeuw, J. and Nolet, B.A. 2012. Net energy intake rate as a common currency to explain swan spatial distribution in a shallow lake. *Wetlands*, 32:119–127.

Hawlena, D., Saltz, D., Abramsky, Z. and Buskila, A. 2010. Ecological trap for desert lizards caused by anthropogenic changes to habitat structure that favor predator activity. *Conservation Biology*, 24:803–809.

Hochman, V. and Kotler, B.P. 2006. Effects of food quality, diet preference and water on patch use by Nubian ibex. *Oikos*, 112:547–554.

Hoffmeister, D.F. 1989. *Mammals of Illinois*. Champaign: University of Illinois Press.

Houston, A.I. and McNamara, J.M. 1999. *Models of Adaptive Behavior*. Cambridge: Cambridge University Press.

Huntly, N.J. 1987. Influence of refuging consumers (pikas – *ochotona-princeps*) on sub-alpine meadow vegetation. *Ecology*, 68:12–26.

Iribarren, C. and Kotler, B.P. 2012a. Patch use and vigilance behavior by Nubian ibex: the role of the effectiveness of vigilance. *Evolutionary Ecology Research*, 14:223–234.

Iribarren, C. and Kotler, B.P. 2012b. Foraging patterns of habitat use reveal Nubian ibex' landscape of fear. *Wildlife Biology*, 18:194–201.

Kjellander, P. and Nordström, J. 2003. Cyclic voles, prey switching in red fox, and roe deer dynamics – a test of the alternative prey hypothesis. *Oikos*, 101:338–344.

Kokko, H. and Sutherland, W.J. 2001. Ecological traps in changing environments: ecological and evolutionary consequences of a behaviorally mediated Allee effect. *Evolutionary Ecology Research*, 3:537–551.

Kotler, B.P. 1985. Owl predation on desert rodents which differ in morphology and behavior. *Journal of Mammalogy*, 66:824–828.

Kotler, B.P. and Brown, J.S. 1990. Rates of seed harvest by two species of gerbilline rodents. *Journal of Mammalogy*, 71:591–596.

Kotler, B.P., Brown, J.S. and Hasson, O. 1991. Factors affecting gerbil foraging behavior and rates of owl predation. *Ecology*, 72:2249–2260.

Kotler, B.P., Brown, J.S and Mitchell, W.A. 1993. Environmental factors affecting patch use in gerbilline rodents. *Journal of Mammalogy*, 74:614–620.

Kotler, B.P., Brown, J.S. and Subach, A. 1993. Mechanisms of coexistence of optimal foragers: temporal partitioning in two species of sand dune dwelling gerbils. *Oikos*, 67:548–556.

Kotler, B.P., Gross, J.E. and Mitchell, W.A. 1994. Applying patch use in Nubian ibex to measure resource assessment ability, diet selection, indirect interactions between food plants, and predatory risk. *Journal of Wildlife Management*, 58:300–308.

Kotler, B.P., Dickman, C.R. and Brown, J.S. 1998. The effects of water on patch use in arid-zone birds and rodents in the Simpson Desert, central Australia. *Australian Journal of Ecology*, 23:574–578.

Kotler, B.P., Brown, J.S., Oldfield, A., Thorson, J. and Cohen, D. 2001. Patch use in three species of gerbils in a risky environment: the role of escape substrate and foraging substrate. *Ecology*, 82:1781–1790.

Kotler, B.P., Brown, J.S., Dall, S.R.X., Gresser, S., Ganey, D. and Bouskila, A. 2002. Foraging games between owls and gerbils: temporal dynamics of resource depletion and apprehension in gerbils. *Evolutionary Ecology Research*, 4:495–518.

Kotler, B.P. and Brown, J.S. 2007. Community Ecology. In Stephens D.W., Brown, J.S. and Ydenberg, R.C. (eds.) *Foraging: Behavior and Ecology*, pp. 397–436. Chicago: University of Chicago Press.

Kotler, B.P., Morris, D.W. and Brown, J.S. 2007. Behavioral indicators and conservation: wielding "the biologist's tricorder." *Israel Journal of Ecology and Evolution*, 53:237–244.

Kotler, B.P., Brown, J.S., Mukherjee, S., Berger-Tal, O. and A. Bouskila. 2010. Moonlight avoidance in gerbils reveals a sophisticated interplay among time allocation, vigilance, and state-dependent foraging. *Proceeding of the Royal Society of London B*, 277:1469–1474.

Kramer, A.M. and Drake, J.M. 2010. Experimental demonstration of population extinction due to a predator-driven Allee effect. *Journal of Animal Ecology*, 79:633–639.

Laundre, J.W., Hernandez, L. and Altendorf, K.B. 2001. Wolves, elk, and bison: re-establishing the "landscape of fear" in Yellowstone National Park, USA. *Canadian Journal of Zoology*, 79:1401–1409.

Laundre, J.W. 2010. Behavioral response races, predator–prey shell games, ecology of fear, and patch use of pumas and their ungulate prey. *Ecology*, 91:2995–3007.

Lucas, J.R. and Grafen, A. 1985. Partial prey consumption by ambush predators. *Journal of Theoretical Biology*, 113:455–473.

Mabry, K.E. and Stamps, J.A. 2008. Dispersing brush mice prefer habitat like home. *Proceedings of the Royal Society B*, 275:543–548.

Manly, B.F. J., McDonald, L.L., Thomas, D.L., McDonald, T.L. and Wallace P.E. 2002. *Resource Selection by Animals: Statistical Design and Analysis for Field Studies*, 2nd Edition. Boston, MA: Kluwer Academic Publishers.

McArthur, C., Orlando, P., Banks, P.B. and Brown. J.S. 2012. The foraging tight rope between predation risk and plant toxins: a matter of concentration. *Functional Ecology*, 26:74–83.

McLoughlin, P.D., Morris, D.W., Fortin, D., VanderWal, E. and Contasti, A.L. 2010. Considering ecological dynamics in resource selection functions. *Journal of Animal Ecology*, 79:4–12.

Messier, F., Virgl, J.A. and Marinelli, L. 1990. Density-dependent habitat selection in muskrats – a test of the ideal free distribution model. *Oecologia*, 84:380–385.

Mitchell, W.A. and Valone, T.J. 1990. The optimization research program: studying adaptations by their functions. *Quarterly Review of Biology*, 65:43–52

Morris, D.W. 1987. Tests of density dependent habitat selection in a patchy environment. *Ecological Monographs*, 57:269281.

Morris, D.W. 1988. Habitat dependent population regulation and community structure. *Evolutionary Ecology*, 2:253269.

Morris, D.W. 1999. Has the ghost of competition passed? *Evolutionary Ecology Research*, 1:3–20.

Morris, D.W. 2002. Measuring the Allee effect: positive density dependence in small mammals. *Ecology*, 83: 14–20.

Morris, D.W. 2003a. Toward an ecological synthesis: a case for habitat selection. *Oecologia*, 136: 1–13.

Morris, D.W. 2003b. How can we apply theories of habitat selection to wildlife conservation and management? *Wildlife Research*, 30:303–319.

Morris, D.W. 2011. Adaptation, habitat selection, and the eco-evolutionary process. *Proceedings of the Royal Society B*, 278:2401–2411.

Morris, D.W. 2014. Can foraging behaviour reveal the eco-evolutionary dynamics of habitat selection? *Evolutionary Ecology Research*, 16:1–18.

Morris, D.W. and Mukherjee, S. 2007. Can we measure carrying capacity with foraging behavior? *Ecology*, 88:597–604.

Morris, D.W., Kotler, B.P., Brown, J.S., Ale S.B. and Sundararaj, V. 2009. Behavioral indicators for conserving mammal diversity. The Year in Ecology and Conservation Biology, *Annals of The New York Academy of Sciences*, 1162:334–356.

Morris, D.W., Moore, D. E., Ale, S. B. and Dupuch, A. 2011. Forecasting ecological and evolutionary strategies to global change: an example from habitat selection by lemmings. *Global Change Biology*, 17:1266–1276.

Morris, D.W. and Dupuch A. 2012. Habitat change and the scale of habitat selection: shifting gradients used by coexisting Arctic rodents. *Oikos*, 121:975–984.

Morris, D.W., Dupuch, A. and Halliday, W.D. 2012. Climate induced habitat selection predicts future evolutionary strategies of lemmings. *Evolutionary Ecology Research*, 14:689–705.

Nelson, R., Keller, C. and Ratnaswamy, M. 2005. Estimating the extent of Delmarva fox squirrel habitat using an airborne LiDAR profiler. *Remote Sensing of Environment*, 96:292–301.

Nolet, B.A., 2006. The use of a flexible patch leaving rule under exploitative competition: a field test with swans. *Oikos*, 112: 342–352.

Nolet, B.A., Gyimesi, A. and Klaassen, R.H. 2006. Prediction of bird-day carrying capacity on a staging site: a test of depletion models. *Journal of Animal Ecology*, 75:1285–1292.

Nolet, B.A. and Klaassen, M. 2009. Retrodicting patch use by foraging swans in a heterogeneous environment using a set of functional responses. *Oikos*, 118:431–439.

Olsson, O., Wiktander, U., Holmgren, N.M.A. and Nilsson, S.G. 1999. Gaining ecological information about Bayesian foragers through their behaviour. II. A field test with woodpeckers.*Oikos*, 87:264–276.

Olsson, O., Molokwyu, M. and Ngozi, M. 2007. On the missed opportunity cost, GUD, and estimating environmental quality. *Israel Journal of Ecology and Evolution*, 53:263–278.

Olsson, O., Brown, J.S. and Heft, K.L. 2008. A guide to central place effects in foraging. *Theoretical Population Biology*, 74:22–33.

Oyugi, J.O., Brown, J.S. and Whelan, C.J. 2012. Foraging behavior and coexistence of two sunbird species in a Kenyan woodland. *Biotropica*, 44:262–269.

Patten, M.A. and Kelly, J.F. 2010. Habitat selection and the perceptual trap. *Ecological Applications*, 20: 2148–2156.

Perrin, M.R. and Kotler, B.P. 2005. A test of five mechanisms of species coexistence between rodents in a southern African savanna. *African Zoology*, 40:55–61.

Pulliam, H.R. 1974. On the theory of optimal diets. *American Naturalist*, 108:59–74.

Reid, C. 2004. Habitat suitability and behavior of springbok (*Antidorcas marsupialis*) at Augrabies Falls National Park, South Africa. University of Port Elizabeth, M. Sc.thesis. pp.

Rieuceu, G., Vickery, W.L. and Doucet, G.J. 2009. A patch use model to separate effects of foraging costs on giving-up densities: an experiment with white-tailed deer (*Odocoileus virginianus*). *Behavioral Ecology and Sociobiology*, 63:891–897.

Rosenzweig, M.L. 1979. Optimal habitat selection in two-species competitive systems. *Fortschritte der Zoologie*, 25:283–293.

Rosenzweig, M.L. 1981. A theory of habitat selection. *Ecology*, 62:327–335.

Sanchez, F. 2006. Harvest rates and patch-use strategy of Egyptian fruit bats in artificial food patches. *Journal of Mammalogy*, 87:1140–1144.

Sanchez, F., Kotler, B.P., Korine, C. and Pinshow, B. 2008a. Sugars are complementary resources to ethanol in foods consumed by Egyptian fruit bats. *Journal of Experimental Biology*, 211: 1475–1481.

Sanchez, F, Kotler, B.P., Korine, C. and Pinshow, B. 2008b. Ethanol and the foraging behavior of Egyptian fruit bats, *Rousettus aegyptiacus*. *Naturwissenschaften*, 95: 561–567.

Schlaepfer, M.A., Runge, M.C. and Sherman, P.W. 2002. Ecological and evolutionary traps. *Trends in Ecology & Evolution*, 17:474–480.

Schmidt, K.A., Brown, J.S. and Morgan, R.A. 1998. Plant defenses as complementary resources: a test with squirrels. *Oikos*, 81:130–142.

Shrader, A.M, Kotler, B.P., Brown, J.S. and Kerley. 2008. Providing water for goats in arid landscapes: effects of feeding effort with regards to time period, herd size, and secondary compounds. *Oikos*, 117:446–472.

Shochat, E., Patten, M.A., Morris, D.W., Reinking, D.L., Wolfe, D.H. and Sherrod, S. K. 2005. Ecological traps in isodars: effects of tallgrass prairie management on bird nest success. *Oikos*, 111:159–169.

Sih, A. 1980. Optimal behavior – can foragers balance 2 conflicting demands? *Science*, 210:1041–1043.

Speziale, K.L., Lambertucci, S.O. and Olsson, O. 2008. Disturbance from roads negatively affects Andean condor habitat use. *Biological Conservation*, 141:1765–1772.

Stamps, J.A. 2001. Habitat selection by dispersers: integrating proximate and ultimate approaches. In Clober, J., Danchin, E., Dhondt, A.A. and Nichols, J.D. (eds.) *Dispersal*, pp. 230–242, New York: Oxford University Press.

Stephens, D.W. and Krebs, J.R. 1986. *Foraging Theory*. Princeton: Princeton University Press.

Sutherland, W.J. 1996. *From Individual Behavior to Population Ecology*. Oxford: Oxford University Press.

Tadesse, S.A. and Kotler, B.P. 2012. Impact of tourism on Nubian ibex revealed through assessment of behavioral indicators. *Behavioral Ecology*, 23:1257–1262.

Tadesse, S.A. and Kotler, B.P. 2013. Habitat use by mountain nyala *Tragelaphus buxtoni* determined using stem bite diameters at point of browse, bite rates, and time budgets in the Bale Mountains National Park, Ethiopia. *Current Zoology*, 59:707–717.

Tilman, T. 1988. *Plant Strategies and the Dynamics and Structure of Plant Communities*. Princeton: Princeton University Press.

van der Merwe, M. and Brown, J.S. 2007. Foraging ecology of North American tree squirrels on cacheable and less cacheable foods: a comparison of two urban habitats. *Evolutionary Ecology Research*, 9:705–716.

van der Merwe, M. and Brown, J.S. 2008. Mapping the landscape of fear of the cape ground squirrel (*Xerus inauris*). *Journal of Mammalogy*, 89:1162–1169.

van Gils, J.A., Edelaar, P., Escudero, G. and Pierma, T. 2004. Carrying capacity models should not use fixed prey density thresholds: a plea for using more tools of behavioral ecology. *Oikos*, 104:197–204.

van Gils, J.A., Kraan, C., Dekinga, A., Drent, J., de Goeij, P. and Pierma, T. 2009. Reversed optimality and predictive ecology: burrowing depth forecasts population change in a bivalve. *Biology Letters*, 5:5–8.

van Horne, B. 1983. Density as a misleading indicator of habitat quality. *Journal of Wildlife Management*, 47:893–901.

Vickery, W.L., Rieucau, G. and Doucet, G.J. 2010. Comparing habitat quality within and between environments using giving up densities: an example based on the winter habitat of white-tailed deer *Odocoileus virginianus*. *Oikos*, 120:999–1004.

Watkinson, A.R. and Sutherland, W.J. 1995. Sources, sinks and pseudo-sinks. *Journal of Animal Ecology*, **64**:126–130.

Wecker, S.C. 1963. The role of early experience in habitat selection by the prairie deer mouse, *Peromyscus maniculatus baridi*. *Ecological Monographs*, **33**:307–325.

Yip, S.J.S., Rich, M.A. and Dickman, C.R. 2015. Diet of the feral cat, *Felis catus*, in central Australian grassland habitats during population cycles of its principal prey. *Mammal Research*, **60**:39–50.

Ziv, Y., Abramsky, Z., Kotelr, B.P. and Subach, A. 1993. Interference competition and temporal and habitat partitioning in two gerbil species. *Oikos*, **66**:237–246.

# Indirect behavioral indicators and their uses in conservation and management

ODED BERGER-TAL AND DAVID SALTZ

## 12.1 INTRODUCTION

Animals inhabit environments that are rapidly changing due to anthropogenic activities, such as the destruction and fragmentation of habitats, the introduction of exotic species and the alteration of local and global climate regimes. These changes are stretching the capacity of animals to cope, with conditions potentially outside the bounds of those experienced over the recent evolutionary history of the species. For managers of protected areas and endangered populations to respond in time to the threats posed by changing environments, these threats must be recognized early on, when still relatively benign, so as to be able to mitigate or counteract the adverse consequences of the altered environment. Coping takes place most immediately through behavioral responses, perhaps followed at a later stage by adjustments in physiology, and maybe over generational scales by shifts in morphology. This means that changes in animal behavior are potentially sensitive indicators that can provide the necessary early warning. Such behaviors must be documented in such a way that the changes will be revealed. In Chapter 11, Kotler *et al.* describe in detail how managers can use the behavior of animals as an indicator for their population's status and as a monitoring tool for the success of management programs aimed to assist these populations. However, behavioral indicators can tell us even more.

Animals in the wild do not live their life in isolation. They are a part of a complex ecosystem that includes many different species as well as various abiotic features. All species in a given system interact, to some extent, either directly or indirectly. There are many types of biological interactions between species: The most common ones are competition (either direct competition through interference or indirect through the utilization of shared resources), predation, parasitism and mutualism (e.g. pollination or seed dispersal). In addition, animals interact with their abiotic environment. The environment provides resources such as food and shelter, as well as constraints that may limit the behavior of animals (e.g. barriers that limit movement, the chemical composition of the water that can affect the behavior of aquatic species or noise that can limit communication between individuals). There is a constant feedback between animals and their environment, which, depending on existing conditions, may be either positive or negative. Thus, the behavior of animals may not only reflect their own state, but can also shed light on their interactions with other species and their interactions with their abiotic environment. By understanding these interactions and how they impact the behavior of animals, managers can get an imperative "head start" in assessing changes in both the biotic and the abiotic conditions of an animal's surroundings.

In this chapter we focus on what we term "indirect behavioral indicators." These behaviors are not used to evaluate the state of the individual depicting them or the state of its population, but are rather used as indicators to the state of its community, ecosystem and even to large-scaled global processes. The structure of this chapter is top-down, beginning with looking at how behavior can be used as an indicator for global changes such as climate change or large-scale pollution. We then scale-down to ecosystem management and the use of behavioral indicators to assess ecosystem health and as an early warning protocol for ecosystem collapse. Lastly, we consider the role of behavioral indicators in predicting community shifts. Indirect behavioral indicators have the potential to be an extremely powerful management tool with far-reaching consequences. Despite this, they are still rarely used in most systems. We hope this chapter will encourage managers to incorporate indirect behavioral indicators into their management plans when possible.

## 12.2 BEHAVIORAL INDICATORS OF GLOBAL CHANGES

### 12.2.1 Why behavioral indicators?

The use of indicators to monitor large-scale ecological phenomena or processes was first suggested almost a century ago (Hall & Grinnel 1919). The logic behind the concept is simple – large or multi-scale phenomena (e.g. biodiversity) are very hard to measure or predict due to their scale, complexity and rate of progression. Indicators provide measurable surrogates for such processes that enable cost- and time-efficient assessments (Noss 1990). As such, the use of indicators has frequently been incorporated into management policies and regulations (Carignan & Villard 2001). As the magnitude and consequences of global climate change became apparent in the last few decades (Pachauri & Reisinger 2008), realization of the need to monitor the rate and effects of these changes grew. Indicators, and in particular, indicator species, became an important tool for effective environmental monitoring, with several publications offering guidelines for selecting species that are efficient indicators of climate change (Hughes 2003). Butterflies, for example, are a very popular indicator group, and long-term data on the distribution and abundance of many butterfly species or assemblies have been used to generate specific, testable predictions related to climate change (Hellmann 2002, Parmesan 2003). Other examples include the arrival of a migratory moth species, *Plutella xylostella*, to certain arctic islands where these species have never been seen before, which has been suggested as an indicator of changing wind regimes in the arctic as a result of global warming (Coulson *et al.* 2002); and changes to the plant species composition in the Bavarian Forest National Park in Germany that were used to identify climate-sensitive zones that require the attention of managers (Bassler *et al.* 2010).

In all of the examples above, the species' occurrence, abundance or distribution are used as indicators for climate change and its effects. However, there are many cases in which the behavior of animals can serve as indicators of climate change just as effectively as non-behavioral indicators, or even surpass them. To effectively use an indicator in a management program, the choice of the indicator is critical but often very difficult (Noss *et al.* 1997). A good indicator should possess as many as possible from the following characteristics: It should provide early warning for environmental impacts; it should directly indicate the cause of change and not only the existence of a change; it should provide continuous assessment over a wide range of stresses; it should be reliable and create a minimum of "false alarms"; and it should be simple and cost-effective to measure (Noss 1990, Carignan & Villard 2001 and references within). Animal behavior can serve as an excellent indicator, as many behaviors possess most or all of the above requirements. First and foremost, behaviors are in many cases much easier to monitor than the distribution or abundance of species. Animals will usually exhibit a behavioral reaction to environmental changes long before demographic responses are evident, providing a much earlier warning of the change. Most behaviors can be measured on a continuous scale, either within individuals (e.g. the amount of food left in a foraging patch) or within the populations (e.g. the percentage of individuals choosing a certain habitat), allowing for a continuous assessment of change. Knowledge and understanding of a species' life history and behavior can serve to minimize the occurrences of "false alarms." Finally, the nature of the behavioral change can in many cases give much more information than simply the occurrence or absence of individuals, which makes behavioral indicators more likely to suggest the cause of the change and not just its existence.

It is also important to be aware of the limitations of using behavioral indicators. In most cases, there is a high variability among individuals in their behavioral responses, as well as variability within individuals as a function of their age, their reproductive stage and season, making precise measurements challenging. For many species we lack baseline data (i.e. control measurements) and data that is obtained in the lab may be irrelevant in the field. Lastly, there are cases in which the behavioral measurements are actually the more difficult and expensive methods (Scherer 1992, Zala & Penn 2004). Thus, while behavior is an important addition to the conservation biologist's toolbox, it should not be preferred blindly over other methods.

### 12.2.2 Which behaviors?

Since we want to gain insight on environmental changes on very large spatial and temporal scales, many behaviors may be unsuitable to serve as indicators. While an animal's foraging or anti-predatory behavior can tell us a lot about changes to the environment it currently inhabits, it can be extremely difficult to determine whether these changes to the environment are the result of some local disturbance, or whether they are caused by a global process. Therefore, in order to learn of large-scale environmental changes, we need to look for changes in large-scale behavioral *patterns*, or behaviors that can be compared along a large-scale spatial gradient. Large-scale spatial behavioral patterns usually refer to seasonal or annual large-scale movements, while temporal patterns concern the time allocated to specific activities and scheduling of these activities during the passage of days, seasons, years or an individual's lifetime. In many cases, these patterns encompass most or all members of the population (as opposed to the great variation among individuals in response to a local change in conditions). For example, Taylor *et al.* (2007) studied the behavior of migrating sandpipers, *Calidris mauri*, at stopover sites, and showed that choice of stopover site and foraging intensity within a site can serve as indicators for

Box 12.1:   Movement ecology as an indicator of global changes: indicators of stressful environmental conditions provided by animal movements tracked using GPS telemetry

NORMAN OWEN-SMITH

As a consequence of rising concentrations of greenhouse gases in the atmosphere, global environments are expected to become generally warmer, with precipitation as rain or snow more widely variable between and within years (Easterling *et al.* 2000). At the same time, large mammal movements are becoming increasingly restricted by expanding human settlements, with many populations now largely confined within the bounds of national parks and other protected areas. The viability of these populations depends on how effectively animals can cope with extreme conditions by adjusting their behavior. Behavioral responses can buy time for species until genetic changes occur, adapting animals physiologically or morphologically to the altered environmental conditions. Herbivores, for example, may need to spend more time searching for sparse food resources remaining during the most arid period of the year toward the end of winter or the dry season. But movement is costly, and exposes animals to heightened risks of predation, especially if foraging activity needs to be shifted into the night to avoid high temperature conditions during the day. Hence, movement patterns can serve as an indicator of how successfully particular species of large herbivore are responding to seasonal and annual variation in food availability and thermal conditions. Underlying these responses are changes in the growth responses of the plants that they depend on to changing ecosystem regimes in temperature and precipitation.

   With the advent of high-resolution and high-volume movement data-collecting techniques using GPS animal tracking, the ecology of movement can be studied in a far more rigorous manner than it has been in the past (Cagnacci *et al.* 2010). The wealth of data on animal movements provided by GPS telemetry opens new opportunities to derive early-warning indicators of stressful conditions, before the population consequences become manifested. Changes in movement patterns may occur at various spatio-temporal scales, from shortened stays in feeding sites to greater time spent seeking out places retaining food, and wandering excursions beyond usual home ranges. GPS technology enables movement responses to be documented without the direct involvement of human observers over annual seasonal cycles and through multiple years. The challenge is how to analyze and interpret the voluminous data provided. Candidate indicators could measure the proportion of time spent feeding relative to relocation moves between feeding stations (Owen-Smith 1979), daily amount of time spent foraging in relation to prevailing temperature conditions (Owen-Smith 1998), time lost to foraging as a result of travel to and from remaining water sources (Cain *et al.* 2012), day-to-day shifts in the foraging areas exploited (Owen-Smith 2013), and frequency of excursions beyond usual home range limits (Owen-Smith & Cain 2007). These responses must be interpreted in the context of

Box 12.1:   (cont.)

seasonal rhythms of food abundance and scarcity, governed in African savanna ecosystems largely by rainfall variation.

While much research has been done on methods for analyzing the movement sequences, less attention has been given to ecological inter-pretations of these data. A particular challenge was posed by the need to establish the causes of the extreme population declines shown by several less common antelope species in one of Africa's largest and most inten-sely managed protected areas, the Kruger National Park in South Africa (Ogutu & Owen-Smith 2003). The leading question was whether these population trends were a reflection of changing climate, or simply a result of mis-directed management interventions. A concern was that excessive provision of water points had benefited the more common ungulate species and their predators at the expense of the rarer species, and hence at a cost to overall biodiversity the park was intended to conserve. The associated need was to investigate whether extreme food shortages as a result of grassland responses to climatic variation could underlie the apparent vulnerability of the rarer species to predation.

Our study was focused on sable antelope (*Hippotragus niger*), the most widely distributed of the rarer antelope species, which enables us to investigate their movement patterns across the rainfall gradient that existed between the dry northern end of the Kruger Park and the relatively wet south-eastern region. Rainfall was strongly seasonal, with about 80% received during the summer months from October or November through March or April. Grass growth lagged slightly behind rainfall, with peak biomass attained in January or February, and grasses becoming progres-sively dry and deficient nutritionally as the dry season progressed. GPS collars were placed on female sable representing the herds with which they were associated in three widely separated regions of the Kruger National Park. Although the period covered was limited by the working life of the collars, it encompassed quite wide variation in annual rainfall totals.

The challenge was to identify a feature of the movement sequences documen-ted by the GPS collars that could indicate how adequately the animals coped with seasonal and annual variation in food stress. Indicators initially considered included (1) diel (24-h) displacement distances between times of the day when the animals were likely to be foraging, (2) frequency of excursions beyond the core home range, and (3) extent of movement at night when the risk of predation was heightened (Owen-Smith & Cain 2007). Most promising were diel displacement distances, provided excursions to water sources were excluded. Compared with zebra and buffalo herds in the same region, the northern-most sable herd showed earlier and substantially greater increases in diel displacement distances during the dry season month, especially during a year with below average rainfall (Owen-Smith 2013). This sable herd also moved further between successive days during the dry season than did sable herds occupying the wetter southern region of the park.

**Box 12.1:** (cont.)

Assembling the data from the sable herds in the three regions across the different years covered provides supporting insights into how differences in annual rainfall (shown in Figure 12.1) affected movement patterns. Diel displacement distances generally increased earlier in dryer years and reached a higher peak in the late dry season of the driest years (Figure 12.2). This pattern is clearly evident when the mean diel displacement during the dry season months is plotted against the annual rainfall total over the preceding wet season (Figure 12.3). However, this finding does not negate other evidence indicating that increased predation was the primary mechanism generating the sable population decline, because sable had also declined substantially in abundance in the wetter south-west despite the more favorable rainfall conditions (Owen-Smith *et al.* 2012). Sable herds inhabiting this region showed a lesser increase in diel displacement distances during the dry season than the sable herd in the north, under the same rainfall conditions, indicating that they were under less nutritional stress. However, our GPS tracking study was conducted after the sable population had declined on the herds that remained, meaning that we cannot establish the actual conditions that had led to the earlier demise of numerous herds (Owen-Smith *et al.*

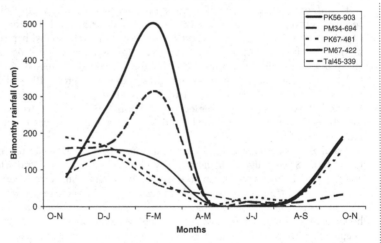

**Figure 12.1:** Rainfall patterns comparing study areas and years covered by the movement data over seasonal cycle extending from October at the beginning of the wet season to November of the following year. Legend indicates the study area (from north to south PM, Tal and PK), specific years represented in each (2003/4, 2004/5, 2005/6 and 2006/7, shown by numbers following the study area acronyms) and annual rainfall totals in these years (in mm, at the end of each label).

Box 12.1:   (cont.)

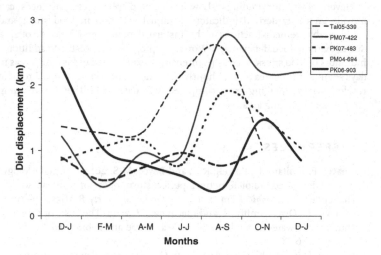

**Figure 12.2:**   Diel displacement distances (bimonthly medians) recorded for sable herds bearing GPS collars in the study areas and years corresponding with the rainfall data shown in Figure 12.1.

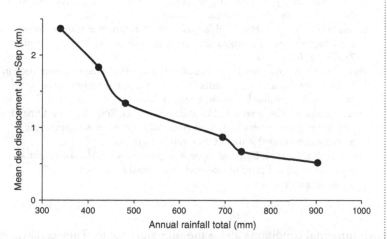

**Figure 12.3:**   Diel displacement distances shown by the sable herds through the dry season months (June–September) in these study areas and years plotted against the annual rainfall totals.

Box 12.1: (cont.)

2012). GPS tracking needs to be operating pre-emptively in order to provide effectual indicators of impending animal losses.

Nevertheless, our findings show that diel displacement distances can provide an easily derived indicator of annual variation in food stress and could enable remedial actions to be instigated before populations collapse due to extreme food deficiencies. Work is in progress to investigate additional indicators of food stress, such as (1) intensity of exploitation of foraging areas in days spent per unit area and (2) hourly movement rates while animals remain resident within foraging areas compared with days when they are roaming seeking new foraging areas.

## REFERENCES

Cagnacci, F., Boitani, L., Powell, R.A. and Boyce, M.S. 2010. Animal ecology meets GPS-based radiotelemetry: a perfect storm of opportunities and challenges. *Philosophical Transactions of the Royal Society B*, **365**:2157–2162.

Cain, J.W. III, Owen-Smith, N. and Macandza, V. 2012. The costs of drinking: comparative water dependency of sable antelope and zebra. *Journal of Zoology*, **286**:58–67.

Easterling, D.R., Meehl, G.A., Parmesan, C. *et al.* 2000. Climate extremes: observations, modelling, and impacts. *Science*, **289**:2068–2074.

Ogutu, J. and Owen-Smith, N. 2003. ENSO, rainfall and temperature influences on extreme population declines among African savanna ungulates. *Ecology Letters*, **6**:412–419.

Owen-Smith, N. 1979. Assessing the foraging efficiency of a large herbivore, the kudu. *South African Journal of Wildlife Research*, **9**:102110.

Owen-Smith, N. 1998. How high ambient temperature affects the daily activity and foraging time of a subtropical ungulate, the greater kudu. *Journal of Zoology*, **246**:183–192.

Owen-Smith, N. and Cain III, J.W. 2007. Indicators of adaptive responses in home range use and movements by a large mammalian herbivore. *Israel Journal of Ecology and Evolution*, **53**:423–438.

Owen-Smith, N., Chirima, G.J., Macandza, V. and Le Roux, E. 2012. Shrinking sable antelope numbers in Kruger National Park: what is suppressing population recovery? *Animal Conservation*, **15**:195–204.

Owen-Smith, N. 2013. Daily movement responses by an African savanna ungulate as an indicator of seasonal and annual food stress. *Wildlife Research*, **40**:232–240.

environmental conditions along the migratory route. They constructed a model showing that the mass action of many individual migrants, each optimizing its own migration timing and route, leads to the emergence of distinctive patterns of behavior and site choice, where a flyway-wide reduction in the amount of available food causes increased foraging intensity in stopover sites and a flyway-wide increase in predation risk causes a preference for larger stopover sites. An additional example is given in Box 12.1, where Norman Owen-Smith describes the use of GPS telemetry to discern the movement patterns of terrestrial animals, and how these patterns can serve as indicators for environmental consequences of global climate change.

Climate change may affect species associations with different habitats. For example, polar bears have become less associated with ice and more associated with land and open water, and this can be used to assess which areas are more affected by climate change and which are less affected (Gleason & Rode 2009). In such cases, however, it is important to distinguish between habitat association, which may simply reflect the change in the spatial availability and distribution of resources, and habitat selection, which reflects a behavioral response to a changing environment, increasing the attractiveness of one type of habitat (perhaps as it provides better conditions in an altered climate) and decreasing the attractiveness of others.

Climate change has also brought about various alterations to animals' phenology (i.e. life history temporal patterns). These include the timing of departure from wintering grounds and arrival to migratory stopovers or breeding areas (Butler 2003, Ptaszyk et al. 2003, Lehikoinen et al. 2010), and the timing of various aspects of breeding (such as territorial formation, courting, mating, egg laying, incubating and hatching/giving birth; Dunn & Winkler 1999, Cresswell & McCleery 2003, Sergio 2003). The scope of the effects of climate change on the phenology of species worldwide is enormous, making phenology a preferred indicator for the occurrence of global warming and for its effects (Menzel et al. 2006). The exact effect of climate change on a species' phenology may vary, depending on the ecology and life history of a species (Jenni & Kery 2003), highlighting the importance of applying knowledge of animal behavior (theory and empirical data) and life history in order to make the right predictions regarding the rate and the implications of climate change.

## 12.3 BEHAVIORAL INDICATORS FOR ECOSYSTEM HEALTH

There is now wide acceptance of the need for ecosystem-based assessment tools to monitor, protect and restore ecosystems, with an emphasis on "early warning" tools that will give managers enough time to react to deteriorating ecosystems before they collapse (Borja *et al.* 2008, Hellou 2011). There is virtually no ecosystem on our planet that is not affected by anthropogenic stressors, whether through direct anthropogenic interference (land use, harvesting, introduction of exotic species), or through anthropogenically created pollutants that are transported by air, water and with sediments to the most remote regions of our planet where animals have been found to carry significant amounts of industrial chemicals in their tissues (e.g. polar bears and albatrosses; Kannan *et al.* 2001a,b). Since ecosystem collapse is a process that is nearly impossible to reverse (Folke *et al.* 2005), maintaining ecosystems in a healthy condition and employing a reliable early warning system is a top priority in conservation biology. The study of animal behavior provides valuable opportunities for such early warning tools.

In order to quantify the effects of pollutants, scientists normally use two standardized measurements – $LD_{50}$ and LOEC (Clotfelter *et al.* 2003). $LD_{50}$ is the dosage that is lethal to 50% of exposed organisms ($LC_{50}$ is a similar test which refers to the concentration of chemicals in the air or in the water), whereas LOEC is the lowest concentration that produces observable effects, traditionally morphological deformities and their likes. While these measurements are very useful in generating guidelines to prevent mass mortality, they are sometimes very limited in their ability to provide an early warning compared to behavioral indicators. First, lethality tests ignore all cases of "ecological death" that may occur following toxicant exposure at much lower levels. In these cases, animals may not die as a direct cause of the pollutant, but may be unable to maintain their functions within the ecological system (Scott & Sloman 2004). Second, behavior is 10–1000 times more sensitive to pollutants than conventional $LC_{50}$ tests (Hellou *et al.* 2008, Robinson 2009). Lastly, behavioral tests are in many cases fast, simple to perform, noninvasive, cheap and of high ecological relevance (Hellou 2011). However, it is important to remember that using behavior as an indicator may also have several disadvantages, depending on the system in question (see previous section).

One of the earlier cases in which behavioral observations were used as an indicator for the presence of pollutants in the ecosystem is Broley's observations on the American bald eagle (Broley 1958). Broley started banding and monitoring populations of bald eagles along the west coast of Florida in 1939. Between the years 1952–1957 he reported changes in the reproductive behavior of the eagles that were followed by a severe decline in their reproductive success. This was one of the first reports alarming people to the dangers of DDT and other chemical pollutants (Zala & Penn 2004). In

1966, Warner *et al.* were among the first to suggest that behavioral measures have great potential as bioindicators for ecosystem health, and more specifically they suggested behavioral observations as a way to monitor the effects of chemical contaminants on aquatic systems. As can be understood from all of these examples, the use of behavior as an indicator for ecosystem health and integrity is currently most developed within the field of ecotoxicology.

### 12.3.1  Behavioral ecotoxicology

Ecotoxicology is the combination of the fields of Ecology (the study of the relationship between animals and their environment) and Toxicology (the study of toxic agents). It aims to determine whether detrimental substances are influencing the state of organisms and to assess the health of ecosystems in order to promote the sustainability of ecosystems and to stop early progressions of environmental degradation whilst conditions are still reversible (Hellou 2011). Behavioral ecotoxicology, as can be easily deducted from its name, is the use of animal behavior and behavioral ecology to achieve the same goals (Dell'Omo 2002). This follows the many observations confirming that organisms change their behavior after exposure to contaminants, and that these behavioral changes are far more sensitive and rapid than physiological changes or mortality of the organisms. The field has grown substantially in the last couple of decades, yielding several review papers (e.g. Clotfelter *et al.* 2003, Scott & Sloman 2004, Hellou 2011) and a comprehensive book on the subject (Dell'Omo 2002). Within the field, there is a great emphasis on fish in aquatic ecosystems, since fish species do not only play a role as organismic monitors of water quality, but are also part of intricate food webs that include humans (Scherer 1992).

The range of behaviors that have been used so far in behavioral ecotoxicology is very wide and encompasses behaviors from all behavioral domains including levels of activity, avoidance and escape behaviors, migration, homing, foraging efficiency, vigilance and response to predator cues, communication, courtship, mating, parental and social behaviors. The review papers mentioned above give well over a hundred examples on how these behaviors (and many more) change when animals are exposed to the presence of metals, oils and a large variety of endocrine-disrupting chemicals, many of which have been found to harm humans as well (Clotfelter *et al.* 2003). These examples span across species and taxa and include unnecessary schooling behavior and hyperactivity as a response to copper exposure in Atlantic silverside, *Menidia menidia* (Koltes 1985), delayed spawning onset in fathead minnows, *Pimephales promelas*, exposed to dietary methylmercury (Hammerschmidt *et al.* 2002), reduced courtship and nesting behaviors in male ringed turtle doves, *Streptopelia risoria*, that were fed with DDE, the principle DDT metabolite (Haegele & Hudson 1997), smaller and lower-quality nests of tree swallows, *Tachycineta bicolor*, living near PBC-contaminated sources compared to birds living in cleaner areas (McCarty & Secord 1999), increased agonistic behaviors in gold fish, *Carassius auratus*, exposed to carbofuran, a carbamate insecticide (Saglio *et al.* 1996), increased aggressiveness in male house mice, *Mus musculus*, and wild deer mice, *Peromyscus maniculatus*, exposed to a mixture of chemicals that are common in drinking water (Porter *et al.* 1999), and changes in jumping behavior of springtail, *Folsomia candida* (a species used as a "standard" test organism for estimating the effects of pesticides and environmental pollutants on non-target soil arthropods; Fountain & Hopkin 2005), in copper- and nickel-polluted soils (Kim & An 2014).

The use of behavioral indicators as an early warning system for environmental pollution has been successfully integrated into environmental monitoring procedures and innovative automatic behavioral monitoring instruments have been developed specifically for this. Today, online continuous bio-monitoring instruments that automatically measure changes to pre-defined behaviors of test organisms are available. For example, the Multispecies Freshwater Biomonitor™ uses various behaviors (e.g. movement, ventilation, foraging) from a wide range of test organisms to monitor the quality of water, soil or sediment in real time. The system is composed of a number of test chambers in which the relevant organisms are placed. The system then automatically converts their movements to measurements of electrical currency (Gerhardt *et al.* 1998). When coupled with previous knowledge of the organisms' response to various contaminates, these systems provide constant monitoring and an early warning system for a variety of ecosystems (e.g. Gerhardt & Schmidt 2002, Mohti *et al.* 2012).

### 12.3.2 Early warnings of ecosystem collapse

Outside the field of ecological toxicology there are but a few examples for studies exploring the use of behavioral indicators in the monitoring of ecosystem health. Couvillon *et al.* (2014) demonstrated how the waggle dance of honey bees, *Apis mellifera*, in which a forager bee informs her fellow bees of the most profitable foraging locations, can be used as an indicator for the quality of their environment. In another study, Searle *et al.* (2007) used various foraging metrics such as bite size, bite rate and rumination time as reliable behavioral indicators for assessing the state of the vegetation in a given landscape from the perspective of different foraging species. By assessing vegetation state "through the eyes" of the different foragers, the authors wielded a much more sensitive vegetation index than using biomass measurements alone, which allowed them to make better predictions regarding the response of different species to environmental changes.

Persson and Nilsson (2007) have demonstrated that the foraging behavior of benthivorous fish in shallow lakes can provide first-hand information about the ecosystem state. Shallow lakes can exist in one of two states – a clear state dominated by submerged macrophytes and a turbid state dominated by phytoplankton (Sceffer & van Nes 2007). Each of these states may be maintained by various feedback mechanisms, making it extremely difficult to shift between the states. Depleting benthic prey resources can be used as an early indication that the system is approaching a threshold level that will induce regime shift (Persson & Stenberg 2006), but may be difficult to monitor using traditional sampling techniques. Monitoring the feeding behavior of benthivorous predators using standardized feeding patches (giving-up densities technique, see Chapter 11) provides one possible alternative approach that is easy to implement and can provide early indication that the system is approaching its shifting threshold.

Samways (2005) reported that changes in the territorial behavior (e.g. patrolling) of different species of butterflyfish in the Seychelles islands in the western Indian Ocean can be used as an early indication of massive coral bleaching events. While it is not entirely clear whether the fish were reacting to the dying corals, or to the same abiotic factors that induced the bleaching event, it is clear that the change in fish behavior preceded the coral bleaching event and could be used as an early indicator for this system collapse.

The last two examples make a compelling case for the use of behavioral indicators as an early warning system for ecosystem collapse. Ecosystems are constantly subjected to a multitude of stressors, of both natural and anthropogenic origins (Rapport & Whitford 1999). Eventually, these stressors can cause a regime shift in which complex ecosystems undergo rapid transformation to very simple systems containing very few species, and even fewer interactions. This process is termed ecosystem collapse. Examples include the desertification of semi-arid systems, the eutrophication of aquatic systems, the bleaching of coral reefs, and bush encroachment in grasslands systems. Once a system has collapsed, it is highly resistant to further transformations, and it is therefore extremely difficult to restore it to its previous state (Folke et al. 2005). Consequently, the most effective way to treat system collapse is to prevent it from happening in the first place. The problem is that such regime shifts occur swiftly and their onset is very difficult to predict or anticipate (Clark *et al.* 2001). This is due to the synergism among the components of the ecosystem, so in many cases ecosystems react to stressors in a non-linear fashion (Rapport & Whitford 1999). A system might show no noticeable changes, or the change in the system's state may seem very gradual, only to be suddenly interrupted by a drastic shift, rapidly collapsing to a simplified state (Scheffer *et al.* 2001). In order to prevent systems from collapsing we have to be able to perceive that the system is heading toward a regime shift while it is still in its gradual, hard-to-notice, state of decline. Theory suggests that some variables within ecosystems do change during this period, and these variables may serve as early indicators of the coming change (Carpenter & Brock 2006, Carpenter *et al.* 2008). Identifying these indicators, and monitoring them, may alert us that the system is heading toward collapse and give us a crucial management head start that may prevent the collapse.

Many behaviors of organisms are potentially excellent candidates for early warning indicators of ecosystem collapse. Because we generally know which systems are prone to collapse (e.g. coral reefs, open grasslands), efforts should be made to identify indicative behaviors in species that are common in these systems (preferably keystone species), even under lab conditions (when possible). Finding a good indicator of ecosystem collapse (Figure 12.4) is not an easy task and requires extensive knowledge on the behaviors and life-history of the relevant species. In the case of species of which we still know little, developing a behavioral indicator may require years of research, defeating the purpose of obtaining a quick, cost-effective and easy-to-use indicator. However, behavioral ecologists, ethologists and psychologists have already gathered a vast amount of information regarding the life history and behaviors of a great many species. The benefits of applying this knowledge to creating easy to monitor and reliable early warning systems for ecosystem collapse are immense.

## 12.4  BEHAVIORAL INDICATORS OF COMMUNITY SHIFTS

Species constantly interact with each other. They prey upon, compete with, cooperate with or parasitize each other. This means that any change to the distribution, abundance or behavior of one species can affect the behavior of other species in the community. In other words, a change in the behavior of

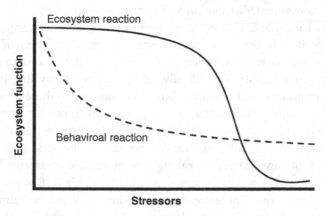

**Figure 12.4:** A schematic representation of the effects of natural and anthropogenic stressors on ecosystem functions in most ecosystems (solid line). This response is usually non-linear due to the synergetic relationship among the components of the ecosystems, and thus most systems will not show any significant changes to their functions while the stressors accumulate, until some threshold will be crossed and the system will be abruptly reduced to a much simplified state. The dashed line represents the response of a useful behavioral indicator to ecosystem collapse. Because the behavioral reaction to the environmental stressors is much more rapid than the ecosystem response, it can serve as an early detection system, warning managers that the system in question may be approaching the above-mentioned collapse threshold.

any species can reflect shifts in the composition or structure of its community.

Classic models of interactions between species (among and within trophic levels) assume that these interactions are density mediated (i.e. governed by changes in populations' densities). However, it is now widely accepted that much of the interactions between species are driven by changes to the traits of individuals within populations. These are called trait-mediated indirect interactions (TMII; Abrams 1995, Werner & Peacor 2003). TMIIs can represent any trophic or nontrophic interaction between species. When TMIIs are strong enough to structure ecosystems through the indirect top-down effects of predators on their entire food web (Shurin et al. 2002, Schmitz et al. 2004), they are often referred to as behaviorally mediated trophic cascades (BMTCs, Abrams 1984, Beckerman et al. 1997).

One of the most famous examples of BMTC is the release of wolves, *Canis lupus*, to the greater Yellowstone area in 1995–1996 (White & Garrott 2005). The reintroduction of wolves into the ecosystem altered the vigilance, movement, distribution and group size of elk, *Cervus Canadensis* (Ripple & Larsen 2000, Ripple *et al.* 2001, Creel & Winnie 2005). These changes to the behavior of the elk as a response to the distribution of wolves are the main driver of a major trophic cascade that is reshaping the Yellowstone region. More specifically, elk do not avoid traveling in areas that are commonly used by wolves, but when traveling in these areas, they switch their habitat preferences from aspen stands to conifer forests and open areas (Fortin *et al.* 2005). This change of preferences seems to have led to a dramatic change in the species composition in the park, with aspen, willow and cottonwood trees increasing in numbers and landscape-shaping colonies of beavers becoming more abundant (Ripple & Beschta 2012), although the relationship between elk behavior and aspen recruitment is still a subject of controversy (Kauffman *et al.* 2010, Winnie 2012). Either way, the Yellowstone case study serves as a compelling reminder of how a behavioral change in one species can cascade through the system and drastically alter entire communities.

Many other examples of trait-mediated cascading effects exist in aquatic (e.g. Peckarsky & McIntosh 1998, Bernot & Turner 2001), marine (reviewed in Dill *et al.* 2003, Heithaus *et al.* 2008) and terrestrial (e.g., Beckerman *et al.* 1997, Rudgers *et al.* 2003) systems. The species changing its behavior as a response to predation can be a predator species itself (e.g. fish – salamanders – isopods system; Huang & Sih 1991), and a behavioral response could also be initiated as a response to individuals of the same species (e.g. juvenile perch changing their behavior in response to the distribution of adult perch; Diehl & Eklov 1995). In addition, TMIIs may be strongly mediated by the hunting mode of the predator (Schmitz *et al.* 2004, Schmitz 2005), where sit-and-wait predators may elicit the strongest effects on the behavior of prey species, while actively searching predators have the weakest effects.

So how can we use the accumulating knowledge on TMIIs and BMTCs for conservation and management? We simply reverse our point of view. Similarly to reverse engineering, in which the purpose of an object is deducted from observing its structure and operation, usually with no knowledge on the procedures that have led to the production of the object (Eilam 2005), what we should aim to do is to deduct the adaptive value of a behavioral shift (and through it, the cause for the shift) from observing a change in the behavior of individuals within a population. If we take the Yellowstone system as an example, and imagine that wolves were not intentionally introduced to the park, but rather found their way into the park on their own, then the change in the behavior of the elk, and the dramatic changes to the ecosystem that followed, should have alerted us to the fact that there likely was a change to the community structure in the Yellowstone region. Moreover, if we would have been regularly monitoring the behavior of elk, we probably could have detected the change to the community structure well ahead of the ensuing ecosystem changes, which could have allowed us to prevent these changes (if they were something we wished to avoid). Thus, behavioral shifts in keystone species can be used to detect the arrival and establishment of alien species into a community, which can be a valuable management tool when the invasive species is cryptic (and therefore difficult to detect through more conventional means), or when it is difficult to estimate its abundance. Other changes to the community structure that can be detected through the monitoring of the behaviors of key or focal species in the community are changes to the abundance or distribution of key predators in the system, and changes to the abundance or distribution of important competitor or collaborator species.

## 12.5 MANAGEMENT IMPLICATIONS

(1) To set up an early warning system for large- and medium-scale environmental changes to ecosystems or communities, focal key species should be chosen for behavioral monitoring. These species should be numerous in the system, easy to monitor, and connected with many other species in the system by ways of competition, cooperation, predator prey interactions, etc.
(2) For a behavior to be a good indicator it should be easy to monitor, have the potential to rapidly change, be measured on a continuous scale and be directly connected to the fitness of the organism displaying it. In addition, the behavior should be well understood in terms of its driving mechanisms since it will increase the chances of rapidly finding the cause for any change to the behavior.
(3) Sensitive ecosystems that are prone to collapse should be given a high priority in finding and monitoring the behaviors of key species within them. These ecosystems include shallow aquatic systems (prone to eutrophication), coral reefs (prone to coral bleaching), semi-arid areas (prone to desertification), grasslands (prone to bush encroachment) and marine systems that are prone to overfishing.

## 12.6 SUMMARY AND PROSPECTS

Ecosystems are inherently complex, and each species within an ecosystem is connected to other species by numerous direct and indirect interactions. Furthermore, the abiotic features of an ecosystem are also tightly connected to the existence of the organisms that comprise it. This means that any changes to the biotic and abiotic processes within the ecosystem, whether it is a global-scale change or a shift in the distribution of just one species, is likely to be reflected in the behavior of species within the ecosystem. Indirect behavioral indicators represent an extremely valuable and relatively underused management tool for the early detection of global changes, ecosystem collapses and community shifts.

The complexity of ecosystems may also be one of the main obstacles in the use of indirect behavioral indicators, and may have contributed to the fact that such indicators are still rarely used in most types of ecosystems. Observing a behavioral change in a focal species does not immediately reveal its cause; the change may be due to alteration of global processes, ecosystem processes, community processes or it can reflect a change in the state of the focal population itself (i.e. direct behavioral indicators, Chapter 11). As wildlife managers, how can we tell the difference? While this can indeed be a challenging feat, by identifying key species and key behaviors *a priori*, and monitoring these behaviors (in many cases an easier management practice than monitoring the abundance of populations), we can get an early indication that something in the system has changed. Once alerted to such a change, looking at the behaviors of other species, as well as taking additional abiotic measurements, can put the behavioral shift into the correct context. The fact that the behavioral responses of organisms to changes in the environment can be extremely rapid "buys" us important time in which we can properly investigate the source of this change before a negative demographic response takes place or before the system undergoes an irreversible shift in its condition.

In some systems the relevant behavioral context can be conjectured in advance through behavioral experimentation. This allows, for example, the creation of online bio-monitoring instruments that automatically measure changes to pre-defined behaviors of test organisms to continuously monitor water quality in aquatic systems (Gerhardt & Schmidt 2002). The type of behavior that is changing can also be used in many cases to delineate the scale of the environmental change that is causing the behavioral shift: behaviors from the foraging and vigilance domain are probably most appropriate as direct behavioral indicators, as in most cases they represent an organism's landscape of fear (Brown *et al.* 1999), which mostly reflects its own state and the state of its population (although landscape of fear indicators may also sometimes be used indirectly to monitor changes to predator populations; Schmidt & Schauber 2007). Behaviors from the movement and space-use domains are much more likely to act as indirect indicators, although this very much depends on the scale of the behavioral shift. Intuitively, the larger the scale of the behavioral shift (e.g. changing local movement patterns vs. changing migration routes, or switching between patches vs. selecting different habitats), the larger the scale of the disturbance this shift indicates. Within the reproductive and social behaviors domain, changes in the phenology of species are usually indicative of large-scale environmental changes rather than of changes in the organism's local surroundings.

Indirect behavioral indicators stem from the realization that the different scales of ecological processes are all linked, and the effects of global changes trickle down until they can be reflected in the behavior of a single individual. This offers us exciting new opportunities to increase the efficiency in which we manage communities and ecosystems, and can provide valuable conservation tools that may sometimes have far-reaching consequences. It is up to us now to seize these opportunities and revolutionize the way we practice community and ecosystem management.

## REFERENCES

Abrams, P.A. 1995. Implications of dynamically variable traits for identifying, classifying and measuring direct and indirect effects in ecological communities. *American Naturalist*, 146:112–134.

Bassler, C., Muller, J. and Dziock, F. 2010. Detection of climate sensitive zones and identification of climate change indicators: a case study from the Bavarian Forest National Park. *Folia Geobotanica*, 45:163–182.

Beckerman, A.P., Uriarte, M. and Schmitz, O.J. 1997. Experimental evidence for a behavior-mediated trophic cascade in a terrestrial food chain. *Proceedings of the National Academy of Sciences (USA)*, 94:10735–10738.

Bernot, R.J. and Turner, A.M. 2001. Predator identity and trait-mediated indirect effects in a littoral food web. *Oecologia*, 129:139–146.

Borja, A., Bricker, S.B., Dauer, D.M. *et al.* 2008. Overview of integrative tools and methods in assessing ecological integrity in estuarine and coastal systems worldwide. *Marine Pollution Bulletin*, 56:1519–1537.

Broley, C. 1958. The plight of the American bald eagle. *Audubon Magazine*, 60:162–163, 171.

Brown, J.S., Laundre, J.W. and Gurung, M. 1999. The ecology of fear: optimal foraging, game theory, and trophic interactions. *Journal of Mammalogy*, 80:385–399.

Butler, C.J. 2003. The disproportionate effect of global warming on the arrival dates of short-distance migratory birds in North America. *Ibis*, **145**:484–495.

Carignan, V. and Villard, M. 2001. Selecting indicator species to monitor ecological integrity: a review. *Environmental Monitoring and Assessment*, **78**:45–61.

Carpenter, S.R. and Brock, W.A. 2006. Rising variance: a leading indicator of ecological transition. *Ecology Letters*, **9**:311–318.

Carpenter, S.R., Brock, W.A., Cole, J.J., Kitchell, J.F. and Pace, M.L. 2008. Leading indicators of trophic cascades. *Ecology Letters*, **11**:128–138.

Clark, J.S., Carpenter, S.R., Barber, M. *et al.* 2001. Ecological forecasts: an emerging imperative. *Science*, **293**:657–660.

Clotfelter, E.D., Bell, A.M. and Levering, K.R. 2003. The role of animal behaviour in the study of endocrine-disrupting chemicals. *Animal Behaviour*, **68**:665–676.

Coulson, S.J., Hodkinson, I.D., Webb, N.R. *et al.* 2002. Aerial colonization of high Arctic islands by invertebrates: the diamondback moth *Plutella xylostella* (Lepidoptera: Yponomeutidae) as a potential indicator species. *Diversity and Distributions*, **8**:327–334.

Couvillon, M.J., Schurch, R. and Ratnieks, F.L.W. 2014. Dancing bees communicate a foraging preference for rural lands in high-level agri-environment schemes. *Current Biology*, **24**:1212–1215.

Creel, S. and Winnie Jr., J.A. 2005. Responses of elf herd size to fine-scale spatial and temporal variation in the risk of predation by wolves. *Animal Behaviour*, **69**:1181–1189.

Cresswell, W. and McCleery, R.H. 2003. How great tits maintain synchronization of their hatch date with food supply in response to long-term variability in temperature. *Journal of Animal Ecology*, **72**:356–366.

Dell'Omo, G. (ed.) 2002. *Behavioural Ecotoxicology*. West Sussex: Wiley.

Diehl, S. and Eklov, P. 1995. Effects of piscivore-mediated habitat use on resources, diet, and growth of perch. *Ecology*, **76**:1712–1726.

Dunn, P.O. and Winkler, D.W. 1999. Climate change has affected the breeding date of tree swallows throughout North America. *Proceedings of the Royal Society of London. Series B. Biological Sciences*, **266**:2487–2490.

Eilam, E. 2005. *Reversing: Secrets of Reverse Engineering*. Indianapolis: Wiley Publishing.

Folke, C., Carpenter, S.R., Walker, B. *et al.* 2005. Regime shifts, resilience, and biodiversity in ecosystem management. *Annual Review of Ecology, Evolution, and Systematics*, **35**:557–581.

Fortin, D., Beyer, H.L., Boyce, M.S. *et al.* 2005. Wolves influence elk movements: behavior shapes a trophic cascade in Yellowstone National Park. *Ecology*, **86**:1320–1330.

Fountain M.T. and Hopkin, S.P. 2005. *Folsomia candida* (Collembola): a "standard" soil arthropod. *Annual Review of Entomology*, **50**:201–22.

Gerhardt, A., Carlsson, A., Ressenmann, C. and Stich, K.P. 1998. A new online biomonitoring system for *Gammarus pulex* (L.) (Crustacea): *in situ* test below a copper effluent in south Sweden. *Environmental Science & Technology*, **32**:150–156.

Gerhardt, A. and Schmidt, S. 2002. The multispecies freshwater biomonitor: a potential new tool for sediment biotests and biomonitoring. *Journal of Soils and Sediments*, **2**:67–70.

Gleason, J.S. and Rode, K.D. 2009. Polar bear distribution and habitat association reflect long-term changes in fall sea ice conditions in the Alaskan Beaufort Sea. *Arctic*, **62**:405–417.

Hall, H.M. and Grinnell, J. 1919. Life-zone indicators in California. *Proceedings of the California Academy of Sciences*, 9:37–67.

Hellmann, J.J. 2002. Butterflies as model systems for understanding and predicting climate change. In Schneider, S.H. and Root, T.L.(eds.) *Wildlife Responses to Climate Change*. Washington, DC: Island Press. pp. 93–126.

Hellou, J., Cheeseman, K., Desnoyers, E. *et al.* 2008. A non-lethal chemically based approach to investigate the quality of harbor sediments. *Science of the Total Environment*, 389:178–187.

Hellou, J. 2011. Behavioural ecotoxicology, an "early warning" signal to assess environmental quality. *Environmental Science and Pollution Research*, 18:1–11.

Huang, C. and Sih, A. 1991. Experimental studies on direct and indirect interactions in a three trophic-level stream system. *Oecologia*, 85:530–536.

Hughes, L. 2003. Climate change and Australia: trends, projections and impacts. *Austral Ecology*, 28:423–443.

Jenni, L. and Kery, M. 2003. Timing of autumn bird migration under climate change: advances in long-distance migrants, delays in short-distance migrants. *Proceedings of the Royal Society of London. Series B. Biological Sciences*, 270:1467–1471.

Kannan, K., Koistinen, J., Beckmen, K. *et al.* 2001a. Accumulation of perfluorooctane sulfonate in marine mammals. *Environmental Science and Technology*, 35:1593–1598.

Kannan, K., Franson, J.C., Bowerman, W.W. *et al.* 2001b. Perfluorooctane sulfonate in fish-eating water birds including bald eagles and albatrosses. *Environmental Science and Technology*, 35:3065–3070.

Kim, S.W. and An, Y.J. 2014. Jumping behavior of the springtail *Folsomia candida* as a novel soil quality indicator in metal-contaminated soils. *Ecological Indicators*, 38:67–71.

Lehikoinen, A., Saurola, P., Byholm, P., Linden, A. and Valkama, J. 2010. Life history events of the Eurasian sparrowhawk *Accipiter nisus* in a changing climate. *Journal of Avian Biology*, 41:627–636.

Menzel, A., Sparks, T.H., Estrella, N. *et al.* 2006. European phonological response to climate change matches the warming pattern. *Global Change Biology*, 12:1969–1976.

Mohti, A., Shuhaimi-Othman, M. and Gerhardt, A. 2012. Use of the Multispecies Freshwater Biomonitor to assess behavioral changes of *Poecilia reticulate* (Cyprinodontiformes: Poeciliidae) and *Macrobrachium lanchesteri* (Decapoda: Palaemonidae) in response to acud mine drainage: laboratory exposure. *Journal of Environmental Monitoring*, 14:2505–2511.

Noss, R.F. 1990. Indicators for monitoring biodiversity: a hierarchical approach. *Conservation Biology*, 4:355–364.

Noss, R.F., O'Connel, M.A. and Murphy, D.D. (eds.) 1997. *The Science of Conservation Planning: Habitat Conservation under the Endangered Species Act*. Washington, DC: Island Press.

Pachauri, R.K. and Reisinger, A. (eds.) 2008. *Climate Change 2007*. Synthesis report. Contribution of working groups I, II, and III to the fourth assessment report. Intergovernmental Panel on Climate Change, Geneva, Switzerland.

Parmesan, C. 2003. Butterflies as bioindicators for climate change effects. In Boggs, C. L., Watt, W.B. and Ehrlich, P.R.(eds.) *Butterflies: Ecology and Evolution Taking Flight*. Chicago: University of Chicago Press. pp. 541–560.

Peckarsky, B.L. and McIntosh, A.R. 1998. Fitness and community consequences of avoiding multiple predators. *Oecologia*, 113:565–576.

Persson, A. and Stenberg, M. 2006. Linking patch use behaviour, resource density and growth expectations in fish. *Ecology*, **87**:1953–1959.

Persson, A. and Nilsson, E. 2007. Foraging behavior of benthic fish as an indicator of ecosystem state in shallow lakes. *Israel Journal of Ecology and Evolution*, **53**:407–421.

Ptaszyk, J., Kosicki, J., Sparks, T.H. and Tryjanowski, P. 2003. Changes in the timing and pattern of arrival of the white stork (*Ciconia ciconia*) in western Poland. *Journal fur Ornithologie*, **144**:323–329.

Rapport, D.J. and Whitford, W.G. 1999. How ecosystems respond to stress. *Bioscience*, **49**:193–203.

Ripple, W.J. and Larsen, E.J. 2000. Historic aspen recruitment, elk, and wolves in northern Yellowstone National Park, USA. *Biological Conservation*, **95**:361–370.

Ripple, W.J., Larsen, E.J., Renkin, R.A. and Smith, D.W. 2001. Trophic cascades among wolves, elk and aspen on Yellowstone National Park's northern range. *Biological Conservation*, **102**:227–234.

Ripple, W.J. and Beschta, R.L. 2012. Trophic cascades in Yellowstone: the first 15 years after wolf reintroduction. *Biological Conservation*, **145**:205–213.

Robinson, P.D. 2009. Behavioural toxicity of organic chemical contaminants in fish: application to ecological risk assessments (ERAs). *Canadian Journal of Fisheries and Aquatic Sciences*, **66**:1179–1188.

Rudgers, J.A., Hodgen, J.G. and White, J.W. 2003. Behavioral mechanisms underlie an ant–plant mutualism. *Oecologia*, **135**:51–59.

Samways, M.J. 2005. Breakdown of butterflyfish (*Chaetodontidae*) territories associated with the onset of a mass coral bleaching event. *Aquatic Conservation: Marine and Freshwater Ecosystems*, **15**:S101–S107.

Scheffer, M., Carpenter, S., Foley, J.A., Folke, C. and Walker, B. 2001. Catastrophic shifts in ecosystems. *Nature*, **413**:591–596.

Scheffer, M. and van Nes, E.H. 2007. Shallow lakes theory revisited: various alternative regimes driven by climate, nutrients, depth and lake size. *Hydrobiologia* **584**:455–466.

Scherer, E. 1992. Behavioural responses as indicators of environmental alterations: approaches, results, developments. *Journal of Applied Ichthyology*, **8**:122–131.

Schmidt, K.A. and Schauber, E.M. 2007. Behavioral indicators of predator space use: studying species interactions through the behavior of predators. *Israel Journal of Ecology and Evolution*, **53**:389–406.

Schmitz, O.J., Krivan, V. and Ovadia, O. 2004. Trophic cascades: the primacy of trait-mediated indirect interactions. *Ecology Letters*, **7**:153–163.

Scott, G.R. and Sloman, K.A. 2004. The effects of environmental pollutants on complex fish behaviour: integrating behavioural and physiological indicators of toxicity. *Aquatic Toxicology*, **68**:369–392.

Searle, K.R, Hobbs, N.T. and Gordon, I.J. 2007. It's the "foodscape," not the landscape: using foraging behavior to make functional assessments of landscape condition. *Israel Journal of Ecology & Evolution*, **53**:297–316.

Sergio, F. 2003. From individual behaviour to population pattern: weather-dependent foraging and breeding performance in black kites. *Animal Behaviour*, **66**:1109–1117.

Shurin, J.B., Borer, E.T., Seabloom, E.W. *et al.* 2002. A cross-ecosystem comparison of the strength of trophic cascades. *Ecology Letters,* 5:785–791.

Warner, R.E., Peterson, K.K. and Borgman, L. 1966. Behavioural pathology in fish: a quantitative study of sublethal pesticide toxication. *Journal of Applied Ecology,* 3:223–247.

Werner, E.E. and Peacor, S.D. 2003. A review of trait-mediated indirect interactions in ecological communities. *Ecology,* 84:1083–1100.

White, P.J. and Garrot, R.A. 2005. Yellowstone's ungulates after wolves – expectations, realizations, and predictions. *Biological Conservation,* 125:141–152.

Zala, S.M. and Penn, D.J. 2004. Abnormal behaviours induced by chemical pollution: a review of the evidence and new challenges. *Animal Behaviour,* 68:649–664.

# Index

Printed in the United States
by Baker & Taylor Publisher Services